Advances in
Bacterial Paracrystalline
Surface Layers

NATO ASI Series

Advanced Science Institutes Series

A series presenting the results of activities sponsored by the NATO Science Committee, which aims at the dissemination of advanced scientific and technological knowledge, with a view to strengthening links between scientific communities.

The series is published by an international board of publishers in conjunction with the NATO Scientific Affairs Division

A	**Life Sciences**	Plenum Publishing Corporation
B	**Physics**	New York and London
C	**Mathematical and Physical Sciences**	Kluwer Academic Publishers
D	**Behavioral and Social Sciences**	Dordrecht, Boston, and London
E	**Applied Sciences**	
F	**Computer and Systems Sciences**	Springer-Verlag
G	**Ecological Sciences**	Berlin, Heidelberg, New York, London,
H	**Cell Biology**	Paris, Tokyo, Hong Kong, and Barcelona
I	**Global Environmental Change**	

Recent Volumes in this Series

Volume 246 —New Developments in Lipid–Protein Interactions and Receptor Function
edited by K. W. A. Wirtz, L. Packer, J. Å. Gustafsson, A. E. Evangelopoulos, and J. P. Changeux

Volume 247 —Bone Circulation and Vascularization in Normal and Pathological Conditions
edited by A. Schoutens, J. Arlet, J. W. M. Gardeniers, and S. P. F. Hughes

Volume 248 —Genetic Conservation of Salmonid Fishes
edited by Joseph G. Cloud and Gary H. Thorgaard

Volume 249 —Progress in Electrodermal Research
edited by Jean-Claude Roy, Wolfram Boucsein, Don C. Fowles, and John H. Gruzelier

Volume 250 —Use of Biomarkers in Assessing Health and Environmental Impacts of Chemical Pollutants
edited by Curtis C. Travis

Volume 251 —Ion Flux in Pulmonary Vascular Control
edited by E. Kenneth Weir, Joseph R. Hume, and John T. Reeves

Volume 252 —Advances in Bacterial Paracrystalline Surface Layers
edited by Terry J. Beveridge and Susan F. Koval

Series A: Life Sciences

Advances in Bacterial Paracrystalline Surface Layers

Edited by

Terry J. Beveridge

University of Guelph
Guelph, Ontario, Canada

and

Susan F. Koval

University of Western Ontario
London, Ontario, Canada

Plenum Press
New York and London
Published in cooperation with NATO Scientific Affairs Division

Proceedings of a NATO Advanced Research Workshop on
Advances in Bacterial Paracrystalline Surface Layers,
held September 27–30, 1992,
in London, Ontario, Canada

NATO-PCO-DATA BASE

The electronic index to the NATO ASI Series provides full bibliographical references (with keywords and/or abstracts) to more than 30,000 contributions from international scientists published in all sections of the NATO ASI Series. Access to the NATO-PCO-DATA BASE is possible in two ways:

—via online FILE 128 (NATO-PCO-DATA BASE) hosted by ESRIN, Via Galileo Galilei, I-00044 Frascati, Italy

—via CD-ROM "NATO Science and Technology Disk" with user-friendly retrieval software in English, French, and German (©WTV GmbH and DATAWARE Technologies, Inc. 1989). The CD-ROM also contains the AGARD Aerospace Database.

The CD-ROM can be ordered through any member of the Board of Publishers or through NATO-PCO, Overijse, Belgium.

Library of Congress Cataloging-in-Publication Data

Advances in bacterial paracrystalline surface layers / edited by Terry
  J. Beveridge and Susan F. Koval.
      p.   cm. -- (NATO ASI series. Series A, Life sciences ; v.
  252)
    Includes bibliographical references and index.
    ISBN 0-306-44582-4
    1. Bacterial cell surfaces--Congresses.   I. Beveridge, Terrance
  J. II. Koval, Susan F. III. NATO Advanced Research Workshop on
  Advances in Bacterial Paracrystalline Surface Layers (1992 : London,
  Ontario, Canada) IV. Series.
  QR77.35.A39 1993
  589.9'087'5--dc20   93-5613
                                                              CIP

ISBN 0-306-44582-4

PREFACE

This book is a compilation of the research which was presented during the NATO-Advanced Research Workshop (ARW) entitled "Advances in Bacterial Paracrystalline Surface Layers" held in London, Ontario, Canada during September 27 to 30, 1992. The organizing committee consisted of the two Workshop directors, S.F. Koval and T.J. Beveridge, and H. König, U.B. Sleytr and T.J. Trust; their summary statements about the significance and success of the NATO-ARW are in Chapter 37 of this book.

This was the third international workshop on bacterial S-layers and it demonstrated unequivocally how rapidly research is progressing. The Workshop was made possible by financial support from the North Atlantic Treaty Organization (NATO), the Medical Research Council of Canada (MRC), the Natural Sciences and Engineering Research Council of Canada (NSERC), and the Canadian Bacterial Diseases Network (CBDN) which is a Canadian National Centre of Excellence (NCE). We are very grateful for the support from all of these agencies since their financial aid made it possible to bring to London, Canada a truly international group of S-layer experts. We encouraged the attendance and participation of graduate students, postdoctoral fellows and research associates, and their presentations constitute the "Poster" section of this book. The NATO-ARW was an intense three day workshop held at a delightful secluded location (Spencer Hall) so that the delegates had both formal and informal occasions to interact and evolve new ideas.

Many individuals helped in the compilation of this book, In particular Elizabeth Copland, Beverley Sharpe and Susanne Schultze-Lam of the Department of Microbiology, University of Guelph, worked many hours word processing and formatting the book into camera-ready form. Ms Copland and Anita Evans (Department of Microbiology and Immunology, University of Western Ontario) were instrumental in helping us organize the Workshop.

- Terry J. Beveridge
Susan F. Koval, March, 1993

CONTENTS

I. INTRODUCTION

II. STRUCTURAL ANALYSIS OF S-LAYERS

III. S-LAYERS OF AGRICULTURAL AND ENVIRONMENTAL IMPORTANCE

VI. POSTER PRESENTATIONS

VII. SUMMARY STATEMENTS

I. INTRODUCTION

Chapter 1

A PERSPECTIVE ON S-LAYER RESEARCH

R.G.E. Murray

Department of Microbiology and Immunology
University of Western Ontario
London, Ontario, Canada N6A 5C1

It is almost forty years since Houwink (1953) demonstrated by electron microscopy a monolayer of macromolecules in a regular array, which we would now call an S-layer, on the external surface of the cell wall of an unidentified *Spirillum*. It was an elegant oddity for only a short time; within three years shadow-cast preparations displayed arrays on representatives of four more distinct genera. An ever increasing number of genera and species have been added to the list and continue to be recognized as having hexagonal (p6), tetragonal (p4), trimeric (p3), and oblique (p2) lattice forms; to these must be added some with layer upon layer, even with different symmetries expressed in each layer. Messner and Sleytr (1992) have tabulated from the literature more than 300 strains representing some 93 genera belonging to all the major phylogenetic groups of prokaryotes including *Archaea*. Forty years of research have generated a substantial body of knowledge which is available in several recent reviews on S-layers (Messner and Sleytr, 1992; Koval, 1988; Smit, 1986) as well as on the range of surface components of bacterial walls (Beveridge and Graham, 1991). More specialized areas of study include the self-assembly of units to form S-layers (Sleytr and Messner, 1989), participation of carbohydrates in the macromolecules (Messner and Sleytr, 1991), and the analysis of computer-enhanced micrographic images with generation of 3-D models (Hovmöller et al., 1988). The status of work in hand at the time of the previous workshop is published in a book (Sleytr et al., 1988) and this present volume is for current observations and concepts.

It is hard to describe today our feelings on first seeing an S-layer array and to explain how big a challenge it was to find the means to learn more about these compelling structures. Studies of cellular components were still primitive and micrographs were only a stimulus to speculation, while the analytical methods were cumbersome and peptide separation was at the level of paper chromatography. At the beginning, when the examples of S-layers were few, we derived pleasure and motivation because they seemed to be unique and beautiful. The excitements were yet to come after increasingly elegant images provided by electron microscopy impelled deeper research concerning the nature of the structures. Now, the variety of sophisticated approaches to analysis test our technological capability; they involve a shifting but selective marriage of microscopy, biochemistry, biophysical analysis, physiology, molecular genetics, crystallography, serology, pathology, and taxonomy, all of which developed at different rates. Interdisciplinary research has been applied to S-layers from the beginning and there are more questions than answers arising from our researches.

The form of research that will provide the next quantum-leap forward is not yet apparent, but it is certain that the qualities of creative curiosity, perceptiveness, patience, and readiness to cross disciplinary boundaries will still reward diligent research. This was true in past times when biological studies flowered in the decade 1945-1955; when electron microscopy was a prime stimulus; when bacteriologists were in process of realizing that bacteria needed to be thought of and described as cells (Stanier and van Niel, 1962), and when we learned that cells were susceptible to fractionation for the proper study of structure and function. This does not seem very revolutionary now, but, as we see by the later association with genetics and molecular techniques, this was a powerful if deceptively simple understanding that carried through to all the coming aspects of cell biology. We learned that diverse technologies must be applied if we were learn more about walls and their S-layers.

To us, at the time, the most crucial early development (allowing that there were many simultaneous discoveries) for what we were about to do was provided by the fractionation of bacterial cell walls assisted by the control of procedure provided by electron microscopy. The seminal studies of Salton and Horne (1951) and Salton (1953) showed that electron microscopy complemented biochemistry and that gram-positive and gram-negative cell walls were distinguishable in both biochemical and structural terms. These observations were soon elaborated by electron microscopy of sections showing varied complexity of layering (Kellenberger and Ryter, 1958) and by chemical and enzymatic fractional analysis of walls such as was accomplished by Weidel et al. (1960). Microbiologists were persuaded that significant developments needed the association of biochemistry and electron microscopy and we were gaining experience in our laboratory from a pioneer study on fractionation of bacterial endospores by P.C. Fitz-James (1953) combining biochemistry and structure. Finally, the development of the technique of negative staining (Brenner and Horne, 1959) and its application to electron microscopy of bacteria, viruses, and cell fractions was crucial to the display of macromolecules and their associations, both in _situ_ and in fractions. The studies increased in number and depth in the 1960's; both the philosophy and the techniques were ready. We are still sailing along with the favourable winds they provided. However, we must be prepared to recognize now that the world of biological wonders of today is seductive and productive, but it is becoming excessively repetitive. We should be encouraged to look deeply into what surface science and physical chemistry can provide for new ways to think about how the microenvironments in cells operate and identify emerging problems in our sort of biology. A new intellectual and technical preparedness will be needed now, as it was 40 years ago, in the study of S-layers and bacterial cell walls, which still should form "model systems" for understanding aspects of structure and morphogenesis, regulation, macromolecular interactions in differentiation, and dynamic features of growth.

There are persistent and recurring deficiencies in our understanding of the biological significance of S-layers, some of which will be examined in this volume, and these should be kept in mind even while pursuing research that does not appear to be related to function. As Smit (1986) points out, structures that look alike do not have to function alike. It is tempting to think that location on a surface means that the structure has evolved in response to environmental and, by association, ecological forces (Messner and Sleytr, 1992); this may be true in many cases. However, it is also clear that S-layers are a structural and even an integral component of the cell wall, which serves a complex of functions including being a chaperon to the somewhat sensitive requirements of the periplasm and the plasma membrane. Clearly, this is crucial to the S-layer-walled species of archaeobacteria. So functions may include both outward and inward concerns of the bacterial cell and of its dynamic cell wall. Some species of gram-negative S^+ eubacteria have provided S^- mutants which have not been shown to have other deficiencies.

The great number of species with S-layers probably conserve that structure by necessity in natural environments because S^- strains of S^+ species are seldom found in nature, despite the energy required to synthesize the protein. It seems likely that a lost S-layer has to be replaced if the clone is to avoid serious or fatal consequences. There should be alternative proteins with that potential, but not necessarily competent as an alternate, to fill the function. We should try to identify the sorts of basic properties that these molecules must have to be of service. There are some indications that this supposition might be worth considering. Strains of a single phenotypic and genetic species can show several distinctive lattice forms among their number, e.g. *Bacillus stearothermophilus* (Messner et al., 1984) and *Aquaspirillum serpens* (Boivin et al., 1985). A number of cellular enzymes form aggregates that display regular multimeric units in the electron microscope and at least one of them, when over-produced, forms tubular intracellular assemblies (Elmes et al., 1986) not unlike those formed by some S-layers when shed into the growth medium. Gas vacuoles (vesicles), flotation devices in many pelagic bacteria (Walsby, 1972), are probably the most extreme example of an intracellular assembly system inherent in a single species of protein and, like S-layers, having no interactive function. If species can control their density equilibrium by synthesizing or collapsing their gas vacuoles, it is not too much of a stretch of imagination to consider a response (to light, or pressure, or nutrients) that leads to synthesis or depolymerization of some other protein, even an S-layer. It seems unlikely that S-layers are derived from a single genetically conserved protein from archaean times and the variety of them suggests many independent adaptations. Therefore, the biological reasons for the existence of these structures need to be sought along with any related proteins from that lineage of bacteria. In addition, field studies should be extended to include the broadest range of ecosystems.

The relationships of the S-layer proteins in the families of bacterial proteins have yet to be worked out, and the techniques of molecular biology are likely to be of service. As pointed out long ago by Sleytr and Plohberger (1980), any protein capable of assembling a continuous sheet from anisotropic units (e.g. hydrophobic on one side) could form specific associations with other macromolecules to form complex membranes by cooperating with lipids or together with other macromolecular associations to form structurally and physiologically significant cell components. Certainly, there are S-layers integrated with the outer membrane of gram-negative bacteria to the degree that the membrane must be dissolved away from the protein layer in order to isolate it (Smith and Murray, 1990; Thompson et al., 1982). Of course, these examples in *Aquaspirillum sinuosum* and *Deinococcus radiodurans* may be recent evolutionary adaptations, but it is suggested (Sleytr and Plohberger, 1980) that a simple protein membrane capable of dynamic growth could have initiated a barrier membrane in the early stages of biological evolution. An idea that should not be forgotten.

Our laboratory has studied several bacteria that form multiple S-layers on their surfaces, and it is not clear whether or not this confers some functional significance. One might think that a single layer formed by a single protein species should be sufficient for the needs of a bacterium; an understandable solution to some environmental problem. Even the formation of two overlaying layers, each formed by single but distinct proteins (Kist and Murray, 1984), might not be excessive because nature does not mind expending energy as long as the products work. However, it is hard to understand the formation of complex units requiring the assembly of several distinct proteins, some glycosylated and some not, for a functional layer a distance outside of the cell. This is the case for the punctate layer of *Lampropedia hyalina* (Austin and Murray, 1990). That such complexity evolves is the more mysterious because that layer combines with an underlying perforate layer to enclose the sheets of cells and plays a dynamic role in the enclosure and separation (division) of the square tablets of cells (Chapman et al., 1963) despite being some distance outside of the essential cell wall. So there are deeper mysteries.

S-layers provide for some of a number of bacterial strategies utilizing wall components to reduce the impact of dire environmental influences (Koval and Murray, 1986). Their diverse functions (see Messner and Sleytr, 1992 for an extensive analysis), varying among taxa of bacteria, must have some strong selective value if the S-layer is to persist. Some recognized functions may co-exist in many bacteria to form a barrier against predators, a sieve that retains useful and excludes damaging macromolecules, and a promoter of cell associations or adhesion to surfaces, but it is equally likely that the real basis for evolutionary selection is cryptic in each species and appropriate to their natural environment. It would have to be actively sought in their natural environments. Observations made on the efficacy of the barrier function against *Bdellovibrio bacteriovorus* as a bacterial predator, grazing protozoa, and some bacteriophages (S.F. Koval, Chapter 9) support this thesis as do the studies on roles in virulence and pathogenicity (M.J. Blaser and W.W. Kay, Chapters 15 and 17). The possible functions are many, varied, incompletely surveyed, and only partially understood. Clarification of the nature of functions is an important goal because of the remarkable ubiquity of these structured surfaces.

The study of the roles of S-layers as surface components having functions should be integrated with studies on a wider range of other functional structures associated with cell walls. Some of them are structured assemblies of proteins (such as pili, fimbriae, spinae, and flagella); others may be less ordered surface proteins of high molecular mass, or integral proteins for enzymatic or transport functions but with uncertain physical associations in wall structure. It may be too limiting to expect all the complex S-layers to be crystalline, and there may be some with a mosaic or fractal ordering. We have observed an arrangement that might be termed "regular irregularity" in *Leptotrichia buccalis* (Listgarten and Lai, 1975; R.G.E. Murray, unpublished observations). This particular surface is specifically reactive, e.g., recognizing and adhering to a specific sugar moiety, for the sole purpose of close physical association with other cells while the main barrier/filter function remains and is essentially non-reactive. Although S-layers may serve generally to cover-up and protect a functional macromolecular (and antigenic) mosaic that lies underneath, some components, e.g., side-chains of lipopolysaccharide, may protrude through the meshwork (Chart et al., 1984) to confuse observations but add to function in the disease state.

Our interest in diseases affecting our lives, directly or indirectly, ensures continued research on the importance of surface components in terms of pathogenicity and virulence. Whether or not some of the antigenic variations involved in human disease involve S-layers is not established. However, some of the pathogens have high molecular weight protein superficial components suspected of complicity even though an S-layer is not present. None so far give the clear example of a role for S-layers provided by *Aeromonas salmonicida* and relatives in causing disease in salmonid fish (see T.J. Trust and W.W. Kay, Chapters 16 and 15), but S-layers have been identified on a number of pathogens (e.g., species of *Campylobacter, Cardiobacterium, Clostridium,* and *Bacillus*) and roles in disease may be defined. There are opportunities, perhaps, that need further exploration. An example is the recurrent fevers due to species of *Borrelia* which, like the S-layer equipped trypanosome parasites, must shift their surface antigens to evade the effects of antibody. There are other spirochaetes, e.g., *Leptospira interrogans*, which have highly host-specific serovars. In the plant pathogenic bacteria, there are highly host-specific pathovars of genetically defined species, whose surface components are likely to be involved. So there is still much to be done in this field that is not so very different from understanding what is involved in a defined species managing to produce variants with a different S-layer protein.

It is my prejudice that all research on bacteria and their surfaces must retain an awareness of the biological significance. The techniques of molecular biology and genetics will continue to be productive and may well provide indications of otherwise inaccessible relationships among genes and their products. However, productive molecular studies need a precise biological phenomenon, a special circumstance, or a product or process identified as significant to provide the essential basis for the research. It is probable that S-layers will be recognized on more and more species of bacteria and most of these will be unexciting variations on a theme. Further understanding needs concentration on exemplary bacteria and increased awareness and concern for the physiology and interactions of bacteria in their ecological niches. Such a breadth of interests and involvements (happily this exists despite the tendency to focus on S-layer structure) is essential to the recognition of stimulating novelties and fresh associations which will stimulate our researches.

Despite the incredible amount of work done on model systems concerning the regulation and behaviour, in sickness and in health, of the murein and some other components of bacterial walls, there is still much to learn about how the integrated and complex structure works. The S-layer forms a defined and productive part of this overall study and has some remarkable properties: elegance, multiple functions, influential associations, accessibility and variety. Diverse disciplines come into play, increase our fascination, and provide for an exemplary study in biology.

It is a particular pleasure to me and to my microbiological colleagues from the University of Western Ontario and University of Guelph, the organizers of this meeting, to hold the meeting here because of the years spent in our Departments of Microbiology studying S-layers together with a number of graduate students, post-doctoral fellows, and visiting scientists. Now we expand our contacts in the family of scientists by bringing together an international group experienced in the broadest range of research on S-layers. We are all the beneficiaries of this exchange of information and understanding derived from research and thought.

REFERENCES

Austin, J.W., and Murray, R.G.E., 1990, Isolation and *in vitro* assembly of the components of the outer S-layer of *Lampropedia hyalina, J. Bacteriol.* 173:3681.
Beveridge, T.J., and Graham, L., 1991, Surface layers of bacteria, *Microbiol. Rev.* 55:684.
Boivin, M.F., Morris, V.L., Lee-Chan, E.C.M., and Murray, R.G.E., 1985, Deoxyribonucleic acid relatedness between selected members of the genus *Aquaspirillum* by slot/blot hybridization: *Aquaspirillum serpens* (Mueller 1786) Hylemon, Wells, Krieg, and Jannasch 1973 emended to included *Aquaspirillum*

bengal as a subjective synonym, *Int. J. System. Bacteriol.* 35:512.

Brenner, S., and Horne, R.W., 1959, A negative staining method for high resolution electron microscopy of viruses, *Biochim. Biophys. Acta* 34:103.

Chapman, J.A., Murray, R.G.E., and Salton, M.R.J., 1963, The surface anatomy of *Lampropedia hyalina*, *Proc. Roy. Soc. London Ser. B.*, 158:498.

Chart, H., Shaw, D.H., Ishiguro, E.E., and Trust, T.J., 1984, Structural and immunochemical homogeneity of *Aeromonas salmonicida* lipopolysaccharide, *J. Bacteriol.* 158:16.

Elmes, M.L., Scraba, D.G., and Weiner, J.H., 1986, Isolation and characterization of the tubular organelles induced by fumarate reductase overproduction in *Escherichia coli*, *J. Gen. Microbiol.* 132:1429.

Fitz-James, P.C., 1953, The structure of spores as revealed by mechanical disruption, *J. Bacteriol.* 66:312.

Houwink, A.L., 1953, A macromolecular mono-layer in the cell wall of *Spirillum* spec., *Biochim. Biophys. Acta* 10:360.

Hovmöller, S., Sjögren, A., and Wang, D.N., 1988, The structure of crystalline bacterial surface layers, *Prog. Biophys. Molec. Biol.* 51:131.

Kellenberger, E., and Ryter, A., 1958, Cell wall and cytoplasmic membranes of *Escherichia coli*, *J. Biophys. Biochem. Cytol.* 4:323.

Kist, M.L., and Murray, R.G.E., 1981, Cell wall proteins of *Aquaspirillum serpens*, *J. Bacteriol.* 146:1083.

Koval, S.F., and Murray, R.G.E., 1986, The superficial protein arrays on bacteria, *Microbiol. Sci.* 3:357.

Koval, S.F., 1988, Paracrystalline protein surface arrays on bacteria, *Can. J. Microbiol.* 34:407.

Listgarten, M.A., and Lai, C.H., 1975, Unusual cell wall ultrastructure of *Leptotrichia buccalis*, *J. Bacteriol.* 123:747.

Messner, P., Hollaus, F., and Sleytr, U.B., 1984, Paracrystalline cell wall surface layers of different *Bacillus stearothermophilus* strains, *Int. J. Syst. Bacteriol.* 34:202.

Messner, P., and Sleytr, U.B., 1991, Bacterial surface layer glycoproteins, *Glycobiol.* 1:545.

Messner, P., and Sleytr, U.B., 1992, Crystalline bacterial cell-surface layers, *Adv. Microbial. Physiol.* 33:213.

Murray, R.G.E., 1963, Role of superficial structures in the characteristic morphology of *Lampropedia hyalina*, *Can. J. Microbiol.* 9:593.

Salton, M.R.J., and Horne, R.W., 1951, Studies on the bacterial cell wall. II. Methods of preparation and some properties of cell wall, *Biochim. Biophys. Acta* 7:177.

Salton, M.R.J., 1953, Studies of the bacterial cell wall. IV. The composition of the cell walls of some gram-positive and gram-negative bacteria, *Biochim. Biophys. Acta* 10:512.

Sleytr, U.B., and Plohberger, R., 1980, The dynamic process of assembly of two dimensional arrays of macromolecules on bacterial cell walls, *in*: "Electron Microscopy at Molecular Dimensions", W. Baumeister and W. Vogell, ed., pp.36-47, Springer-Verlag, Berlin.

Sleytr, U.B., Messner, P., Pum, D., and Sára, M. (ed.), 1988, "Crystalline Bacterial Cell Surface Layers", Springer-Verlag, Berlin.

Sleytr, U.B., and Messner, P., 1989, Self-assembly of crystalline bacterial cell surface layers (S-layers), *in*: "Electron Microscopy of Subcellular Dynamics", H. Plattner, ed., pp. 13-31. CRC Press, Boca Raton.

Smit, J., 1986, Protein surface layers, *in*: "Bacterial Outer Membranes as Model Systems", M. Inouye, ed., pp. 343-376, John Wiley and Sons, Chichester.

Smith, S.H., and Murray, R.G.E., 1990, The structure and associations of the double

S-layer on the cell wall of *Aquaspirillum sinuosum, Can. J. Microbiol.* 36:327.

Stanier, R.Y., and van Niel, C.B., 1962, The concept of a bacterium. *Arch. Microbiol.* 42:17.

Thompson, B.G., Murray, R.G.E., and Boyce, J.F., 1982, The association of the surface array and the outer membrane of *Deinococcus radiodurans, Can. J. Microbiol.* 28:1081.

Walsby, A.E., 1972, Structure and function of gas vacuoles, Bacteriol. Rev. 36:1.

Weidel, W., Frank, H., and Martin, H.H., 1960, The rigid layer of the cell wall of *Escherichia coli* strain B, *J. Gen. Microbiol.* 22:158.

II. STRUCTURAL ANALYSIS OF S-LAYERS

Chapter 2

CRYSTALLOGRAPHIC IMAGE PROCESSING APPLICATIONS FOR S-LAYERS

Sven Hovmöller

Structural Chemistry
Stockholm University
Stockholm, Sweden

INTRODUCTION

The three-dimensional structures of many crystalline bacterial surface layers (S-layers) have been studied by electron microscopy. Protein molecules are composed of light elements (mainly C,O,N and H) which give very little contrast and are so sensitive to radiation damage in the electron microscope that it is impossible to observe many details in a protein molecule before it has been destroyed by the electron beam. There are two ways to overcome this dilemma. One is to embed the protein in a staining medium which both increases the contrast and resists radiation damage. The other way is to take micrographs using very low electron doses, before the protein molecules are completely destroyed. Images of hundreds of these "noisy" protein molecules can then be averaged into one image. These two methods can be used individually or in combination with one another.

The process of adding the diffuse information obtained from many individual protein molecules into one image has to be performed with great care. The relative orientations and positions of the individual molecules must be known with very high accuracy. In the case of crystalline proteins this procedure is greatly facilitated because all protein monomers are oriented the same way or are related by a known crystallographic symmetry such as, for example, a 4- or 6-fold axis. The positions of the different protein molecules are also easily determined; they are regularly spaced by one unit cell. The procedure for structure determination of crystalline objects by applying crystallographic techniques to electron microscopy data is called Crystallographic Image Processing (CIP). It has been used to determine the 3D structure of several S-layers to intermediate resolution (Baumeister and Engelhardt, 1987; Hovmöller et al., 1988). Recent advances in computer hardware and software has made CIP considerably easier, faster and cheaper to apply (Hovmöller, 1992).

Advances in Bacterial Paracrystalline Surface Layers
Edited by T.J. Beveridge and S.F. Koval, Plenum Press, New York, 1993

CRYSTALLINE NATURE OF S-LAYERS

A crystal is an object with components that repeat periodically in an ordered way.Usually when we think about crystals, we think about 3D crystals, but crystals may also be oriented into two-dimensional planes or even one-dimensional lines. The S-layers found on many bacteria are crystalline since the components, the protein monomers, are repeated in a periodic fashion. Many molecules, including proteins, crystallize if they are obtained in a pure and concentrated form. All the protein molecules in a pure sample are equal in terms of sequence and environment, and so will have identical folding structure. In an S-layer the protein molecules are in contact with two very different environments; on the inside with the bacterial membrane or wall and on the outside with the exterior solution. There will be a strong tendency for the protein molecules to orient with one specific surface towards the bacterium with an opposite surface facing outwards. The only remaining degree of freedom for the S-layer monomers is the rotation around an axis perpendicular to the membrane. When the protein molecules are packed together there will be a few ways of contact between adjacent monomers which are favoured over others for steric and electrostatic reasons. If new monomers are added at a pace which is slow enough to give the monomers time to find the optimal contact, a crystalline order will result.

Temperature is an important factor for crystallization. The most well-ordered crystals are formed if crystal growth is performed near the melting point of the crystal. At higher temperatures the thermal motions of the molecules dominate over the attractive forces between molecules, and the components move about in a liquid fashion. If, on the other hand, the temperature is much lower and the thermal motion is so small that a protein monomer which has come in contact with another monomer in an orientation which is not the optimal in terms of strong interaction, then the monomer will remain at that suboptimal position. The result is an aggregate without long-range order.

CRYSTAL STRUCTURE DETERMINATION BY X-RAY CRYSTALLOGRAPHY

The classical method for structure determination of molecules, including proteins, is X-ray crystallography. With X-ray crystallography very accurate results can be obtained. The atomic positions of thousands of atoms is determined in proteins with an accuracy of about 0.02 nm, i.e., a tenth of a typical interatomic distance. The only problem is the strict requirement on the sample; it has to be a 3D crystal with all three dimensions being about 100 μm or bigger. This is very much larger than the size of an S-layer, which is limited in two dimensions by the size of the bacterium to one to a few micrometers. In the third dimension, an S-layer is just a single layer of molecules rather than ~10000 which would have been needed to reach a thickness of 100 μm. In principle this problem can be overcome by purifying the S-layer protein and then recrystallizing it in vitro, as has been done with some 400 water soluble proteins (Brändén and Tooze, 1991) and even with a few membrane proteins (Deisenhofer et al., 1984).

Unfortunately no S-layer protein crystals suitable for X-ray diffraction studies have yet been made. There may be specific problems in obtaining such crystals, because it can be suspected that the S-layer proteins are tailor-made to make only single layers with strong interactions within the plane of the crystal. However, hydrophobic and amphiphilic membrane proteins pose a similar problem and that has been overcome by trying out specific conditions for crystallization, e.g., detergents can be incorporated into the crystallization medium. There is no reason why it should not be possible to find conditions in which S-layer proteins can also form 3D crystals.

STRUCTURE DETERMINATION BY ELECTRON MICROSCOPY

Electrons interact about 10000 times more strongly with matter than do X-rays. As a consequence an acceptable signal can be obtained even with a very small crystal. The problems of radiation sensitivity and low contrast for proteins studied by electron microscopy are usually overcome by adding a heavy metal stain to the preparation. With this technique, usually performed by negative staining of isolated S-layers, the crystal may be as small as 10 by 10 unit cells. Although this method is used for nearly all structure determinations of S-layers, it has a serious limitation; the protein molecule itself is not visualized, only the staining medium. This means that we see only the outside imprint of the protein in the stain and no internal structure. The resolution is limited to about 2.0 nm, far from the 0.3 nm resolution needed for atomic resolution. Yet there is still a lot of interesting biological information at this resolution. The symmetry and packing arrangement of monomers forming the S-layers as well as the rough shape of the protein molecules can be determined, and it is also easy to make comparisons between S-layers of different bacterial species (Baumeister and Engelhardt, 1987; Hovmöller et al., 1988).

It is possible to reach atomic resolution of 2D protein crystals with electron microscopy, as has been shown for bacteriorhodopsin (Henderson et al., 1990). For such high resolution quite large and well ordered 2D crystals are needed. The crystals need to be several micrometers in diameter, i.e., containing tens of thousands of protein monomers. In order to see the internal structure of the proteins, no stain can be added to the specimen. The crystals may be protected from drying out and disrupting in the vacuum of the electron microscope either by the addition of a glucose solution (which can be thought of as a polymeric water) or by freezing the sample. Over the last years, cryotechniques have become the most widely used. When unstained proteins are viewed in the electron microscope only very low electron doses can be used before the protein is destroyed. The typical limiting dose is about 1 electron per 0.1nm^2. When such a small dose is applied, the film looks almost unexposed, with an optical density of about 0.1 O.D. units compared to 1.0 which is the density for normally exposed films. Furthermore the signal is less than 10% when compared to the noise in these images. When metal staining is used the protein is destroyed at the same rate as it is for unstained specimens. This means that the protein has disappeared before the focus has been adjusted, and what remains is only the mould of very stable metal oxide formed from the stain.

The structure of the outer membrane-bound trimeric porin protein has been determined to high resolution independently by electron crystallography (Jap et al., 1990) and X-ray crystallography (Weiss et al., 1990). It is comforting to note the striking similarity of the two structure determinations. From the structure determination by X-ray crystallography, it was shown that the monomers in the trimer are mainly made up of a 16-stranded β-barrel within the membrane where every second amino acid is hydrophobic and pointing out towards the lipids of the membrane. The other amino acids are generally hydrophilic and point inwards to the porin channel. Porin is not considered to be an S-layer protein, since S-layers are not integral membrane proteins, but it is still important for the S-layer field of research that the porin structure has been determined. From a functional point of view, porins are related to S-layer proteins in that both form channels. Porins are the most abundant proteins in the outer membranes of gram-negative bacteria. Quite often porins are so concentrated that they crystallize in vivo and can easily be mistaken for S-layers. For this reason tightly membrane-bound "S-layers" with p3 symmetry and a unit cell around 8 nm on gram-negative bacteria should always be suspected to be porins. Finally, the molecular architecture of porin represents an interesting new way of protein folding, completely different from the large group of integral membrane proteins which have only α-helices transversing the membranes.

CRYSTALLOGRAPHIC IMAGE PROCESSING

Electron micrographs of both stained and unstained protein crystals are very noisy. The signal-to-noise ratio can be improved considerably by averaging over many unit cells. The periodic information is the same in each unit cell and so increases linearly with the number of unit cells used for the averaging. The non-periodic noise only increases by the square root of the number of unit cells used. If 100 unit cells are averaged the random noise is reduced to one tenth. This procedure is called lattice averaging and may be quite useful, but it is not sufficient. There are still effects of non-random noise in the image, and there is also a fall-off of information at higher resolution.

The non-random noise in electron micrographs is caused by electron optical effects such as misalignment of the electron optics, wrong focus and astigmatism. Since these effects are the same over the whole area of the micrograph, they will not cancel out when many unit cells are averaged. There are different ways of eliminating these problems. The first, of course, is to improve the imaging conditions by making sure the microscope is perfectly aligned. There are now available procedures for automatic alignment of electron microscopes. A further improvement can be obtained if the crystal has some symmetry. Microscope misalignment because of beam tilt and astigmatism will have different effects along different directions of the image. If the crystal symmetry is known and imposed, a large part of these distortions can be eliminated. This is the main power of CIP. In many cases the imposition of crystal symmetry adds greater improvement to the image than that which can be gained by lattice averaging alone.

Crystallographic image processing has been used for over 25 years as a method for improving the quality of information about protein structures obtained by electron microscopy. In the early days, micrographs were processed manually by optical techniques. In fact, one of the first crystalline structures to be investigated by electron microscopy and optical filtering was the S-layer of *Bacillus polymyxa* (Finch et al., 1967). Since 1975 computerized image processing has been applied and there are some advantages to this. The data is obtained in a quantitative form and it becomes possible to impose the crystal symmetry. A drawback with computerized image processing was the need for expensive equipment such as microdensitometers and large computers, but this is no longer so.

CRISP: CRYSTALLOGRAPHIC IMAGE PROCESSING ON A PERSONAL COMPUTER

During the last few years several factors have made crystallographic image processing (CIP) much easier and cheaper to use. Modern personal computers (PC) are as powerful as the much larger computers such as the VAX, which were used in

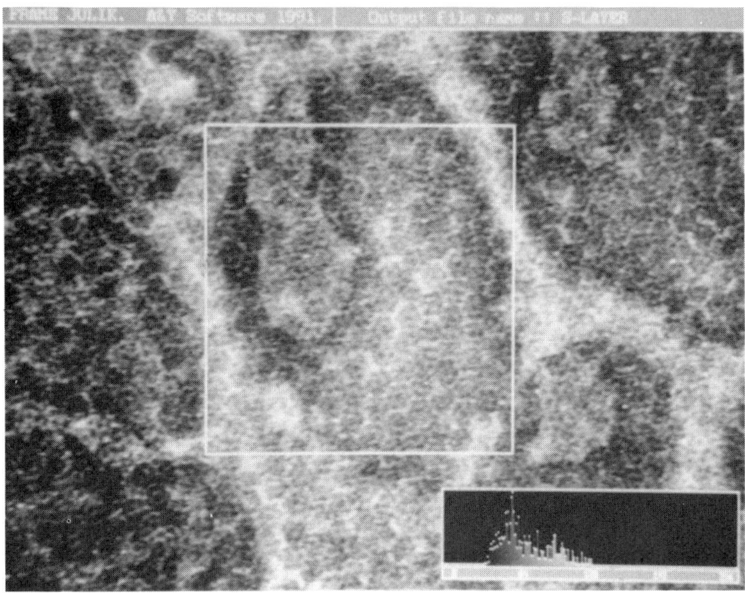

Figure 1. An electron micrograph of a negatively stained S-layer from *Bacteroides buccae* (Sjögren et al., 1985) scanned by a CCD camera, digitized by a frame grabber and displayed on a personal computer.

the 1980s. Yet the price of a PC is only a few percent of what a VAX used to cost. Furthermore, the slow and expensive microdensitometers can now be replaced by small, fast and cheap CCD cameras which provide the PC with a digitized image in a fraction of a second via a frame grabber. Finally, the software for CIP has also now been developed and it is possible to buy all the software needed in a complete package. This eliminates the step of writing or modifying all the computer programs in each laboratory. With the image processing system CRISP (Hovmöller, 1992) the complete processing of a 2D image can be performed in about 15 minutes as compared to 5 hours with the previous generation of software and equipment.

Digitizing electron micrographs with a CCD camera

With a CCD camera (CCD = Charge Coupled Device), images are digitized in a fraction of a second. While microdensitometers may give 2000 grey levels or more, the standard (8-bit) CCD cameras only have 256 grey levels and only about half are in the linear range. Surprisingly, this is not a serious limitation, at least for images with high contrast, such as negatively stained S-layers. An example of the capturing of an EM image of a bacterial surface layer by CCD camera and frame grabber is shown in Fig. 1. The image has a size of 640 x 480 pixels and from it a region of 256 x 256 pixels is selected for image processing by CRISP. In the present version of CRISP this is the maximum image size that can be handled, but a new version is being developed for larger areas.

Once inside CRISP, the images can be trimmed. Areas outside the S-layer or disordered areas are easily cut away. These and practically all other operations in CRISP are carried out by clicking a button on the computer mouse.

Fourier transformation

The Fourier transform is a very powerful mathematical tool for separating periodic and non-periodic features. For an image of an S-layer this means that thesignal from the protein, which is periodic, can be separated out from the random noise, such as contributions from the carbon film, precipitated proteins, scratches on the film, etc. The periodic information will fall on a regular lattice of points in the Fourier transform, while the irregular noise falls everywhere in the transform area (Fig.2).

Only small areas around each lattice point in the Fourier transform are kept, and in this way the random noise is largely eliminated. In the newest version of CRISP (Version 1.2) a Fourier transform of a 256 x 256 image is calculated in only 4 seconds on an IBM-compatible 486 PC.

Symmetry determination

It is very important to know the symmetry of a crystal when CIP is to be applied. It is bad to either underestimate or overestimate the symmetry. If a lower order of symmetry is assigned, we miss the opportunity to average over identical molecules; the result will be noisier than needed and the optical distortions may not be fully eliminated. If too high a symmetry is applied, protein molecules which are not identical will be averaged; the result is a smeared image. For S-layers, the only symmetries we expect to find are p1, p2, p3, p4 and p6 (Hovmöller et al., 1988).

When dealing with different symmetries, different rules apply when identifying and relating the various reflections in the transform. Often the phases are of higher quality than the amplitudes in EM images and symmetry determination is preferentially carried out on phases. All the symmetry rules for all 17 possible plane group symmetries are programmed into CRISP. One very interesting example of an

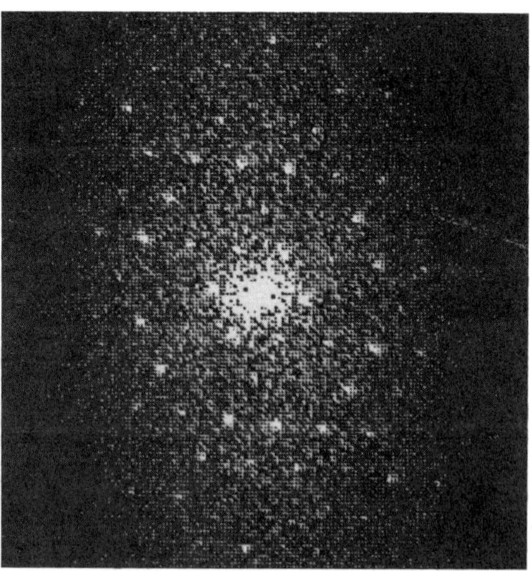

Figure 2. The Fourier transform of the selected area in Figure 1. The information about the regularly arranged S-layer subunits is concentrated into the sharp lattice points.

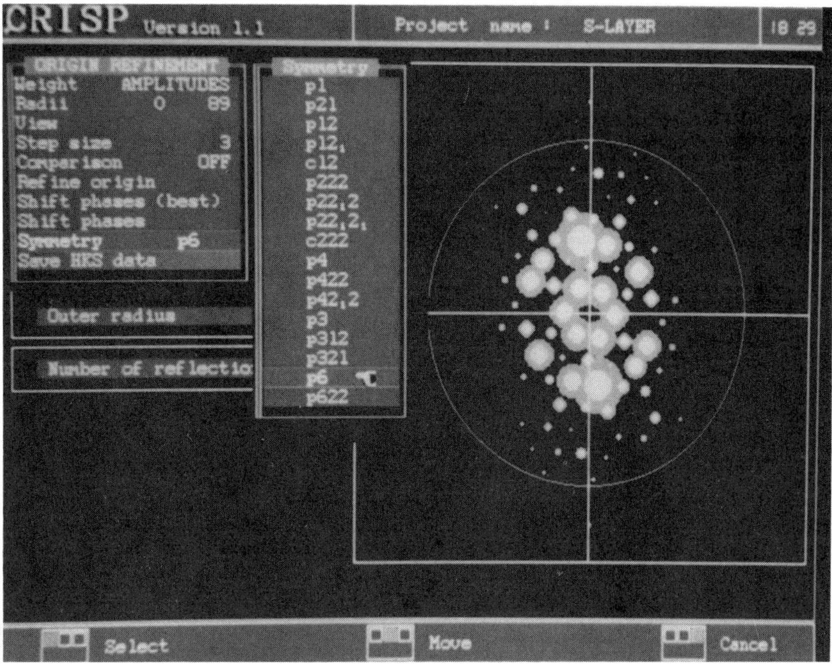

Figure 3. The symmetry of a crystal is determined in CRISP by calculating the phase residual for different plane groups. Here p6 is being selected.

error in the symmetry determination of an S-layer is that of *Sulfolobus acidocaldarius*, which was first assigned p6 symmetry (Deatherage et al., 1983). Later Baumeister and coworkers (1991) showed that this S-layer had only p3 symmetry. The reason for the mistake was that the S-layer is made up of rather small patches of p3 symmetry, and these patches grow together with invisible seams between them. If the processed area covers two or more patches with different orientations, the averaged lattice looks hexagonal with p6 symmetry.

Images can be improved considerably if the correct symmetry has been determined and the rules about amplitude and phase relations have been applied (Fig.3). All these procedures, including finding the exact positions of the symmetry elements in the unit cell (the so-called origin refinement) are started by a click on the mouse in CRISP.

Density map

The final density map after applying the symmetry can be displayed on the computer display, as seen in Fig. 4, or on paper. The scale and colours can be changed, and the density map may be displayed in grey-tones or as contour lines.

We are currently working on including the procedures needed for combining many EM images, taken at different tilt angles, into a single 3D density map into the CRISP image processing system.

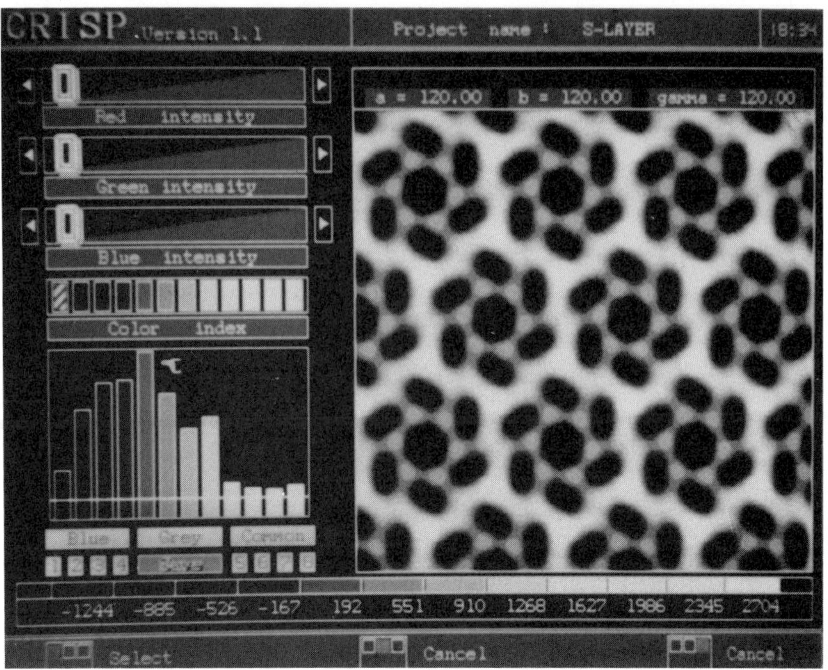

Figure 4. Density map of the S-layer of *B. buccae* after Fourier filtering and imposing the crystal symmetry p6. Protein is white.

ACKNOWLEDGEMENTS

The Swedish Natural Science Research Council (NFR), Sven and Ebba-Christina Hagbergs stiftelse are acknowledged for financial support.

REFERENCES

Baumeister, W., Lembcke, G., Dürr, R. and Phipps, B., 1991, Electron crystallography of bacterial surface proteins, *in*: "Electron Crystallography of Organic Molecules", J.R.Fryer and D.L.Dorset, eds., Kluwer Academic Publishers, Dordrecht.

Baumeister, W. and Engelhardt, H., 1987, Three-dimensional structure of bacterial surface layers, *in*: "Electron Microscopy of Proteins", Vol. 6, R. Harris, ed., Academic Press, London.

Brändén, C.I. and Tooze, J., 1991, "Introduction to Protein Structure", Garland, New York.

Deatherage, J.F., Taylor, K.A. and Amos, L.A., 1983, Three-dimensional arrangement of the cell wall protein of *Sulfolobus acidocaldarius, J. Mol. Biol.* 167:823.

Deisenhofer, J., Epp, O., Miki, K., Huber, R. and Michel, H., 1984, X-ray structure analysis of a membrane protein complex: Electron density map at 3Å resolution and a model of the chromophores of the photosynthetic reaction center from *Rhodopseudomonas viridis, J. Mol. Biol.* 180:385.

Finch, J.T., Klug, A. and Nermut, M.V., 1967, The structure of the macromolecular units on the cell walls of *Bacillus polymxa*, *J. Cell Sci.* 2:587.

Henderson, R., Baldwin, J.M., Ceska, T.A., Zemlin, F., Beckman, E. and Downing, K.H., 1990, Model for the structure of bacteriorhodopsin based on high-resolution electron cryo-microscopy, *J. Mol. Biol.* 213:899.

Hovmöller, S., 1992, CRISP: crystallographic image processing on a personal computer, *Ultramicroscopy* 41:121.

Hovmöller, S., Sjögren, A. and Wang, D.N., 1988, The structure of crystalline bacterial surface layers, *Progr. Biophys. Molec. Biol.* 51:131.

Jap, B.K., Downing, K.H. and Walian, P.J., 1990, Structure of PhoE porin in projection at 3.5 Å resolution, *J. Struc. Biol.* 103:57.

Sjögren, A., Hovmöller, S., Farrants, G., Ranta, H., Haapasalo, M., Ranta, K., and Lounatmaa, K., 1985, Structures of two different surface layers found in six Bacteroides strains, *J. Bacteriol.* 164:1278.

Weiss, M.S., Wacker, T., Weckesser, J., Welte, W. and Schulz, G.E., 1990, The three-dimensional structure of porin from *Rhodobacter capsulatus* at 3 Å resolution, *FEBS Lett.* 267:268.

Chapter 3

STRUCTURES OF PARACRYSTALLINE PROTEIN LAYERS FROM THE HYPERTHERMOPHILIC ARCHAEOBACTERIUM *PYROBACULUM*

Barry M. Phipps

Institute for Biological Sciences
National Research Council
Ottawa, Ontario, Canada

INTRODUCTION

Members of the recently discovered archaeobacterial genus *Pyrobaculum* are sulfur-reducing hyperthermophiles which grow optimally at 100°C (Huber et al., 1987). All known strains were isolated from superheated, sulfurous fresh water springs or from the superheated outflow of a geothermal power plant. The cells are rod-shaped and vary in length up to 8 μm, but their diameter is strikingly uniform. Two strains have been characterized in detail and were found to represent different species, *Pyrobaculum islandicum* (strain GEO3) and *Pyrobaculum organotrophum* (strain H10). *P. islandicum* is a facultative heterotroph capable of chemolitho-autotrophic growth on sulfur, H_2, and CO_2, while *P. organotrophum* is strictly organotrophic. This Chapter reviews the results of investigations into the structure of the cell envelopes of these interesting archaeobacteria (Baumeister et al., 1989, 1990; Phipps et al., 1990, 1991). An element of the cell envelope of most archaeobacteria is a 2D paracrystalline array of protein or glycoprotein subunits, or S-layer, which is often directly adjacent to the cytoplasmic membrane (König, 1988). The array subunit is typically a complex, multi-domain protein and often possesses a putative membrane anchor domain (Baumeister et al., 1989). The unusual structures of *Pyrobaculum* S-layers are described, and the possible functional roles they fulfill are discussed.

ARCHITECTURES OF *PYROBACULUM* ENVELOPES REVEALED BY FREEZE-ETCHING

As one progresses inward from the outside of the cell, freeze-etching of *P. islandicum* GEO3 cells (Fig. 1) reveals an envelope consisting of a fibrillar capsule, an S-layer, and the cytoplasmic membrane. The S-layer exhibits a smooth outer surface, but the inner surface is characterized by distinct projections forming a lattice with hexagonal symmetry and a large lattice constant (Figs. 1B and 1C). Between the membrane and the S-layer is a rather large interspace of uniform width (approx. 17 nm) traversed by regularly spaced projections which appear to be an integral part of the S-layer, and which undoubtedly correspond to those seen on its inner surface (Fig. 1D). The projections seem to make contact with the membrane and may actually insert into it.

A similar architecture is seen in freeze-etched *P. organotrophum* H10 cells (Fig. 2), except that a second S-layer forms the outermost component of the envelope, sandwiching capsular fibrils between it and the inner S-layer. This additional protein array, which displays a kind of pebble-grain texture, also possesses hexagonal symmetry but has a smaller lattice constant (Fig. 2A). In contrast to the

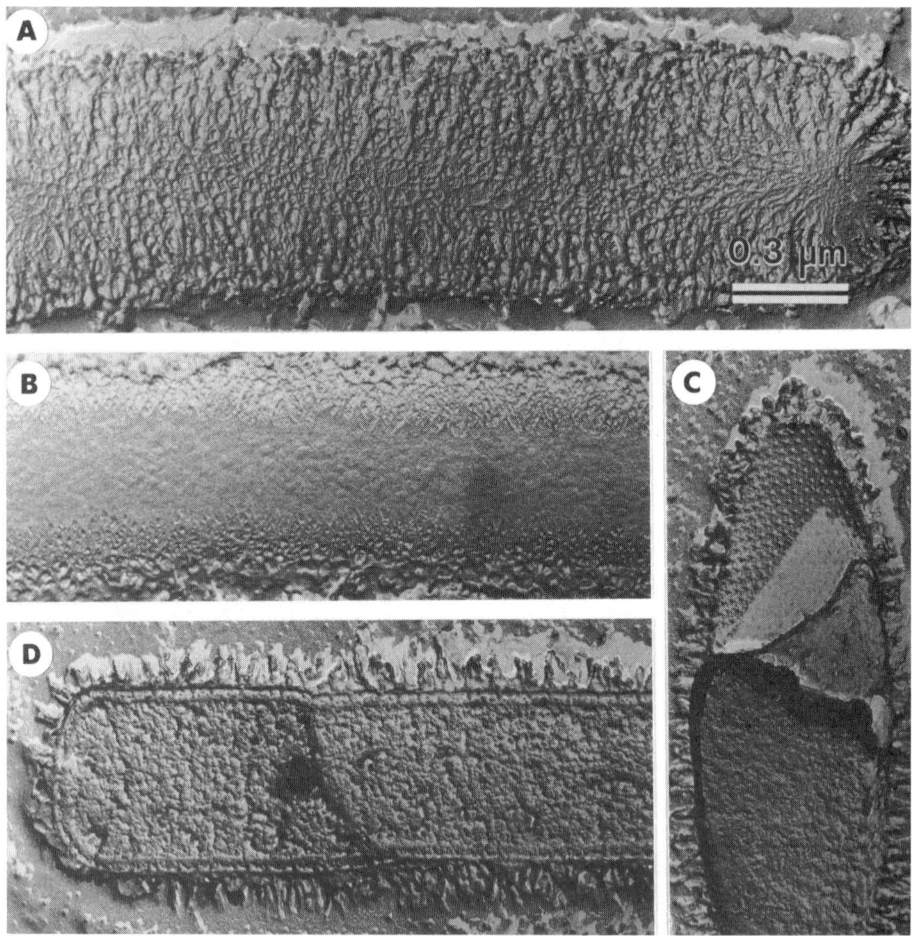

Figure 1. Freeze-etched cells of *P. islandicum* GEO3. (A) Outer surface covered with a fibrous capsule. (B) The smooth outer face of a crystalline protein layer beneath the capsule. (C) Cell in which the cytoplasmic membrane has pulled away, exposing the concave inner surface of the S-layer covered with protrusions forming a p6 lattice. (D) Cross-fractured cell, showing an interspace of uniform width between the membrane and S-layer spanned by regularly spaced elements which seem to be part of the layer and may be inserted into the membrane.

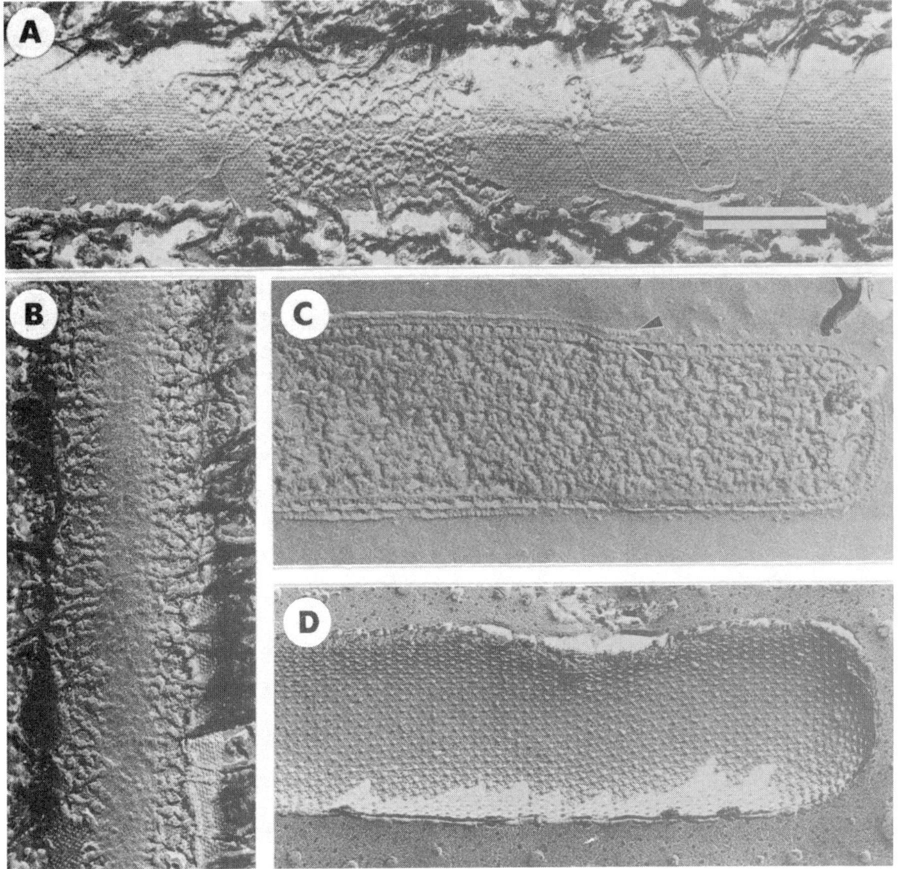

Figure 2. Freeze-etched cells of *P. organotrophum* H10. (A) Outer surface covered with a hexagonal S-layer. Part of the layer is missing, exposing fibrils lying on the smooth outer face of an inner protein array. This face is more clearly visible in (B), where most of the outer S-layer is absent. (C) Cross-fractured cell, showing the inner S-layer with evenly spaced extensions bridging an interspace of uniform width between the layer and the membrane (bottom arrow), and an incomplete outer S-layer, appearing as a row of dots (top arrow). (D) The inner surface of the inner crystalline layer, with hexagonally arranged protrusions. The bar represents 0.3 µm.

inner layer, the outer layer is often incomplete (Figs. 2A-C), suggesting that segments of it tend to be shed, possibly as a result of mechanical stress. The five other known *Pyrobaculum* strains all show a GEO3-type envelope architecture. Thus, the presence of a double S-layer is unique to *P. organotrophum* H10.

ISOLATED S-LAYERS

Pyrobaculum S-layers can be isolated by treatment of envelopes with 2% SDS at 60°C for 1 h. Layers from both strains are very tough, tending to remain as intact sacculi which retain the uniform cylindrical shape of the cell. Metal-shadowed GEO3 sacculi have a smooth outer surface; the inner surface, exposed when a sacculus is torn open, is covered with a p6 arrangement of projections as seen in freeze-etched cells (Fig. 3A). H10 sacculi reveal the presence of two S-layers, an outer layer with a pebble-grain appearance and an inner GEO3-like layer with smooth outer and rough inner surfaces (Figs. 3B and 3C).

STRUCTURE OF *PYROBACULUM ISLANDICUM* GEO3 S-LAYER

When negatively stained, GEO3 sacculi give a Moiré pattern (Fig. 4A) which image analysis shows to result from the superposition of two hexagonal lattices of identical lattice constant which are rotationally misaligned by only a few degrees. These come from the same protein array, which is folded over in the flattened sacculus. Fragments consisting of a single layer, in either of both possible orient-

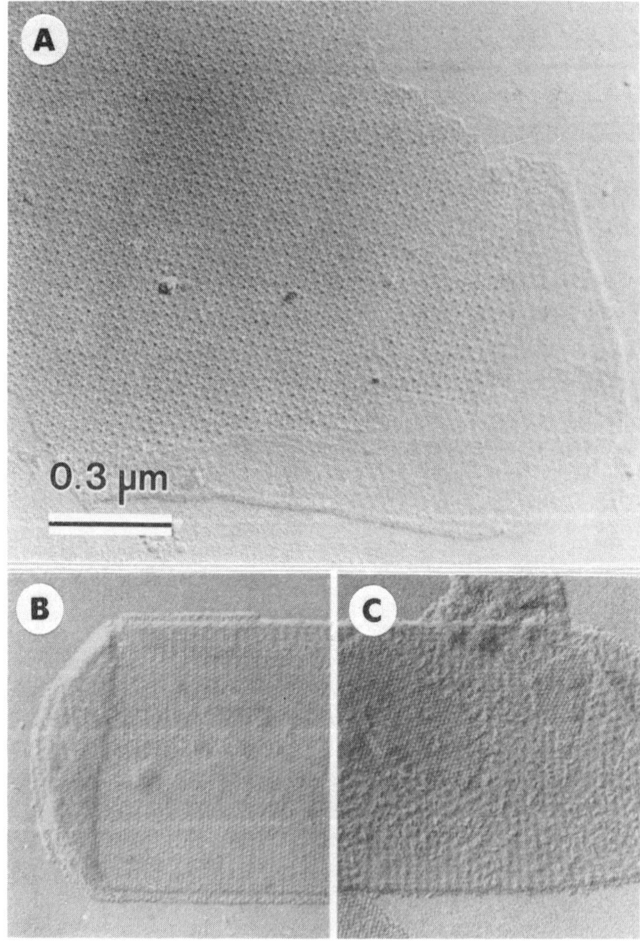

Figure 3. Platinum/carbon-shadowed *Pyrobaculum* S-layers. (A) Large single-layered fragment from *P. islandicum* GEO3. The membrane-directed surface is exposed, revealing a well-ordered p6 lattice characterized by prominent protrusions. The smooth outer face can be seen where the layer is folded over on itself. (B,C) *P. organotrophum* H10 sacculi showing the intact, smooth-surfaced inner layer partially covered by hexagonally arranged outer layer.

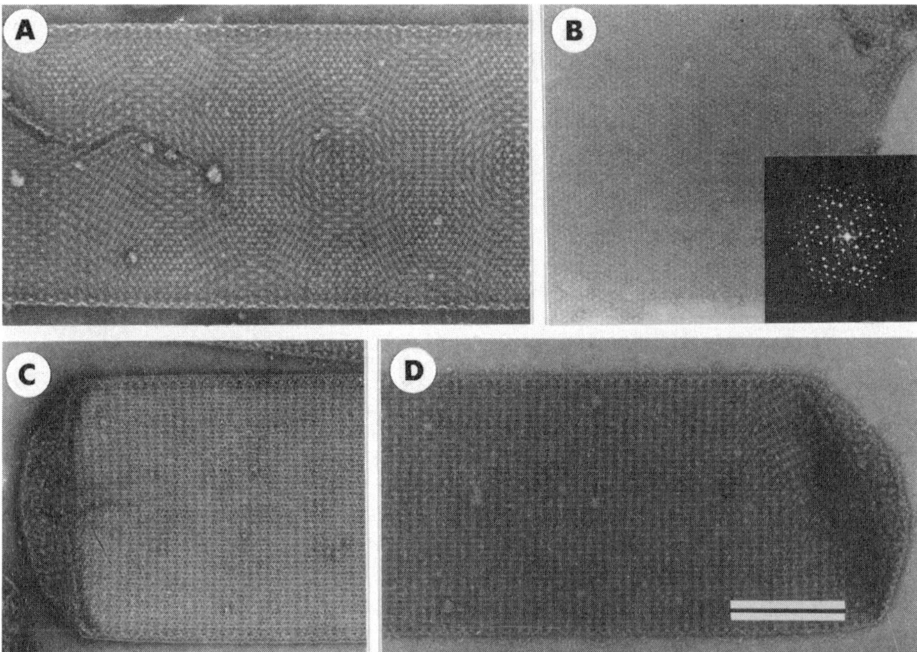

Figure 4. Negatively stained *Pyrobaculum* S-layers. (A) Intact *P. islandicum* GEO3 sacculus. Moiré pattern results from superposition of the top and bottom portions of the same hexagonal array, which are oppositely oriented with respect to the carbon film. (B) Single-layered fragment of GEO3 layer used for the 3D reconstruction shown in Fig. 5, and corresponding diffraction pattern (inset). (C,D) Intact *P. organotrophum* H10 sacculi with (C) and without (D) the outer layer. The rectangular superlattice pattern derives from superposition of the top and bottom portions of the inner array in perfect rotational alignment but translationally out of register, as described in the text. The bar represents 0.3 μm.

ations with respect to the carbon film, can also be found (Fig. 4B). Analysis of lattice positions in single or double layers by cross-correlation indicates that the array is very highly ordered, departing little from perfect crystallinity. From series of micrographs of single-layered fragments, recorded with the specimen at tilt angles ranging from 0° to 80°, two independent 3D reconstructions of the protein array were calculated (Phipps et al., 1990). The combined real space/Fourier space method (Saxton et al., 1984) was employed, in which structural information is extracted from the individual images by correlation averaging (Saxton and Baumeister, 1982), combined in Fourier space, and backtransformed to give a 3D model. One such model is displayed in Fig. 5. It shows a very complex distribution of mass, arranged in a p6 array with an unusually large lattice spacing of 29.9 nm. On the inner surface (facing the cytoplasmic membrane), at the 6-fold symmetry axis, is a solid plug of protein looking rather like a tree stump (Fig. 5B). Radiating out from this are six arms which represent protein monomers. Although the actual outline of the monomer cannot be defined unambiguously, it is clearly composed of several domains and interacts with other monomers across the 6-, 3-, and 2-fold symmetry axes, yielding a highly interconnected structure. We believe each 'stump' represents the base of one of the long projections seen in freeze-etching and shadowing, the rest of the domain disappearing in the reconstruction due to incomplete embedding in stain, variability in its position causing it to be averaged out, or partial collapse of the structure. In contrast, the outer face of the array is relatively smooth (Fig. 5A). A depression at the 6-fold axis could be the beginning of a channel which extends partway or even all the way through the long projection.

The organization and structural features of the GEO3 array are remarkably similar in almost every respect to the S-layer of *Thermoproteus*, another extremely thermophilic sulfur-dependent archaeobacterium with a rod-like shape and constant diameter. Messner et al. (1986) extensively characterized the envelopes of *Thermoproteus tenax* and *T. neutrophilus* by freeze-etching of cells, and metal-shadowing and negative staining of isolated S-layers together with 2D image analysis. Wildhaber and Baumeister (1987) reported a 3D reconstruction of the *T. tenax* S-layer. These studies demonstrated the presence of an interspace crossed by long projections and a hexagonally symmetrical S-layer with the same type of mass distribution and similar lattice spacing to those found for *Pyrobaculum*. In addition, the 3D reconstruction showed a long pillar-like domain extending 22 nm from the inner face of the array. These results clearly suggest a close evolutionary relationship between *Thermoproteus* and *Pyrobaculum*. Although the percentage homology between them determined by DNA-DNA hybridization was judged to be insignificant, it was the same as the percentage homology between two different species of *Thermoproteus* (Huber et al., 1987).

STRUCTURE OF *PYROBACULUM ORGANOTROPHUM* H10 DOUBLE S-LAYER

Negative staining of H10 sacculi typically produces a rectangular superlattice pattern, although the occasional sacculus gives a hexagonal Moiré like that seen for GEO3 (Figs. 4C and 4D). The outer layer, when present, appears as a line of dots or a more or less continuous line at the edge of the sacculus, and a faint triangular

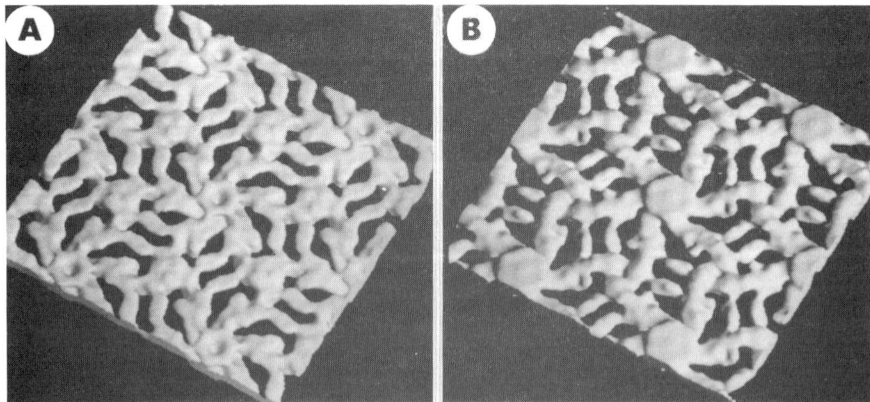

Figure 5. Surface-shaded solid models of the *P. islandicum* GEO3 crystalline protein array, showing the outer surface (A) and inner, membrane-directed surface (B). The lattice centre-to-centre spacing is 29.9 nm.

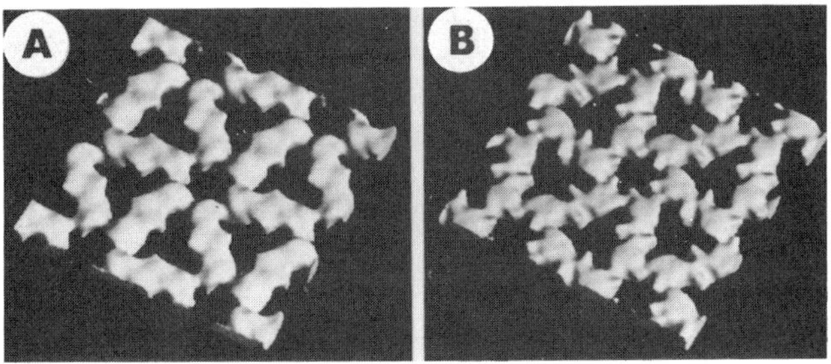

Figure 6. Surface-shaded solid models of the *P. organotrophum* H10 outer protein array, showing the putative outer surface (A) and inner surface (B). The lattice centre-to-centre spacing is 20.6 nm. The models in Fig. 5 and in this figure are shown to the same scale.

pattern over the body of the envelope (Fig. 4C). Patches of outer layer sloughed off from cells exhibit a distinct triangular motif. The image of an intact sacculus is a projection through the superposition of four different protein arrays, the top and bottom lattices of each of the two S-layers. The corresponding Fourier transform contains multiple diffraction patterns, one for each protein array. Using masking methods, it is possible to isolate individual diffraction patterns and analyze the corresponding images by correlation averaging. 2D averages of one of the layers, obtained from fragments and intact sacculi, are very similar to a projection through the 3D reconstruction of the GEO3 array (Figs. 7A and 7B). The arrangement of mass is almost identical and the lattice spacing of 27.9 nm is close to that of GEO3. Thus the H10 inner S-layer and the GEO3 S-layer are very similar at the detailed structural level.

The structure of the outer layer, however, is distinctly different. A 3D reconstruction was obtained from images of an intact sacculus using Fourier space masking (Phipps et al., 1991) and is displayed in Fig. 6. It shows a very simple

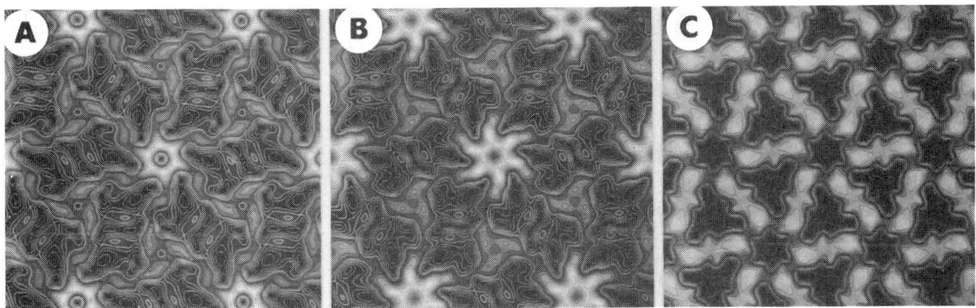

Figure 7. (A) Projection through the 3D reconstruction of the *P. islandicum* GEO3 protein array. (B) Correlation average (six-fold symmetrized) of the *P. organotrophum* H10 inner layer (lattice spacing 27.9 nm). (C) Projection through the 3D reconstruction of the *P. organotrophum* H10 outer layer. Dimensions of each image are 60 × 60 nm.

hexagonal organization of block-like monomers which interact at 6-fold and 2-fold axes, generating a porous triangular network of protein with a lattice spacing of 20.6 nm. An apparently intimate association across the 2-fold axis suggests that the basic organizational unit is a dimer. A side-by-side comparison of projections through the inner and outer layers in Fig. 7 emphasizes the differences between the two structures.

ORGANIZATION OF THE INNER LAYER

Fourier transforms of images of intact H10 sacculi exhibiting the rectangular superlattice pattern contain three diffraction patterns rather than the four expected. This is due to the perfect superposition of diffraction patterns corresponding to the top and bottom lattices of the inner layer, which implies that these arrays are in exact rotational alignment in the sacculus. The fact that a rectangular and not a hexagonal pattern is seen means that there is a relative translational shift between them. Computer simulations in which oppositely oriented inner layer averages are superposed on one another show that the arrays are disposed with one lattice direction exactly parallel to the short axis of the cell, and that they are out of register by one-half of a lattice spacing along that direction (Phipps et al, 1991). This is also shown by the positions of peaks in cross-correlation functions calculated using inner layer averages as references. This means that, over the cylindrical portion of the cell, the inner layer consists of a series of parallel, closed, circular 'hoops' of protein hexamers girdling the cell.

A similar analysis of GEO3 (and H10) sacculi displaying a Moiré pattern shows that one lattice direction is offset a few degrees from the short axis of the cell. Thus one can consider the array to be composed of a small number of parallel helices of hexameric morphological units which wrap around the girth of the cell at a shallow angle. The S-layer of *Thermoproteus* is arranged in a similar fashion. The significance of these modes of organization is not clear. One possibility is that the disposition of one lattice base vector parallel to the short axis of the cell, or nearly so, places the maximum strength of the protein array along the cylindrical radius of the cell. Wildhaber and Baumeister (1987) proposed a 'helical template model' for growth of the surface array in *Thermoproteus* based on the helical arrangement and the demonstration by Messner et al. (1986) that the layer contains a minimal number of lattice faults, all of which are located at the cell poles. According to this hypothesis, new morphological units are added at the ends of the helical strands where they meet the polar caps, at the sites of lattice disclinations. Since new protein is only inserted near the poles, the S-layer and the cell grow in length whilst maintaining a constant diameter.

POSSIBLE FUNCTIONS

The crystalline array of GEO3 and the inner layer of H10 are clearly designed to be mechanically strong and rigid. This is seen in the tendency of isolated layers to remain as sacculi retaining the cylindrical shape of the cell, the order and extensive connectivity of the array, and the positioning of one lattice direction close to the short axis of the cell. It is supported by the finding of a minimal number of lattice faults in the *Thermoproteus* layer, mentioned above. The layer may therefore serve as a kind of exoskeleton to protect the cell from severe mechanical stress. In their biotope, the cells likely encounter powerful shear forces due to violent bubbling of evolved gas, as well as the presence of strong water currents. The rigid layer undoubtedly determines the well-defined cylindrical shape of the cell, as proposed for the *Thermoproteus* layer by Messner et al. (1986).

Given that the long domains projecting from the inner face of the layer probably insert into the cytoplasmic membrane, we suggested that the layer may stabilize the membrane against disintegration at high temperature (Phipps et al., 1990). This may occur in two ways. First, the extensions may physically tether the membrane to the rigid framework of the array. Second, a significant fraction of the membrane lipid will tend to be more highly ordered due to interaction with the hydrophobic protein domains.

The occurrence in *Pyrobaculum* and a wide variety of other archaeobacteria of a well-defined interspace between the cytoplasmic membrane and S-layer, spanned by spacer elements coming from the layer, led us to propose that the delimited volume may serve as the archaeobacterial equivalent of a 'periplasmic space' (Baumeister et al., 1989). In the case of *Pyrobaculum*, the sizes of the holes in the porous S-layer canopy, derived from the 3D reconstructions, suggest a rough molecular weight cut-off for proteins of 20000 Da. Thus the space may house secreted macromolecules involved in the degradation and transport of nutrient molecules or the synthesis and export of surface structures, including the S-layer itself.

Since the layer protein appears to have both transmembrane and surface-exposed domains, another potential function is that of signal transduction between the external milieu and the cytoplasm. In this capacity, it might alert cytoplasmic targets to a variety of environmental signals, such as changes in temperature, pH, ionic strength, or redox state, or the presence of nutrients or noxious substances. The presence of what could be the beginning of a channel in the portion of the long projection furthest from the membrane is reminiscent of connexons in eukaryotic gap junctions, suggesting the possibility of communication between adherent cells, or of regulated ion flow between the cytoplasm and the environment. The formation of 'bacterial connexons' by S-layers was previously hypothesized by Baumeister and Hegerl (1986). Of course, direct interaction of the layers of two GEO3 cells would require some means of bypassing the capsular coat. The presence of the additional protein array in H10 need not pose a problem since patches of it seem to be readily sloughed-off, exposing underlying layers of the cell wall.

The outer array of H10 is likely to engage in one or more surface-related functions, given its location at the outermost surface of the cell (Phipps et al., 1991). One of these is adhesion of cells to inert or biological substrates in their habitat. Another is cell-cell binding or adhesion. H10 cells tend to aggregate into large clumps, while GEO3 cells do not. One occasionally sees what appears to be a specific interaction between outer layers of isolated envelopes in negative stain preparations. Other possible functions include specific binding of ions and organic molecules, as observed for some eubacterial S-layers. Interestingly, in GEO3 a fibrous capsule constitutes the outermost surface, while in H10 this is covered by the outer protein array, suggesting that the role of the outer array may be to provide an alternative to the surface properties supplied by the capsule.

ACKNOWLEDGMENTS

The figures in this article were reproduced with modifications in whole or in part, from Phipps et al. (1990) and Phipps et al. (1991) with permission of Academic Press and Blackwell Scientific Publications, respectively. The author gratefully acknowledges the generous support and encouragement of W. Baumeister (Max-

Planck-Institut für Biochemie, Martinsried, Germany) in whose laboratory this work was carried out.

REFERENCES

Baumeister, W., and Hegerl, R., 1986, Can S-layers make bacterial connexons?, *FEMS Microbiol. Lett.* 36:119.

Baumeister, W., Wildhaber, I., and Phipps, B.M., 1989, Principles of organization in eubacterial and archaebacterial surface proteins, *Can. J. Microbiol.* 35:215.

Baumeister, W., Lembcke, G., Dürr, R., and Phipps, B., 1990, Electron crystallography of bacterial surface proteins, *in*: "Electron Crystallography of Organic Molecules," J.R. Fryer and D.L. Dorset, eds., Kluwer Academic Publishers, Amsterdam.

Huber, R., Kristjansson, J.K., and Stetter, K.O., 1987, *Pyrobaculum* gen. nov., a new genus of neutrophilic, rod-shaped archaebacteria from continental sol-fataras growing optimally at 100°C, *Arch. Microbiol.* 149:95.

König, H., 1988, Archaebacterial cell envelopes, *Can. J. Microbiol.* 34:395.

Messner, P., Pum, D., Sára, M., Stetter, K.O., and Sleytr, U.B., 1986, Ultra-structure of the cell envelope of the archaebacteria *Thermoproteus tenax* and *Thermoproteus neutrophilus*, *J. Bacteriol.* 166:1046.

Phipps, B.M., Engelhardt, H., Huber, R., and Baumeister, W., 1990, Three-dimensional structure of the crystalline protein envelope layer of the hyper-thermophilic archaebacterium *Pyrobaculum islandicum*, *J. Struct. Biol.* 103:152.

Phipps, B.M., Huber, R., and Baumeister, W., 1991, The cell envelope of the hyperthermophilic archaebacterium *Pyrobaculum organotrophum* consists of two regularly arrayed protein layers: three-dimensional structure of the outer layer, *Molec. Microbiol.* 5:253.

Saxton, W.O., and Baumeister, W., 1982, The correlation averaging of a regularly arranged bacterial cell envelope protein, *J. Microsc.* 127:127.

Saxton, W.O., Baumeister, W., and Hahn, M., 1984, Three-dimensional reconstruction of imperfect two-dimensional crystals, *Ultramicroscopy* 13:57.

Wildhaber, I., and Baumeister, W., 1987, The cell envelope of *Thermoproteus tenax*: three-dimensional structure of the surface layer and its role in shape maintenance, *EMBO J.* 6:1475.

Chapter 4

S-LAYERS FOUND ON CLINICAL ISOLATES

Kari Lounatmaa
Academy of Finland, and Department of Electron Microscopy
University of Helsinki

Markus Haapasalo and Eero Kerosuo
Department of Cariology
University of Helsinki

and Hannele Jousimies-Somer
National Public Health Institute
Helsinki, Finland

INTRODUCTION

The cell envelopes of bacteria from clinical infections consist of the cytoplasmic membrane and the cell wall. The cytoplasmic membrane is the site of many important biosynthetic activities of the cell, while the cell wall is mainly responsible for the maintenance of the cell shape. In gram-positive bacteria peptidoglygan surrounds the cytoplasmic membrane, in gram-negative cells a thin peptidoglycan layer is further surrounded by the outer membrane. In both gram-positive and gram-negative species amorphous polymeric material (e.g., capsular polysaccharides) may be present on the cell surface.

Bacterial surface structures are of considerable interest because these structures often have a key role in the interaction of the bacterial cell with its surroundings. Many of the surface structures, such as capsules and fimbriae, are recognized as important virulence factors of various species. During the last ten years, another bacterial surface structure has become important, crystalline proteinaceous surface layers or S-layers. Early S-layer studies concentrated mostly on bacteria of non-human origin. However, recently, several gram-positive and gram-negative species found in clinical infections in humans have been shown to possess an S-layer. *Corynebacterium diphtheriae* and *Bacillus anthracis* are probably the best known human pathogens with an S-layer, but new S-layered species are regularly found, especially from the oral cavity. So far little is known about the role of this layer for bacterial ecology and virulence in opportunistic infections of man. In this article we summarize the results of our studies of the S-layers of bacteria isolated from human clinical infections, and discuss the possible role of the S-layers.

Advances in Bacterial Paracrystalline Surface Layers
Edited by T.J. Beveridge and S.F. Koval, Plenum Press, New York, 1993

METHODS

Collection of Bacteria and Extraction of the S-protein

Bacteria were grown on enriched horse blood agar plates in anaerobic jars at 37°C for three to five days. For thin sections the cells were collected with a glass rod from the agar plates into 1.5 ml of 0.1 M phosphate buffer (pH 7.2) and centrifuged (10 000 x g, 2 min, 20°C). From broth cultures the bacteria were collected by centrifugation and washed as above. The method of extraction and purification used for the S-layer depends on the species studied. For *Eubacterium yurii* we used the following method: freshly cultured bacterial cells were incubated in 20 mM Tris buffer (pH 8.0) in the presence of 6M urea, and 0.1 - 1.0 % sodium dodecyl sulphate (SDS), or in 0.2 M glycine in phosphate buffered saline (PBS). The cells were continuously shaken at 20°C for five and thirty minutes and centrifuged (10 000 xg, 10 min). The supernatant was collected and analyzed by SDS-polyacrylamide gel electrophoresis (PAGE). In some experiments the cells were disrupted in 20 mM Tris-buffer by pulse sonication (3 min; MSE Soniprep 150, MSE Scientific Instruments, Crawley, Sussex, England) in an ice bath. The insoluble fraction consisted mainly of S-layer fragments.

Electron Microscopy

For thin sections the pellet was resuspended in phosphate buffered 3 % glutaraldehyde for 30 min, washed and incubated for 30 min in phosphate buffered tannic acid (4% wt/v), washed again and incubated 30 min with the 1:1 mixture of the glutaraldehyde and tannic acid solutions. For in situ fixations the buffered 3 % glutaraldehyde was pipetted straight onto the agar plate or unbuffered 30 % glutaraldehyde was added to the broth tube to make a final concentration of 3 %. To visualize any acid polysaccharide (i.e., capsular material), ruthenium red (1500 ppm) was added to the prefixative in some samples (Lounatmaa, 1985). For the evaluation of the polymorphonuclear leucocyte (PMN) phagocytosis of S-layer bacteria, a short centrifugation time and slow speed was used (30 sec, 500 xg) to avoid the disintegration of the PMNs. The postfixation (osmium tetroxide), dehydration, embedding, cutting and poststaining (uranyl acetate and lead citrate) were performed as previously described (Lounatmaa, 1985).

For freeze-etching the cells were collected from the plates as described above and centrifuged (10 000 xg, 2 min, 20°C) to form a thick suspension and frozen in liquid Freon 22 cooled by liquid nitrogen. The fracturing in a Balzers freeze-etching unit (model BAF 400T Balzers, Liechtenstein) was performed at -100°C and the platinum shadowed at an angle of 40 degrees after 1 min of etching.

For negative staining, the cell suspension in phosphate buffer was allowed to sediment on a grid with a support film (Pioloform 2295/10, Polaron Equipment Ltd, Watford, England) coated with carbon, and the staining was performed with either 1-2 % (w/v) phosphotungstic acid (pH 6.5) or with 1 % (w/v) ammonium molybdate (pH 7.2).

The electron micrographs were taken with JEM-100CX or JEM-1200EX transmission electron microscopes operating at either 60 or 80 kV.

STRUCTURE AND IMPORTANCE OF S-LAYERS ON HUMAN ISOLATES

While most S-layered bacteria have been found in sources such as soil, hot waters, animal specimens, etc., they have also been isolated from human clinical specimens. The majority of these have been found in infections of the alimentary tract. Nearly all known S-layer species from humans are anaerobes which may partly explain their late detection. Many of the recently detected human S-layer bacteria are from the oral cavity. With the exception of a few bacteria (e.g., *Aeromonas* and *Campylobacter*, see chapters by Blaser, Kay and Trust in this book), there is very little information about the S-layers of human pathogens other than ultrastructural data. Potentially interesting S-layer bacteria isolated from humans are given in Table 1.

Structure and Occurrence of the S-Layers

The S-layers of all gram-negative anaerobes that we have studied have a hexagonal symmetry with the exception of *Bacteroides forsythus* (Table 1). In the *Prevotella* species, *P. buccae* and *P. heparinolytica,* the S-layer is separated from the outer membrane by ca. 20 nm (Haapasalo et al., 1985; Kerosuo et al., 1988a; Kornman and Holt, 1981; Okuda et al., 1985; Ranta et al., 1983). Amorphous polymeric capsule-like substance is regularly seen on both sides of the S-layers of these two species. Contrary to this, the S-layers of *Wolinella recta* and *B. forsythus* are in close connection with the outer membrane (Kerosuo, 1988; Kerosuo et al., 1989). The S-layers of the latter two species have a saw-toothed appearance in thin sections. After repeated subcultures in our laboratory we have noticed that *P. buccae* and *W. recta* can lose their S-layers. In *Treponema denticola* the exact location of the crystalline protein is not clear. It is possible that it is not a true S-layer, but rather a crystalline membrane protein (K. Lounatmaa and M. Haapasalo, unpublished data). In a study of the two-and three-dimensional structures of the S-layer of *P. buccae* we have found two different lattice constants, 7.7 nm and 21.5 nm (Sjögren et al., 1985). The S-layer of *P. buccae* is only 5 nm thick and extremely porous. A rounded central hole is surrounded by six oval-shaped holes. We have also reported a crystalline protein in the outer leaflet of the outer membrane of *P. buccae* with a lattice constant of ca. 8 nm (Kerosuo et al., 1987). The relationship of this protein to the overlying S-layer protein is unclear. Dokland et al. (1988; 1990) studied the 3D structure of *W. recta* using the Fourier transform method and reported a lattice spacing of 21 nm and a thickness of 15 nm. The 3D structure of other gram-negative bacteria from oral infections has not been studied.

Gram-positive, anaerobic *Peptostreptococcus* spp. and *Eubacterium* spp. are routinely found in opportunistic infections of man. Messner and Buckel (1988) reported a hexagonally-arranged S-layer on *P. asaccharolyticus*. We have isolated four different gram-positive anaerobes with an S-layer from a single patient with secretory otitis media (Lounatmaa et al.,1988). These included *P. asaccharolyticus, P. magnus, E. lentum* and *Eubacterium* sp. AHN990. Furthermore, we have isolated *E. yurii,* which is a new species from the oral cavity from several infected dental root canals and it has an S-layer (Kerosuo et al., 1988b). All *E. yurii* strains we have studied so far have the S-layer, however, in the *P. magnus* reference strain DSM20470 did not have an S-layer though our clinical strain did (unpublished data). The S-layers on all these gram-positive bacteria are in close proximity to the peptidoglycan with no visible gap between the two layers in thin sections. Unlike *P. buccae* and *P. heparinolytica,* we have not seen amorphous poly-meric material covering the S-layer in any of the gram-positive anaerobes. 3D structures of *E. yurii* ES4C and *Eubacterium* sp. AHN990 revealed two completely different S-layers (Sjögren et al., 1988). Strain AHN990 has hexagonal periodicity and an open structure with complicated openings at a resolution of 1.8 nm. *E. yurii* ES4C in contrast has a

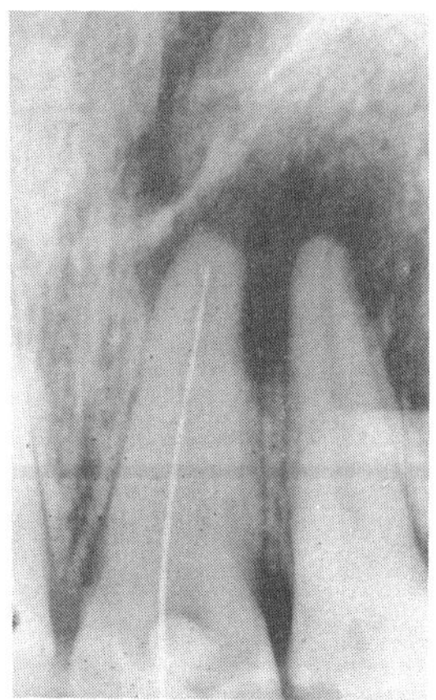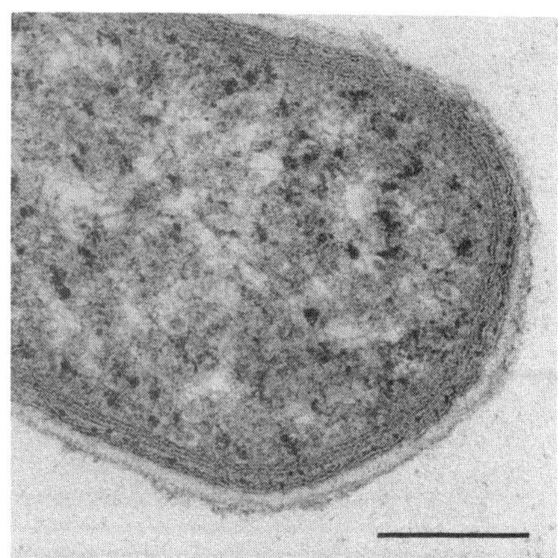

Figure 1. (Left). A radiograph of an acute periapical infection of an upper left maxillary incisor. *P. buccae* was found to be the major infective agent of the polymicrobial flora.
Figure 2. (Right). A thin-sectioned cell of *P. buccae* ES57 showing the cytoplasmic membrane, the outer membrane, and the S-layer. Bar in this and subsequent micrographs represents 0.2 μm.

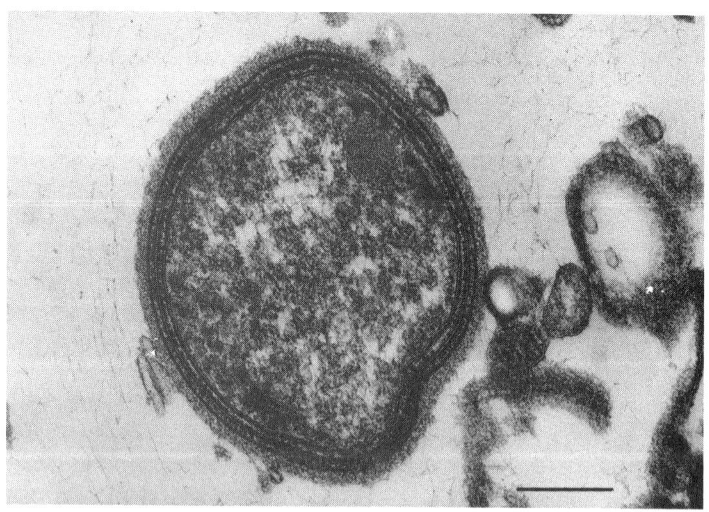

Figure 3. Capsular material is seen on both sides of the S-layer of *P. heparinolytica* ATCC 35895[T]. Notice the ca. 20 nm gap between the OM and the S-layer.

Table 1. Bacteria with an S-layer isolated from human clinical samples

	Typical source	Lattice Symmetry
Gram-negative anaerobes		
P. buccae	EI, PI	H
P. heparinolytica	PI	H
B. forsythus	PI	O
W. recta	EI, PI	H
T. denticola	PI	H
Gram-positive anaerobes		
P. asaccharolyticus	oral infections	H
P. magnus	oral infections	
E. yurii	EI, PI	T
E. lentum	oral infections	
Eubacterium sp.	ear infection	H
C. difficile	PMC	
Gram-positive aerobes and facultatives		
M. bovis		
C. diphtheriae	diphteria	
B. anthracis	anthrax	H
Lactobacillus spp.	normal flora	
Gram-negative aerobes and facultatives		
C. fetus	non-oral infections	H

Abbreviations: EI, endodontic infection; PI, periodontal infection; PMC, pseudomembraneous colitis; H, hexagonal; T, tetragonal; O, oblique.

tetragonal symmetry, and a very closed structure with no openings at a resolution of 2.5 nm.

Role of the S-layer in Bacterial Ecology and Infection

Compared to structural studies, very little information is available about the role of the S-layer in human bacterial ecology and during infection. We compared the occurrence of different bacteria in acute and asymptomatic endodontic infections (infections of human dental root canal) and found that the presence of *P. buccae* and *E. yurii* in the infective flora was regularly accompanied by the presence of acute symptoms. This was not the case with *W. recta* (Haapasalo, 1986; Ranta et al., 1988). Endodontic infections are always polymicrobial infections and some other species, e.g., *Phorphyromonas gingivalis* and *P. endodontalis* (with no S-layer) were also present only in acute cases (Haapasalo et al., 1986). Cell surface hydrophobicity has been suggested to play an important role in bacteria-phagocyte interactions (Absalom, 1988). We have studied the correlation of bacterial surface structures and cell surface hydrophobicity in selected oral bacteria. In *Prevotella* and *Porphyromonas* species with no S-layer, the inter-strain variation in hydrophobicity was very high for each species and was largely dependent on the thickness of the capsule layer. However, in the S-layer species we studied, *E. yurii, W. recta, P. buccae* and *P. heparinolytica,*

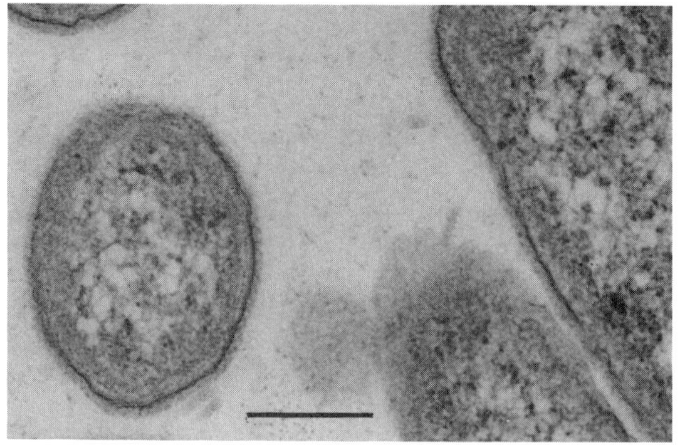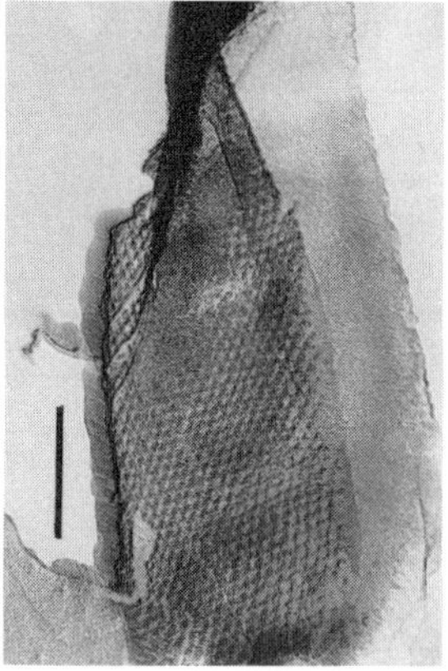

Figure 4. (Left). *W. recta* ES64W from an endodontic infection. The S-layer is seen closely associated with the outer membrane in thin sectioned cells.
Figure 5. (Right). Freeze-etching reveals the hexagonal periodicity of the S-layer of *W. recta* ES64W.

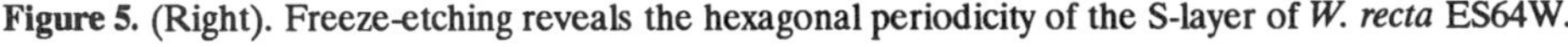
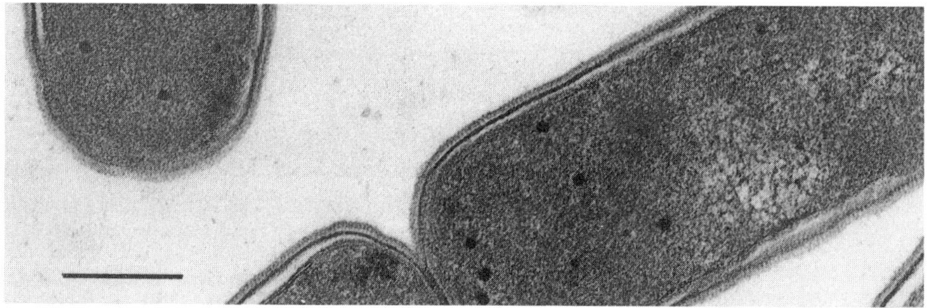

Figure 6. The S-layer of a new gram-positive species from an acute endodontic infection, *E. yurii* strain ES4C, is seen outside the peptidoglycan layer. Note the absence of capsule.

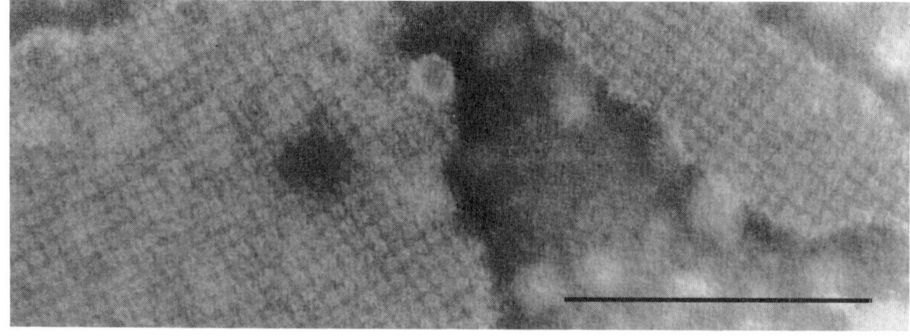

Figure 7. Tetragonal periodicity in the negatively stained, isolated S-layer of *E. yurii* ES4C.

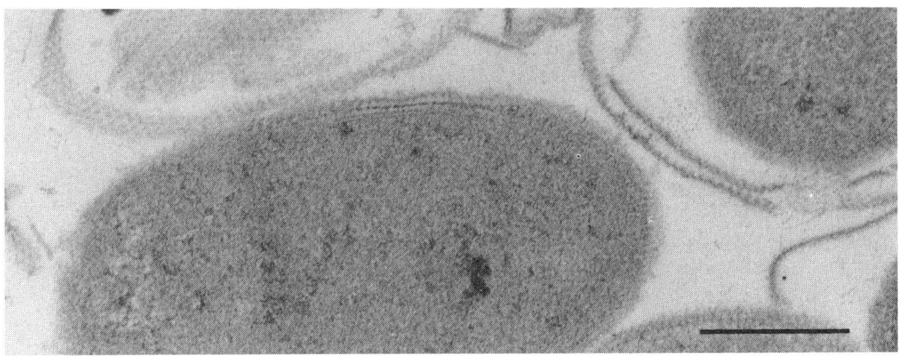

Figure 8. The periodic structure of the S-layer of *Eubacterium* sp. AHN990 isolated from an ear infection can be seen also in thin sectioned cells.

the variation between strains within the same species was extremely low. All strains of the same species were either very hydrophilic or hydrophobic and this, presumably, depended on the properties of the S-protein (Haapasalo et al., 1990). Moreover, our results with *P.buccae* indicated that the crystalline surface protein overrides the importance of the capsule in determining the hydrophobicity of the cell, as the S-layered strains with a thin capsule were equally as hydrophilic as strains with a thick capsule; this was not so with *Prevotella* species without an S-layer (Haapasalo et al., 1990).

We have also studied the non-opsonophagocytosis of some human S-layer species and closely related species without an S-layer. *P. buccae* (all strains were hydrophilic) were totally resistant to phagocytic ingestion by polymorphonuclear leukocytes (PMNs), while the hydrophilic strains of *P. gingivalis* and *F. nucleatum* were not (Kerosuo et al., 1990). Recently we have studied the PMN phagocytosis of *E. yurii*, which is very hydrophobic (Kerosuo et al., manuscript submitted). *E. yurii* subsp. *yurii* subsp. *schtitka* were readily ingested by the PMNs while the strains of subp. *margaretiae* were not ingested when opsonins were not present. However, when *E.yurii* subsp. *margaretiae* was incubated together with monocytes, the bacteria were also rapidly ingested without serum (unpublished). In other studies subsp. *margaretiae* has been shown to be the most pathogenic of the *E. yurii* subspecies (Margaret et al., 1990). Whether the different susceptibility to PMN phagocytosis can be related to differences in the S-layer is not known. When 2% serum was added in the above phagocytosis experiments, both *P. buccae* and *E. yurii* subsp. *margaretiae* were rapidly ingested. The importance in vivo of resistance to non-opsonophagocytosis is not known. In *C. fetus*, which occasionally causes infections in man, the S-layer is regarded as an important factor in making the cells resistant to phagocytosis. Serum resistance of *C. fetus* is due to the inability of complement 3b to bind effectively to the S-protein of this species (Blaser et al., 1988). We have studied the susceptibility of *P. buccae* and other *Prevotella* species to complement-mediated killing. While *Prevotella* species without an S-layer were resistant to 10 % serum, some (but not all) strains of *P. buccae* were susceptible to even 1% serum. We have been unable to relate the different susceptibility of *P. buccae* strains to the S-layer (unpublished data).

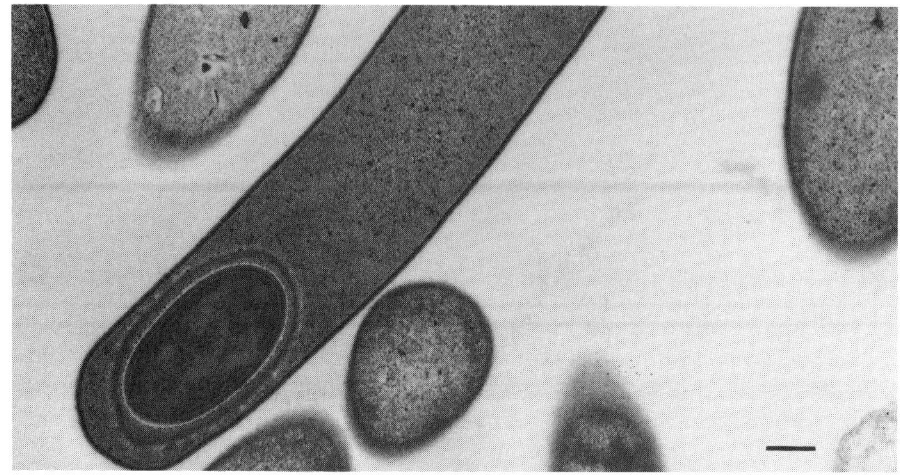

Figure 9. Thin section of *Clostridium difficile* AHS7392.

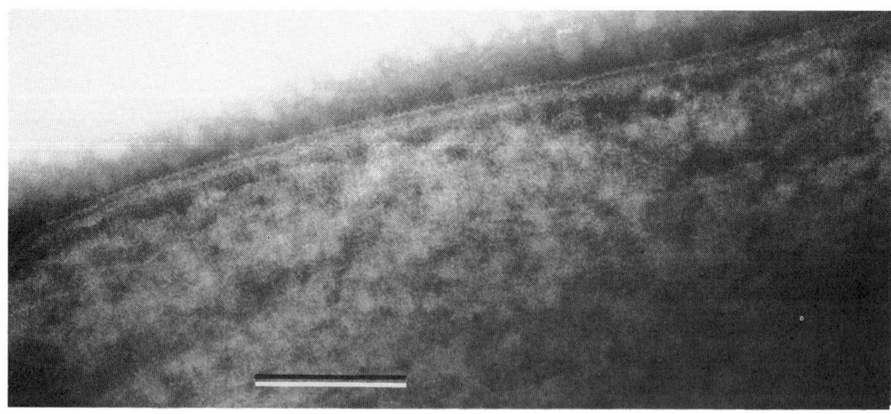

Figure 10. Tetragonal periodicity can be seen on this negatively stained cell of *C. difficile.*

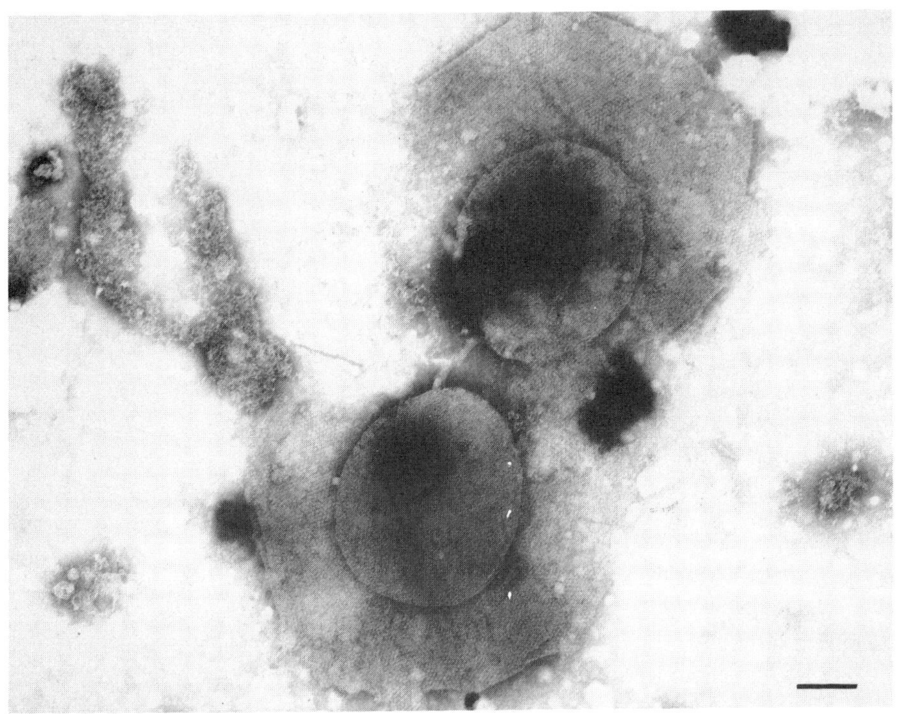

Figure 11. Negative staining of lysed cells of *P. asaccharolyticus* isolated from an ear infection reveals hexagonal periodicity.

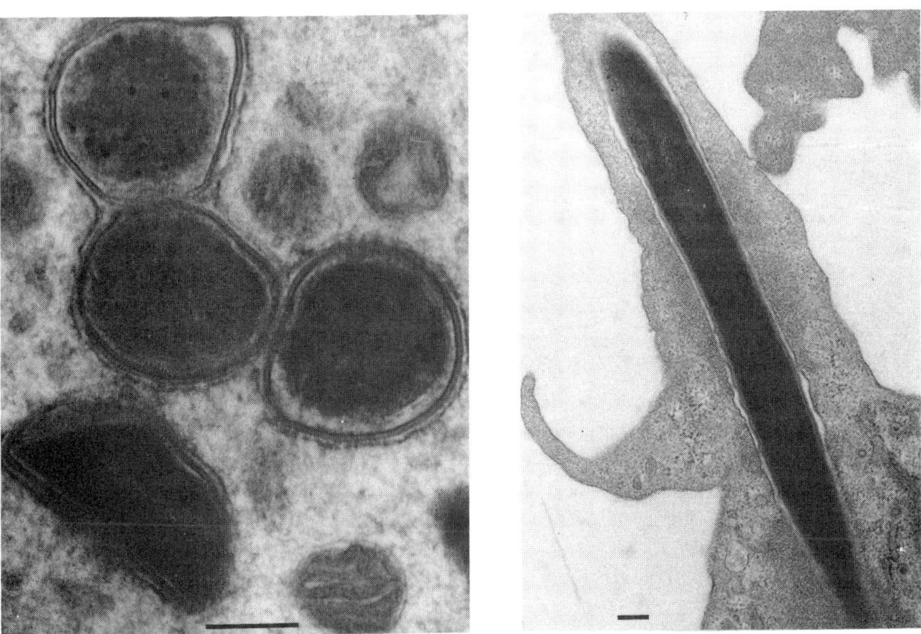

Figure 12. (Left). *E. yurii* ES4C cells phagocytosed by a human monocyte. The S-layer is clearly seen in the phagosome on the bacterial cells.
Figure 13. (Right). A human monocyte has ingested an elongated cell of *E. yurii* ES4C.

ACKNOWLEDGMENTS

This study was financially supported by the Academy of Finland (Grant nr. 1011731, to K.L.) and by the Finnish Dental Society (to M.H.).

REFERENCES

Absalom, D.R., 1988, The role of bacterial hydrophobicity in infection: bacterial adhesion and phagocytic ingestion, *Can. J. Microbiol.* 34:287.

Blaser, M.J., Smith, P.F., Repine, J.E., and Joiner, K.A., 1988, Pathogenesis of *Campylobacter fetus* infections. Failure of encapsulated *C. fetus* to bind C3b explains serum and phagocytosis resistance, *J. Clin. Invest.* 81:1434.

Dokland, T., Johanssen, B.V., Farrants, G.W., and Olsen, I., 1988, Structure of crystalline proteins from the cell envelope of *Wolinella recta, J. Ultrastr. Molec. Struct. Res.* 100:284.

Dokland, T., Olsen, I., Farrants, G., and Johanssen, B.V., 1990, Three-dimentional structure of the surface layer of *Wolinella recta, Oral Microbiol. Immunol.* 5:162.

Haapasalo, M., 1986, *Bacteroides buccae* and related taxa in necrotic root canal infections, *J. Clin. Microbiol.* 24:940.

Haapasalo, M., Kerosuo, E., and Lounatmaa, K., 1990, The hydrophobicities of human polymorphonuclear leukocytes and oral *Bacteroides* and *Porphyromonas* spp., *Wolinella recta* and *Eubacterium yurii* with special reference to bacterial surface structures, *Scand. J. Dent. Res.* 98:472.

Haapasalo, M., Lounatmaa, K., Ranta, H., Shah, H., and Ranta, K., 1985, Ultrastructure of *Bacteroides capillus, B. buccae, B. pentosaceus, B. oris, B. oralis, B. veroralis,* and pentose sugar- fermenting *Bacteroides* sp. from humans with periapical osteitis: occurence of external proteinaceus cell wall layer, *Int. J. Syst. Bacteriol.* 35:65.

Haapasalo M., Ranta H., Ranta K., and Shah H., 1986, Black-pigmented *Bacteroides* spp. in human apical periodontitis. Infect Immun 53:149.

Kerosuo, E., 1988, Ultrastructure of the cell envelope of *Bacteroides forsythus* strain ATCC43037^T, *Oral Microbiol. Immunol.* 3:134.

Kerosuo, E., Haapasalo, M., Ranta, H., and Lounatmaa, K., 1987, Hexagonal periodicity in the outer membrane of *Bacteroides buccae. J. Gen. Microbiol.* 133:2217.

Kerosuo, E., Haapasalo, M., and Lounatmaa, K., 1989, Ultrastructural relationship of cell envelope layers in *Wolinella recta, Scand. J. Dent. Res.* 97:54.

Kerosuo, E., Haapasalo, M., Alli, K., and Lounatmaa, K., 1990, Ingestion of *Bacteroides buccae, B. oris, Porphyromonas gingivalis,* and *Fusobacterium nucleatum* by human polymorphonuclear leukocytes in vitro, *Oral Microbiol. Immunol.* 5:202.

Kerosuo, E., Haapasalo, M., Lounatmaa, K., and Ranta, H., 1988a, Ultrastructural comparison of *Bacteroides heparinolyticus* and *B. buccae, in*: "Crystalline bacterial cell surface layers," U.B. Sleytr, P. Messner, D. Pum, M. Sára, eds., Springer Verlag, Berlin.

Kerosuo. E., Haapasalo, M., Lounatmaa, K., Ranta, H., and Ranta, K., 1988b, Ultrastructure of a novel anaerobic Gram-positive nonsporing rod from dental root canal, *Scand. J. Dent. Res.* 96:50.

Kornman, K.S., and Holt, S.C., 1981, Physiological and ultrastructural characterization of a new *Bacteroides* species (*Bacteroides capillus*) isolated from severe localized periodontitis, *J. Periodont. Res.* 16:542.

Lounatmaa, K., 1985, Electron microscopic methods for the study of bacterial surface structures, *in*: "Enterobacterial surface antigens: methods for molecular characterization," T.K. Korhonen, E.A. Dawes, P.H. Mäkelä, eds., Elsevier Science Publishers, Amsterdam.

Lounatmaa, K., Jousimies-Somer, H., Grenman R., and Rintala, A., 1988, Crystalline surface layers in anaerobic bacteria isolated from a patient with secretory otitis media and a draining ear, *in*: "Crystalline bacterial cell surface layers," U.B. Sleytr, P. Messner, D. Pum, M. Sara, eds., Springer Verlag, Berlin.

Margaret, B.S., Heath, J.R., and Krywolap, G.N., 1990, Pathogenic potential of *Eubacterium yurii* subspecies, *J. Med. Microbiol.* 31:103.

Messner, P., and Buckel, W., 1988, The surface layer of *Peptostreptococcus asaccharolyticus*, *System. App. Microbiol.* 10:226.

Okuda, K., Kato, T., Shiozu, J., Takazoe, I., and Nakamura, T., 1985, *Bacteroides heparinolyticus* sp. nov. isolated from humans with periodontitis, *Int. J. Syst. Bacteriol.* 35:438.

Ranta, H., Lounatmaa, K., Haapasalo, M., and Ranta, K., 1983, Isolation and ultrastructure of *Bacteroides* sp. with external cell wall layer (S-layer) in periapical osteitis. *Scand. J. Dent. Res.* 91:458.

Ranta, H., Haapasalo, M., Ranta, K., Kontiainen, S., Kerosuo, E., Valtonen, V., Suuronen, R., and Hovi, T., 1988, Bacteriology of odontogenic apical periodontitis and effect of penicillin treatment. *Scand. J. Infect. Dis.*, 20:187.

Sjögren, A., Hovmöller, S., Farrants, G., Ranta, H., Haapasalo, M., Ranta, K., and Lounatmaa, K., 1985, Structures of two different surface layers found in six *Bacteroides* strains. *J. Bacteriol.* 164:1278.

Sjögren, A., Wang, D.N., Hovmöller, S., Haapasalo, M., Ranta, H., Kerosuo, E., Jousimies-Somer, H., and Lounatmaa, K., 1988, The three-dimensional structures of S-layers of two novel *Eubacterium* species isolated from inflammatory human processes, *Molecul. Microbiol.* 2:81.

Chapter 5

A COMMON STRUCTURAL PRINCIPLE IN THE SURFACE LAYERS OF THE ARCHAEOBACTERIA *HALOFERAX, HALOBACTERIUM* AND *ARCHAEOGLOBUS*

Martin Kessel and Shlomo Trachtenberg

Department of Membrane and Ultrastructure Research
The Hebrew University-Hadassah Medical School
Jerusalem, Israel

INTRODUCTION

Our present ability to more readily visualise the three dimensional organisation of the bacterial cell surface, especially in the many cases where S-layers are present, has in recent years allowed a comparative approach to the study of this important cell structure. It has also allowed the integration of this structural knowledge with that derived from molecular biology (Baumeister and Engelhardt, 1987; Baumeister and Lembcke, 1992). The bacterial cell surface, whether or not an S-layer is present, is that element of the cell which comes into direct contact with the milieu in which the cell exists. As such, the cell surface is the site of entry for molecules coming into the cell and the site of exit for molecules leaving the cell. This translocation of molecules through the bacterial cell surface involves a combination of both passive and active processes by the components of the various barriers across which molecules enter and leave the cell.

In the specific cases where the S-layer is present, a number of functions have been postulated for this structure, such as protection, recognition, and a passageway for import into and export out of the cell (Baumeister and Hegerl, 1986). In many cases among both gram-positive and gram-negative bacteria the S-layer is not an essential element of the bacterial cell envelope and the same bacterial species may exist with and without the S-layer present depending on growth conditions (Buckmire and Murray, 1976). Among archaeobacteria, however, the S-layer is often the only wall layer and as such performs an essential function in this group of bacteria.

We have examined in detail the S-layers of two species of extreme halophilic archaeobacteria, *Haloferax volcanii* and *Halobacterium halobium* which, despite their different ecological niches, show a very similar S-layer architecture (Kessel et al., 1988b). When these S-layers are compared to the S-layer of the extreme thermophilic archaeobacterium, *Archaeoglobus fulgidus* (Kessel et al., 1990), marked similarities are found in the overall S-layer architecture .

We will first describe the morphological features of the S-layers of these organisms and then discuss the common architectural principles found in these structures in relation to the known features of the S-layers of other archaeobacteria and eubacteria. We will then integrate these structural findings with current knowledge of the biochemical composition and gene structure of the S-layer components as elucidated by molecular biology.

THE S-LAYER OF *HALOFERAX VOLCANII*

H. volcanii was isolated from the Dead Sea and requires in addition to 2.4M NaCl, a high concentration of Mg^{2+} (0.25M) in its growth medium, reflecting the composition of the Dead Sea itself (Mullakhaṇbai and Larsen, 1974). Intact envelopes from exponential growth phase cells of the bacterium, devoid of cytoplasmic contents, are prepared by freezing and thawing followed by incubation with DNAse (Kessel et al., 1988). Fig. 1a is a view of the cell surface of *H. volcanii* contrasted by unidirectional heavy metal shadowing with platinum/carbon. The dominant feature of the surface is a very typical periodic array of morphological units arranged with hexagonal symmetry.

A low resolution surface reconstruction from a freeze-fracture replica of the surface of *H. volcanii*, (Fig. 1b) demonstrates the dome shape of the morphological unit and, by inference, the polarity of the structure. Thin sections of cells of *H. volcanii* also

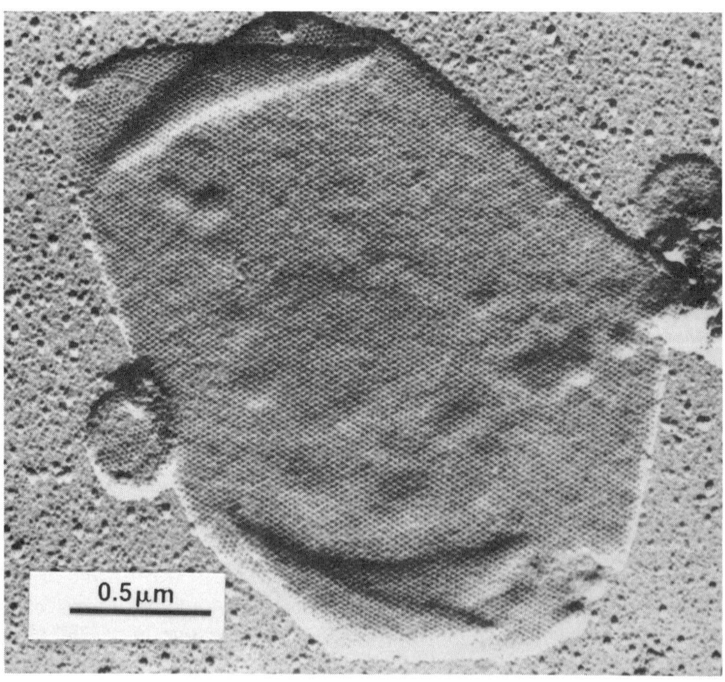

Figure 1a. Freeze-dried shadowed envelope of *Haloferax volcanii* showing the hexagonal array of morphological units. Shadowed with Pt/C at an angle of 30°. Bar = 0.5 μm.

Figure 1b. Shaded surface reconstruction from a freeze-fracture replica of the cell surface of *H. volcanii* showing the dome shape of the morphological unit. Center-to-center spacing is 16nm.

confirmed that the outermost layer of the wall is comprised of dome-shaped protrusions lying exterior to the cell membrane (Kessel et al., 1988a).

Negative staining of the envelopes also revealed the periodic arrangement of morphological units (Kessel et al., 1988a). However, due to uneven stain distribution only certain areas of the envelope showed clear doughnut-shaped morphological units with stain filled centers, whereas in other areas of the envelope the morphological units were seen to be positively stained. Image processing of high resolution micrographs of the negatively stained cell envelope reveals the detail of the morphological unit (Fig. 2a), and its interaction with its neighbours. Each morphological unit, 9nm in diameter, is comprised of six subunits each having two centers of mass: one is close to the hub and the other is located in the arm of the subunit which projects to a three fold axis of symmetry and there meets the arm from an adjacent morphological unit. The center-to-center distance of adjacent morphological units is 16nm.

Tilt series (+75°) from well stained areas of the envelope were collected using an unlimited tilt holder (Chalcroft and Davey, 1984) and three-dimensional reconstruction of the envelopes was carried out using the hybrid real space/reciprocal space method described by Saxton et al. (1984). This reconstruction (Fig. 2b) reveals a dome-shaped structure 9nm in diameter and 4.5nm in height comprised of six radially organised subunits. A small orifice appears to be located at the apex of each dome. When viewed from the inside, the vestibule of the dome is clearly seen with each of the subunits appearing to be lying on top of and presumably connected to the underlying cell membrane. Six radial protrusions emanate from each of the subunits and (presumably) form the points of contact between adjacent morphological units at the three-fold axis of symmetry.

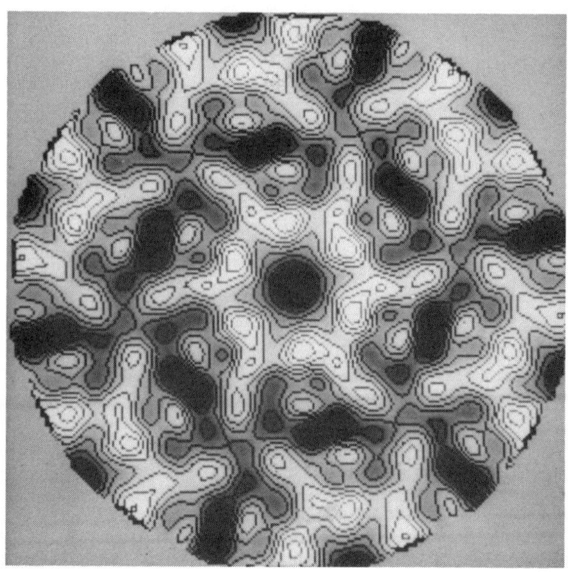

Figure 2a. Averaged two dimensional projection of the morphological unit of *H. volcanii* from a negatively stained envelope showing the detail of the stain-excluding regions of the structure. Note the contact points of neighbouring morphological units at the three-fold axis of symmetry. Center-to-center spacing is 16 nm.

THE S-LAYER OF *H. HALOBIUM*

H. halobium is an extreme halophile which has a salt requirement of 4 to 5M NaCl which is greater than for *H. volcanii*. High resolution electron microscopy of the cell surface was carried out on envelopes prepared by gentle sonication and subsequent treatment with DNAse. Fig. 3a is a cell envelope unidirectionally shadowed showing the periodic hexagonal arrangement of the morphological units. A surface reconstruction from such a preparation (Fig. 3b) reveals the dome-shaped morphological units and their six subunits. It also clearly establishes the polarity with the dome facing outwards from the cell.

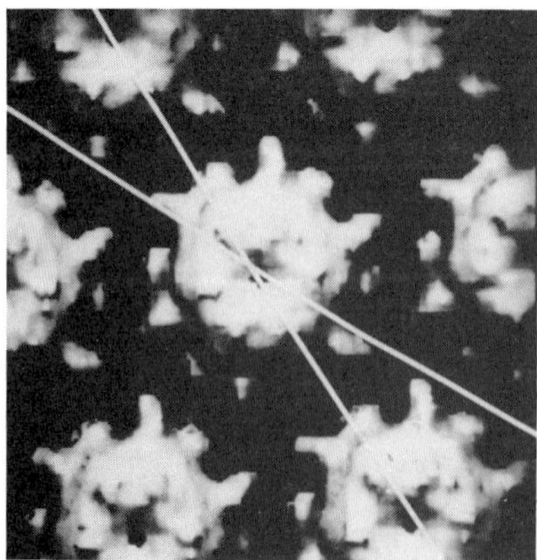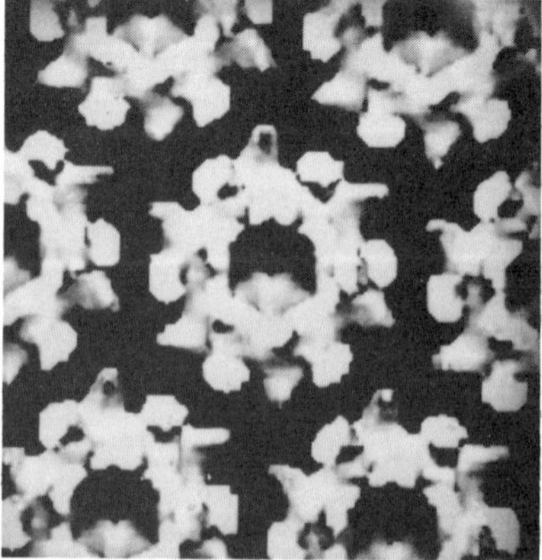

Figure 2b. Three dimensional reconstruction of the morphological unit of *H. volcanii* showing their dome-shaped structure as seen from the outside (left) and the vestibule as seen from the inside (right). Center-to-center spacing is 16 nm.

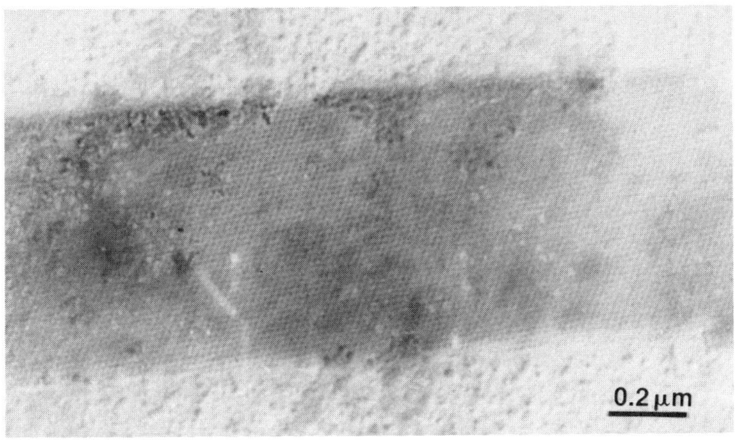

Figure 3a. Shadowed cell envelope of *H. halobium* showing the hexagonal arrangement of the morphological units. Shadowed with Pt/C at an angle of 30°.

The two dimensional projection of negatively stained envelopes of *H. halobium* has been characterised after image processing. The morphological unit, 9nm in diameter, is composed of six subunits showing an inner and outer domain (Fig. 3c). The outer domain forms the projecting arm emanating from each of the subunits which meets the arms from adjacent morphological units at a three-fold axis of symmetry and provides continuity of the S-layer. The center-to-center spacing of adjacent morphological units is 16nm. Thin sections of freeze-substituted *H. halobium* (Fig. 4) (Trachtenberg, 1992) clearly show dome-shaped corrugations in cross section protruding from the cell surface. The relationship of the dome to the underlying cell membrane is also seen.

Figure 3b. Shaded surface reconstruction from a shadowed envelope of *H. halobium* showing the dome-shaped morphological units and their subunits. Center-to-center spacing is 16 nm.

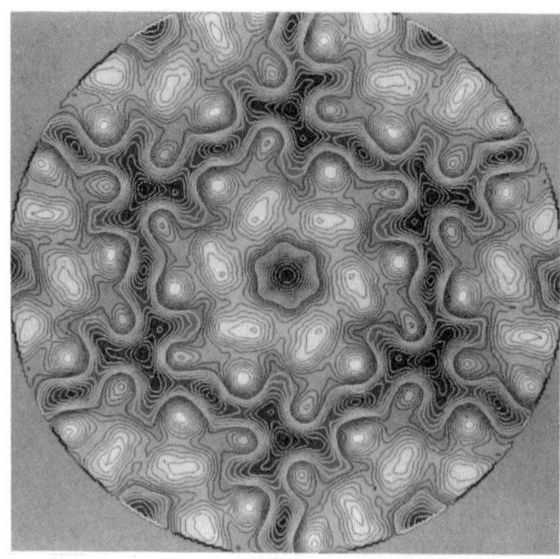

Figure 3c. Averaged two dimensional projection from a negatively stained envelope of *H. halobium* showing the detail of the stain-excluding regions of the structure. Note the strong similarities to the reconstruction from *H. volcanii* (Fig. 2a). Center-to-center spacing is 16 nm.

THE S-LAYER OF *A. FULGIDUS*

A. fulgidus is an extremely thermophilic archaeobacterium isolated from hot sediments of marine hydrothermal systems in Southern Italy (Stetter et al., 1987). Envelopes of the cells, devoid of cytoplasmic contents, were prepared by freezing in liquid nitrogen, thawing the cells, and subsequent treatment with DNAse. In this case, in order to render the periodic surface array visible, it was also necessary to treat the envelopes after deposition on the grid with 0.01M sodium dodecyl sulfate (Kessel et al., 1990).

Analysis of the two dimensional projection of the surface layer of *A. fulgidus* (Fig.5a) shows the morphological unit to be a hexamer with a complex pore. The

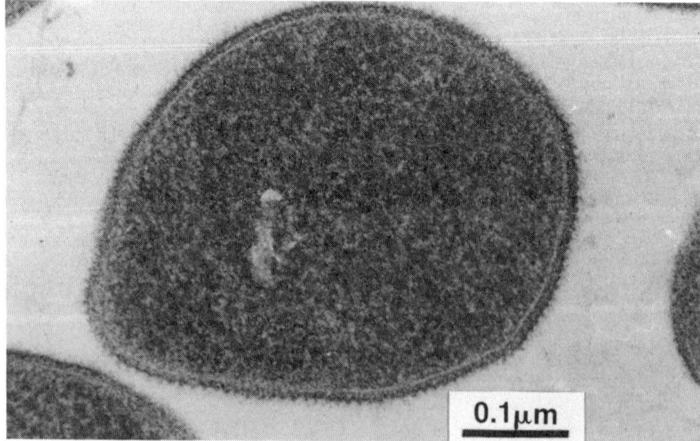

Figure 4. Thin section of a cell of *H. halobium* after freeze-substitution. The outermost layer of the envelope clearly shows the corrugated appearance of the cell surface which is due to the outward projection of the morphological units.

50

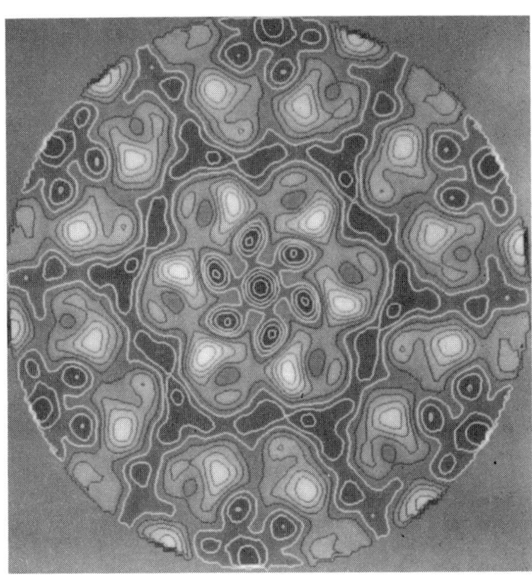

Figure 5a. A two dimensional projection of the morphological unit from a negatively stained envelope of *A. fulgidus* showing the open character of the structure. Center-to-center spacing is 17.5 nm.

central orifice, on the six-fold axis, is connected to six slightly oval satellite channels which appear to have a handedness and are only slightly shallower than the central orifice. The major stain excluding sickle-shaped mass of the monomer is adjacent to the satellite channel. The morphological unit is surrounded by a deep groove of stain separating it from its neighbours, which shows an apparent lack of connectivity between adjacent subunits. The three dimensional reconstruction from a tilt series (+72°) of envelopes (Fig. 5b) reveals a dome shaped structure of the hexamer with a relatively large stain-filled area. Evidence for the dome shape is also obtained from the sculped-out appearance of the monomer as seen from the inside. An analysis of thin sections of *A. fulgidus* demonstrated corrugations which approximate a dome shape projecting from the cell surface (Kessel et al., 1990). Heavy metal shadowing of the cell surface (Stetter et al., 1987) showed the cell surface to be comprised of hexagonally-arrayed morphological units with a center-to-center spacing of 17.5nm.

GENE STRUCTURE AND GLYCOSYLATION OF THE GLYCOPROTEINS FROM *H. HALOBIUM* AND *H.VOLCANII*

Mescher and Strominger (1976) determined that the chemical composition of the cell wall of *Halobacterium* was a glycoprotein. Lechner and Sumper (1987) and Sumper et al. (1990) have studied the glycosylation, and cloned and sequenced the genes of these glycoproteins from both *H. halobium* and *H. volcanii*. This provides the opportunity to integrate the high resolution structural information from three dimensional reconstructions with the polypeptide sequence of the protein. The 90kDa proteins from both of these species show distinct regions of homology, particularly a 21 amino acid hydrophobic domain at the C-terminus which is a presumed membrane anchor. This domain is adjacent to a region of 11 threonine residues, all of which are O-glycosylated. The homology between the two proteins becomes less towards their

N-termini and there are marked differences in the pattern of glycosylation between the two species.

Based on this molecular information, we proposed a model for the S-layers of *H. volcanii* and *H. halobium* (Kessel et al., 1988a) in which the threonine-rich string of amino acids would form a spacer region just above the cell membrane with the bulk of the protein in the region above and closer to the outside of the cell. The suggestion for a spacer region based on X-ray crystallography of envelope pellets has been previously proposed by Blaurock et al. (1976) for *H. halobium*.

COMMON ARCHITECTURAL THEMES

Three dimensional reconstructions of these three S-layers have allowed us to make comparisons between these structures as they appear in different archaeobacteria. The dominant feature is clearly the dome shape of the morphological units. Based on metal-shadowing and reconstruction the domes of the two halobacteria are closed structures and appear similar (cf. Figs 1b and 3b). Details of the dome are seen in the three dimensional reconstruction of the S-layer of *H. volcanii* (Fig. 2b). In the case of *A. fulgidus* the dome is present but in a more porous form as evidenced by higher stain accumulation (Fig. 5b) 3D reconstruction shows the dome structure to be hollow forming a vestibule within which solute can accumulate. The composition of the solute will depend upon the extent of free diffusion between the external medium and this space. So far, we are unable to determine whether or not this is the case for the space created between the S-layer and the cell membrane and are, therefore, unable to refer to this as a periplasmic space similar to that found in gram-negative eubacteria. A distinct difference between the ionic composition of the cytoplasmic compartment and the external medium has clearly been shown in the case of the halobacteria. The information from three dimensional reconstructions allows a determination of the mass density distribution within the S-layer as it extends from the cell membrane to the outermost extremity of the cell.

Besides the three archaeobacteria discussed in detail above, Baumeister and

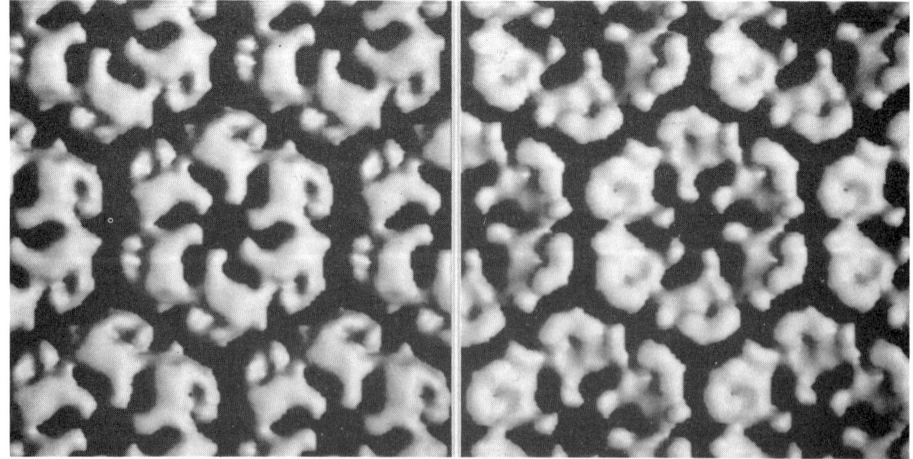

Figure 5b. Three dimensional reconstruction of the morphological unit from an envelope of *A. fulgidus* negatively stained with 1% uranyl acetate. As seen from the outside (left) the dome shape is evident and when viewed from the inside (right) the vestibule is seen. Center-to-centre spacing is 17.5 nm.

Lembcke (1992) have pointed out that a similar architectural principle of a roof-like structure, and by implication, a space between the cell membrane and the S-layer, exists in the archaeobacteria *Sulfolobus shibata*, *Thermoproteus tenax* and *Staphylothermus marinus*. The main difference between the S-layers of these species is the extent of the spacer region between the cell membrane and the S-layer. These organisms may in addition have extra filiform components on the outermost surface of the S-layer. We have also pointed out the similarity between the projection images of *A. fulgidus* and *P. occultum* (Kessel et al., 1990)

It thus appears reasonable to conclude that there is a unifying theme underlying the structure of S-layers of many of the archaeobacteria and as more of these are elucidated at both the structural and molecular level we will approach an understanding of their physiological role as well.

ACKNOWLEDGEMENTS

This research has been supported by a grant from the German-Israeli Foundation for Scientific Research and Development. We would particularly like to thank B. Pinnick and E. Sadovnic for expert technical assistance. Special thanks are due to S. Volker, U. Santarius, I. Wildhaber and W. Baumeister at the Max Planck Institut für Biochemie, Martinsried, Germany, where much of this work was carried out.

REFERENCES

Baumeister, W., and Hegerl, R., 1986, Can S-layers make bacterial connexons? *FEMS Microbiol. Lett.* 36:119.

Baumeister, W., and Engelhardt, H., 1987, Three-dimensional structure of bacterial surface layers *in*: "Electron Microscopy of Proteins Vol.6: Membranous Structures", J.R.Harris and R.W. Horne, eds., pp.109-154, Academic Press, London.

Baumeister, W., and Lembcke, G., 1992, Structural features of archaebacterial envelopes, *J. Bioenerg. and Biomemb.* 24:567.

Blaurock, A.E., Stoeckenius, W., Oesterhelt, D., and Scherphof, G.L., 1976, Structure of the cell envelope of *Halobacterium halobium*, *J. Cell Biol.* 71:1.

Buckmire, F.L.A., and Murray, R.G.E., 1976, Substructure and in vitro assembly of the outer, structured layer of *Spirillum serpens*, *J. Bacteriol.*, 125:290.

Chalcroft, J., and Davey, C.L., 1984, A simply constructed extreme-tilt holder for the Philips eucentric goniometer stage, *J. Microsc.* 134:41.

Kessel, M., Buhle, Jr., E.L., Cohen, S., and Aebi, U., 1988, The cell structure of a magnesium-dependent halobacterium *Halobacterium volcanii* CD-2 from the Dead Sea, *J. Ultrastruc. Mol. Struct. Res.* 100:94.

Kessel, M., Wildhaber, I., Cohen, S., and Baumeister, W., 1988a, Three-dimensional structure of the regular surface glycoprotein layer of *Halobacterium volcanii* from the Dead Sea, *EMBO J.*, 7:1549.

Kessel, M., Volker, S., Santarius, U., Huber, R., and Baumeister, W., 1990, Three-dimensional reconstruction of the surface protein of the extremely thermophilic archaebacterium *Archaeoglobus fulgidus*, *System. Appl. Microbiol.* 13:207.

Lechner, J., and Sumper, M., 1987, The primary structure of a prokaryotic glycoprotein: cloning and sequencing of the cell surface glycoprotein gene of halobacteria, *J. Biol. Chem.* 262:9724.

Mescher, M.F., and Strominger, J.L., 1976, Purification and characterisation of a prokaryotic glycoprotein from the cell envelope of *Halobacterium salinarium*, *J. Biol. Chem.* 251:2005.

Mullakhanbai, M.F., and Larsen H., 1975, *Halobacterium volcanii* spec. nov., a Dead Sea halobacterium with a moderate salt requirement, *Arch. Microbiol.* 104:207.

Saxton, W.O., Baumeister, W. and Hahn, M., 1984, Three-dimensional reconstruction of imperfect two dimensional crystals, *Ultramicroscopy* 13:57.

Sumper, M., Berg, E., Mengele, R., and Strobel, I., 1990, Primary structure and glycosylation of the S-layer protein of *Haloferax volcanii*, *J. Bacteriol.* 172:7111.

Trachtenberg, S., 1993, Programmable freeze-substitution and cryoembedding device, *Microsc. Res. Tech.* 24: 173.

III. S-LAYERS OF AGRICULTURAL AND ENVIRONMENTAL IMPORTANCE

Chapter 6

CRYSTALLINE SURFACE-LAYERS OF THE GENUS LACTOBACILLUS

Sylvie Lortal

I.N.R.A.
Laboratory of Milk Research and Technology
Rennes, France

INTRODUCTION

Lactobacilli are fundamental contributors to food technology and agriculture. Moreover, they are thought to induce beneficial effects to public health as valuable inhabitants of the intestinal and urogenital tract of humans. The envelope of these microorganisms could play an important role (i.e., resistance, adhesion, etc.) in all of these applications. The aim of this review is to summarize the available knowledge about S-layer carrying strains of the genus *Lactobacillus*.

BIOTECHNOLOGICAL USES OF LACTOBACILLI

Lactobacilli decrease the pH of their growth environment to below 4.0 by lactic acid production, thus preventing, or at least efficiently delaying, growth of virtually all others competitors except other lactic acid bacteria and yeasts. Since ancient times they have been involved in the manufacture of a variety of food and beverages; these are dairy products, wine, meat products, vegetables and bakery products as illustrated in Table 1. These applications have been well described (Sharpe, 1979; Sugihara, 1985; Daeschel et al., 1987; McKay and Baldwin, 1990). Lactobacilli are also valuable contributors to agriculture. The inoculation of silage by *Lactobacillus* spp. effectively preserves the food stuff and suppresses the growth of unwanted pathogenic bacteria like clostridia (Seale, 1986). Moreover, various *Lactobacillus* species associated with plants are believed to inhibit the growth of plant pathogens such as *Xanthomonas* and *Erwinia* (Visser et al., 1986).

The importance of *Lactobacillus* as a member of the normal flora of the mouth, intestine and urogenital tract of women is also well recognized. Thus a recent and exciting field of research concerns the potential applications of *Lactobacillus* species for public health (Gilliland, 1990). The term "probiotic" is largely used to name preparations containing viable enteric bacteria, such as lactobacilli (mainly *L. acidophilus*), bifidobacteria and enterococci. The hypothetical beneficial effects of the

Table 1. Biotechnological uses of some Lactobacillus species

Species	Products
L. fermentum	Kefir, and bakery products
L. helveticus	Cheeses (Swiss type cheese)
L. brevis	Kefir, fermented vegetables (sauerkraut, olives and silage), and bakery products
L. bulgaricus	Yoghurts
L. acidophilus	Cheeses and probiotics
L. casei	Cheeses, bakery products, and probiotics
L. plantarum	Cheeses, fermented vegetables (sauerkraut, olives, pickles and silage), bakery products, and wines
L. lactis	Cheeses (Swiss type cheese)

ingestion of probiotics include nutritional benefits, prevention of undesirable intestinal infections (Perdigon et al., 1990), an anticholesteremic effect (Fernandez et al., 1987), and an induction of mucosal and systemic immune response against epitopes associated with these organisms (Perdigon et al., 1988 ; Gerritse et al., 1990).

To exert a positive action, the probiotic microorganisms have to colonize or to multiply in the gut. These conditions imply that they are resistant to the lytic enzymes of the mouth and the acidic conditions and bile salts of the stomach. The outermost envelope of the bacterium could contribute to this resistance and to adhesion onto the intestinal mucosa.

SURFACE LAYERS OF THE GENUS *LACTOBACILLUS*

The presence of surface layers in *Lactobacillus* was first shown in 1974 (Table 2). By freeze-etching and negative staining of *L. fermentum*, Kawata et al. (1974) clearly observed a regular surface array. The lattice had an oblique symmetry (p2) with a 9.6/6.2 nm center-to-center spacing. Biochemical information about the S-layer protein was not provided. Interestingly, in previous work about the cell wall of *L. fermentum* (mainly focused on the peptidoglycan; Wallinder and Neujahr, 1970), a large fraction of the crude cell wall was protein (20% of the dry weight). Nevertheless electron microscopy observations of thin sections revealed an homogeneous (40 nm) cell wall without any separate surface layer.

In 1979 and 1980, Masuda and Kawata characterized a surface layer in *L. brevis*. A tetragonal lattice with a center-to-center spacing of 7.0/4.5 nm was observed using freeze-etching and negative staining. The S-layer protein (M_r = 51000) was extracted by H-bond disrupting agents like urea or guanidine hydrochloride (GHCl) and by ionic detergents such as sodium dodecyl sulphate (SDS). The protein on isolated cell walls was resistant to proteases with the exception of pepsin. Moreover the GHCl-extracted subunits formed regular self-assembly products on dialysis, even in the absence of specific metal cations. The reformation of the S-layer on GHCl-treated cell walls of *L. brevis* was also effective. Some experiments of heterologous re-attachment were successful with *L. fermentum* cell walls and unsuccessful with *L. plantarum* and *L. casei* cell walls.

Table 2. Crystalline surface layers on Lactobacillus species

Species	No of Strains		Characterization of S-layer			References
	S+	S-	M_r in kDa	Glycosylation	Lattice Type	
L. helveticus	2		51.5	--	ND	Masuda and Kawata, 1983
	15		51.0-58.0	--	p2	M. Sára, unpublished data
	10		52.0	ND	ND	S. Lortal, unpublished data
	1		52.0	--	p2	S. Lortal et al., 1992
L. acidophilus	4+/-		41.0	--	ND	Masuda and Kawata, 1983
	2+/-		50.0	--	p2	S. Lortal, unpublished data
	4+/-		46.0	+	ND	Bhowmik et al., 1985
	30+/-		46.0-59.0	ND	ND	Johnson et al., 1987
L. bulgaricus	1		51.5	--	ND	Masuda and Kawata, 1983
		3	ND	ND	ND	S. Lortal, unpublished data
L. fermentum	1		ND	ND	p2	Kawata, 1974
	2		51.5	--	ND	Masuda and Kawata, 1983
L. brevis	1		51.0	ND	p4	Masuda and Kawata, 1979
	2		51.0	--	ND	Masuda and Kawata, 1980
						Masuda and Kawata, 1983
L. casei	1		ND	ND	p6	Backer and Thorne, 1970
		3	ND	ND	ND	Masuda and Kawata, 1983
L. buchneri	1		55.0	ND	p6	Masuda and Kawata, 1981
	2		55.0	--	ND	Masuda and Kawata, 1983

+/-, for *L. acidophilus* the results between S+ and S- are variable and strain dependent.
ND, not determined.

The S-layer of *L. buchneri* was also studied by Masuda and Kawata (1981, 1985). The protein contained no carbohydrate and had a M_r = 55000. The hexagonal periodicity of the lattice was seen by freeze-etching and the center-to-center spacing was 6 nm. In order to identify the nature of the binding of the protein to the other cell wall components, the re-attachment of isolated subunits on chemically modified cell walls was attempted. The authors suggested a linkage with the neutral polysaccharide moiety of the cell wall. Both the positive charges (amino groups) and the negative charges (carboxyl groups) of the S-layer protein were required for the morphogenesis of the regular array and its binding to the underlying neutral polysaccharide.

In 1983, the presence of a surface layer on 13 different species of lactobacilli was investigated more systematically using SDS-PAGE electrophoresis and negative staining (Masuda and Kawata, 1983). *L. fermentum, L. brevis* and *L. buchneri* exhibited a surface layer as well as three species not previously described, *L. helveticus, L. bulgaricus,* and *L. acidophilus.* The M_rs of the S-layer proteins were in the range of 51000 to 55000 except for the *L. acidophilus* strain (M_r=41000), and all were considered to be not glycosylated. The amino acid composition of these S-layers was provided for the first time and, as for other bacterial surface layers, hydrophobic amino acids were in the range of 40-50% and the cysteine and methionine contents were low (Messner and Sleytr, 1992). Using an immunodiffusion assay these authors concluded that there were no common antigenic determinants among the S-layer proteins of the different *Lactobacillus* species.

The first extensive work concerning the S-layer of *L. acidophilus* was done by Bhowmik et al. (1985) and Johnson et al.(1987). The presence of a surface protein in this species has a particular importance because of its use in probiotic products. These authors confirmed that the presence of an S-layer was strain dependent. Chemical extraction suggested that H-bonds were important for S-layer attachment to the cell wall. The M_r (46000) and amino acid composition were slightly different from the previous results of Masuda and Kawata (1983). Moreover, this time, the S-layer of *L. acidophilus* was found to be glycosylated. The growth conditions (temperature, carbon source, atmosphere, and threonine and calcium supplementation) did not significantly influence the amount of S-layer synthesized. As shown by our own unpublished observation (Fig. 1), the lattice has an oblique symmetry with irregularities in some places. Surprisingly, freeze-drying of *L. acidophilus* whole cells induced a significant removal of the S-layer protein and should be done under the protective effect of glycerol (5%; v/v) (Ray and Johnson, 1986). The S-layer protein on whole cells was resistant to proteolytic enzymes with the exception of pronase. It was also sensitive to pepsin when isolated wall fragments were treated with these enzymes. In addition, the removal of the S-layer protein by pronase decreased the surface hydrophobicity to nearly zero (Fig. 2). From these last experiments the authors suggested a protective function for the S-layer as well as a role for binding the bacterium to the intestinal mucosa.

Several recent studies based on DNA-DNA hybridization revealed great genetic heterogeneity between strains that are presently designated as *L. acidophilus* (i.e., the six homology groups are A1 to A4, B1 and B2; Johnson et al., 1980). Since none of the strains in the B1 and B2 groups possess an S-layer protein, this feature could be used to differentiate them from the strains in the groups A1 to A4. All the strains in the A1 homology group showed a M_r = 46000 S-layer protein. The M_rs of the S-layer proteins in the groups A2, A3, A4 seem to be different from this. Future studies (e.g., N-terminal sequence, amino acid composition and antigenic specificity)

Figure 1. S-layer of *L. acidophilus* CNRZ 55 observed by freeze-etching. Bar = 100 nm. (unpublished data).

on the S-layer protein could offer a phenotypical way to differentiate the strains in the four homology groups (A1 to A4) without DNA homology data. This is important because "acidophilus preparations" for use as dietary adjuncts should contain bone fide *L. acidophilus* strains belonging to the A1 homology group (Johnson et al., 1987). Indeed, these strains have the desirable characteristics of rapid hydrolysis of lactose with a concomitant high lactic acid production.

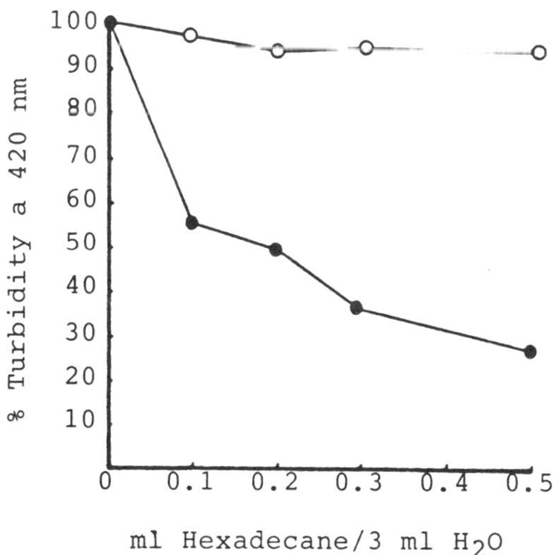

Figure 2. The hydrolysis of the S-layer protein of *L. acidophilus* by pronase drastically decreased the hydrophobicity of the cells. ●, percentage of untreated cells remaining in the aqueous phase ; O, the same for cells digested with pronase. (From Bhowmik et al., 1985 with permission).

The S-layer of *L. helveticus* has recently been studied more extensively (Lortal et al., 1992). The oblique lattice has a center-to-center spacing of 4.5/9.6 nm. A new procedure using lithium chloride was shown to extract the S-layer protein from intact cells efficiently and selectively without a drastic loss of viability. Interestingly, the S-layer reappeared when treated cells were allowed to grow in new medium (Fig. 3). The reappearance of the S-layer subunits was random. Self-assembly products are formed on dialysis of the LiCl crude extract against distilled water. The amino acid composition, M_r (52000) and N-terminal sequence (ATTINADSAINANTNAKYDVDVT) of the purified protein were determined. This surface protein accounts for 14% of the total protein content of *L. helveticus* as estimated from previous work (Lortal et al., 1991). More recently we attempted to obtain the exact molecular weight by mass spectrometry. This was 43533 and will be confirmed in the near future.

The presence of a surface layer on *L. casei* is still uncertain. Barker and Thorne (1970) observed whole cells of *L. casei* ATCC 7469 by negative staining and described an outer layer which was "irregular, amorphous, and having hexagonal subunits". Nevertheless, Masuda and Kawata (1983) did not find a regular array on *L. casei subsp. casei* ATCC 393.

The case of *L. bulgaricus* is also doubtful. The presence of a surface layer was assumed in this species (Masuda and Kawata, 1983). However, we have recently assayed three strains of *L. bulgaricus* without finding any regular array.

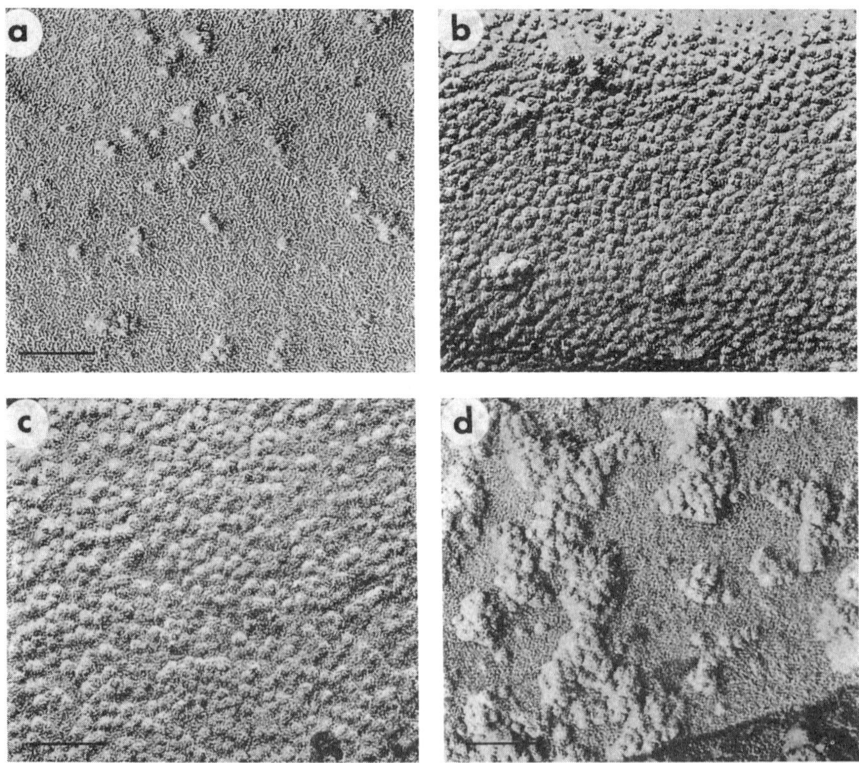

Figure 3. Freeze-dried preparations of *L. helveticus* labelled with polycationic ferritin (PCF), (a) untreated cells, (b) LiCl-treated cells, and (c) and (d) LiCl-treated cells after 8 h and 24 h in a new growth medium respectively. The reappearance of the S-layer is shown by the decrease in the number of bound PCF particles. Bar = 100 nm. (From Lortal et al., 1992).

REMAINING QUESTIONS AND PERSPECTIVES

Distribution of Surface Layers in the Genus *Lactobacillus*

Lactobacillus is a heterogeneous genus and the 50 species in the genus are not all useful for food technology, agriculture or human health. Only these more useful species have been surveyed for S-layers and it is impossible to accurately understand S-layer distribution throughout lactobacilli without a larger screening.

In the few biotechnologically significant species where S-layers have been described, as summarized previously in this chapter, the screening is only preliminary. Except for *L. acidophilus* and *L. helveticus* the number of strains tested is very limited. In addition several ambiguous cases like *L. casei* and *L. bulgaricus* have to be solved. Moreover, all the strains so far studied have come from national or international collections. Do S-layers exist under industrial conditions (e.g., those species used in food technology)? Are they present on the indigeneous strains of the human intestinal and urogenital tract? These questions have to be answered in order to determine the distribution and the stability of S-layers in this genus.

Ultrastructural and Biochemical Characteristics of *Lactobacillus* S-layer Proteins

Depending on the species, the S-layer lattice can have an oblique, square or hexagonal symmetry. The attachment of the S-layer to the neutral polysaccharide moiety of the cell wall in the case of *L. buchneri* has to be verified for other lactobacilli. The M_r of the S-layer proteins is between 41000-60000 and they appear to be non glycosylated (with the exception of *L. acidophilus*, which is still controversial, Table 2). In some cases the amino acid composition is available but there are no features which distinguishes them from other S-layer proteins. The N-terminal sequence plus the exact molecular weight has been determined for only one strain of *L. helveticus*. The accurate determination of the molecular weight by mass spectrometry could be an interesting way to see the variability of S-layer proteins between strains of the same species as well as providing precious information for further genetical approachs. Indeed, to our knowledge, total sequence analysis has never been attempted in lactobacilli, even though genetic systems have been developed (Chassy, 1987), and could be used to rapidly determine sequences or to obtain S^- mutants. The synthetic control of the S-layer proteins also remains unknown. The only observation shows that environmental conditions do not quantitatively influence this S-layer protein synthesis in *L. acidophilus*.

Surface Properties and Roles

Only two experiments can be reported in this part: (i) the S-layer of *L. helveticus* is not an effective barrier against the lytic action of the mutanolysin (Lortal et al., 1992) and (ii) the S-layer of *L. acidophilus* produces an apparent hydrophobic surface (Fig. 2).

Most of the lactobacilli described in this chapter are used in dairy processes. *L. helveticus* could be an interesting model to study the surface properties determined by the presence of an S-layer. Due to non-lethal extraction conditions using LiCl, the comparison of S^+ cells and S^- cells would be possible in this species. This may answer the following questions. Does the S-layer: (i) mask the underlying strong negatively charged peptidoglycan, (ii) protect the cells against the hydrolytic action of milk proteases and lytic enzymes, (iii) allow ionic interactions with milk proteins (caseins) which are then hydrolysed by the cell wall proteases (the proteolysis of milk is of great technological importance), (iv) act as a bacteriophage receptor as suggested by M. Callegari (unpublished data) and, (v) play a role in the interactions between cells? For the latter, spontaneous aggregation or conjugation has previously been observed in *L. helveticus* (Thompson and Collins, 1989).

CONCLUSION

Clearly, basic knowledge about the distribution, stability, ultrastructure and biochemical character of the S-layer has to be extended in lactobacilli. The surface properties of S-layer carrying strains should first be studied (eventually by a comparison between S$^+$ and S$^-$ strains derived by mutagenesis or mild S-layer extraction) in order to answer the most important question: Are S-layers essential for the various biotechnological applications of lactobacilli?

ACKNOWLEDGEMENTS

I wish to thank the colleagues who have critically read the manuscript, particularly J. van Heijenoort. I am deeply grateful to U.B. Sleytr for presenting my paper at the London, Canada NATO-ARW and to C. Hulin for her skillful typing.

REFERENCES

Barker, D.C., and Thorne, K.J., 1970, Spheroplasts of *Lactobacillus casei* and the cellular distribution of bactoprenol, *J. Cell Sci.* 7:755.

Bhowmik, T., Johnson, M.C., and Ray, B., 1985, Isolation and partial characterization of the surface protein of *Lactobacillus acidophilus* strains, *Int. J. Food Microbiol.* 2:311.

Chassy, B.M., 1987, Prospect for the genetic manipulation of lactobacilli, *FEMS Microbiol. Rev.* 46:297.

Daeschel, M.A., Andersson, R.E., and Fleming, H.P., 1987, Microbial ecology of fermenting plant material, *FEMS Microbiol. Rev.* 46:357.

Fernandez, C.F., Shahani, K.M., and Amer, M.A., 1987, Therapeutic role of dietary lactobacilli and lactobacillic fermented dairy products, *FEMS Microbiol. Rev.* 46:343.

Gerritse, K., Posno, M., Schellekens, M.M., Boersma, W.G.A., and Claassen, E., 1990, Oral administration of TNP-*Lactobacillus* conjugates in mice : A model for evaluation of mucosal and systemic immune responses and memory elicited by transformed lactobacilli, *Res. Microbiol.* 141:955.

Gilliland, S.E., 1990, Health and nutritional benefits from lactic bacteria, *FEMS Microbiol. Rev.* 87:175.

Johnson, J.L., Phelps, C.F., Cummins, C.S., London, J., and Gasser, F., 1980, Taxonomy of the *Lactobacillus acidophilus* Group, *Int. J. Syst. Bacteriol.* 30:53.

Johnson, M.C., Ray, B., and Bhowmik, T., 1987, Selection of *Lactobacillus acidophilus* strains for use in "acidophilus products", *Antonie van Leewenhoek* 18:469.

Kawata, T., Masuda, K., Yoskino, K., and Fujimoto, M., 1974, Regular array in the cell wall of *Lactobacillus fermenti* as revealed by freeze-etching and negative staining, *Japan. J. Microbiol.* 18:469.

Lortal, S., Rousseau, M., Boyaval, P., and van Heijenoort, J., 1991, Cell wall and autolytic system of *Lactobacillus helveticus* ATCC 12046, *J. Gen. Microbiol.* 137:549.

Lortal, S., van Heijenoort, J., Gruber, K., and Sleytr, U.B., 1992, S-layer of *Lactobacillus helveticus* ATCC 12046 : isolation, chemical characterization and reformation after extraction with lithium chloride, *J. Gen. Microbiol.* 138:611.

Masuda, K., and Kawata, T., 1979, Ultrastructure and partial characterization of a regular array in the cell wall of *Lactobacillus brevis, Microbiol. Immunol.* 23:941.

Masuda, K., and Kawata, T., 1980, Reassembly of the regularly arranged subunits in the cell wall of *Lactobacillus brevis* and their reattachment to cell walls, *Microbiol. Immunol.* 24:299.

Masuda, K., and Kawata, T., 1981, Characterization of a regular array in the wall of *Lactobacillus buchneri* and its reattachment to the other wall components, *J. Gen. Microbiol.* 124:81.

Masuda, K., and Kawata, T., 1983, Distribution and chemical characterization of regular arrays in the cell walls of strains of the genus *Lactobacillus, FEMS Microbiol. Lett.* 20:145.

Masuda, K., and Kawata, T., 1985, Reassembly of a regularly arrayed protein in the cell wall of *Lactobacillus buchneri* and its reattachment to cell walls : chemical modification studies, *Microbiol. Immunol.* 29:927.

McKay, L.L., and Baldwin, K.A., 1990, Applications of biotechnology : present and future improvements in lactic acid bacteria, *FEMS Microbiol. Rev.* 87:3.

Messner, P., and Sleytr, U.B., 1992, Crystalline bacterial cell-surface layers, *Adv. Microbial Physiol.* 33:213.

Perdigon, M., de Macias, M.E.N., Alvarez, S., Oliver, G., and de Ruiz Holgado, A.P., 1988, Systemic augmentation of the immune response in mice by feeding fermented milks with *Lactobacillus casei* and *Lactobacillus acidophilus, Immunology* 63:17.

Perdigon, G., de Macias, M.E.N., Alvarez, S., Oliver, G., and de Ruiz Holgado A.P., 1990. Prevention of gastrointestinal infection using immunobiological methods with milk fermented with *Lactobacillus casei* and *Lactobacillus acidophilus. J. Dairy Res.* 57:255.

Ray, B., and Johnson, M.C., 1986, Freeze-drying injury of surface layer protein and its protection in *Lactobacillus acidophilus, Cryo-Lett.* 7:210.

Seale, D.R., 1986, Bacterial inoculants as silage additives, *J. Appl. Bacteriol.* Symp. Suppl. 61: 9s.

Sharpe, M.E., 1979, Lactic acid bacteria in the dairy industry, *J. Soc. Dairy Technol.* 32:9.

Sugihara, T.F., 1985, The lactobacilli and streptococci : bakery products, *in*: "Bacterial starter cultures for food", S.E. Gilliland, ed., pp. 119-125, CRC Press, Boca Raton, FL.

Thompson, J.K., and Collins, M.A., 1989. Evidence for the conjugal transfer of a plasmid pVA797::pSA3 co-integrate into strains of *Lactobacillus helveticus.* Lett. Appl. Microbiol. 9:61.

Visser, R., Holzapfel W.H., Bezuidenhout, J.J., and Kotze J.M., 1986, Antagonism of lactic acid bacteria against phytopathogenic bacteria, *Appl. Environ. Microbiol.* 52:552.

Wallinder, I.B., and Neujahr, H.Y., 1971, Cell wall and peptidoglycan from *Lactobacillus fermenti, J. Bacteriol.* 105:918.

Chapter 7

ULTRASTRUCTURAL AND CHEMICAL CHARACTERIZATION OF A CYANOBACTERIAL S-LAYER INVOLVED IN FINE-GRAIN MINERAL FORMATION

Susanne Schultze-Lam and Terry J. Beveridge

Department of Microbiology
College of Biological Science
University of Guelph
Guelph, Ontario, Canada

INTRODUCTION

Cyanobacteria belonging to the *Synechococcus* group are unicellular, marine or freshwater organisms which gain energy from light through oxygenic photosynthesis and carbon from CO_2 (as HCO_3^- in the alkaline waters they usually inhabit). During CO_2 fixation, HCO_3^- is taken into the cell and OH^- is released, leading to alkalization of the microenvironment surrounding each cell (Miller and Colman, 1980). This pH effect has important geochemical implications in the natural environment.

Synechococcus GL24 was isolated from Fayetteville Green Lake, N. Y., a lake in which it has a demonstrated role in large scale geochemical processes (Thompson et al., 1990). These cyanobacteria are the dominant phytoplankton in the lake and are responsible for the formation of extensive calcitic "reefs" or "bioherms" (Fig. 1) where cells are bound to a solid substrate, and marl sediments (formed by a light but constant rain of calcite-encrusted cells to the lake bottom) (Thompson et al., 1990). Calcification is heralded by a whitening of the lake water during the summer months when the cells are most active. The formation of cell-bound calcite ($CaCO_3$) is at least a two-step process which begins with Ca^{2+} binding to the cell surface. This essential first step is the key to the whole process because nucleation sites are provided for subsequent mineral formation (Thompson and Ferris, 1990). SO_4^{2-}, which is abundant in the lake water, joins the bound calcium to build gypsum ($CaSO_4 \cdot 2H_2O$) crystals, a process which occurs readily at the bulk pH of the water (pH 7.9). As the cells become more active, releasing OH^-, the subsequent rise in pH around the cells leads to gypsum instability but encourages calcite formation, which may occur as replacement of SO_4^{2-} by CO_3^{2-} or as de novo calcite formation on the previously formed gypsum template (Thompson and Ferris, 1990). Thus the change in mineral is due to a rise in the microenvironment pH around each cell while the bulk pH of the surrounding water remains fairly constant; the presence of photosynthesizing cells is essential for calcification.

STRUCTURAL AND PHYSICOCHEMICAL STUDIES

Ultrastructural studies undertaken in order to learn where, on the cell surface, the initial mineralization events occur have shown that a hexagonal S-layer with a center-to-center spacing of 22 nm lies above the gram-negative envelope of this cyanobacterium (Schultze-Lam et al., in press). This S-layer shows various patterns in thin-section, depending on the orientation of individual patches to the plane of the cut (Fig. 2). When a cut has been made parallel to an axis of symmetry, the S-layer appears as a row of diamonds. This appearance is likely due to the binding of stain to exposed charged groups on the exterior of the protein while the hydrophobic interior remains unstained. Perpendicular cuts and tangential cuts reveal parallel lines and patterns of dots, respectively (Fig. 3).

Negatively-stained images of the *Synechococcus* S-layer show that the hexagonal pattern is made up of a series of "pinwheel" motifs which give the S-layer

Figure 1. Underwater photograph of the reef in Fayetteville Green Lake. Numerous *Synechococcus* cells have adhered to the solid substratum and been encased in calcite to form this structure on the lake's shore. Bar = 1 m.

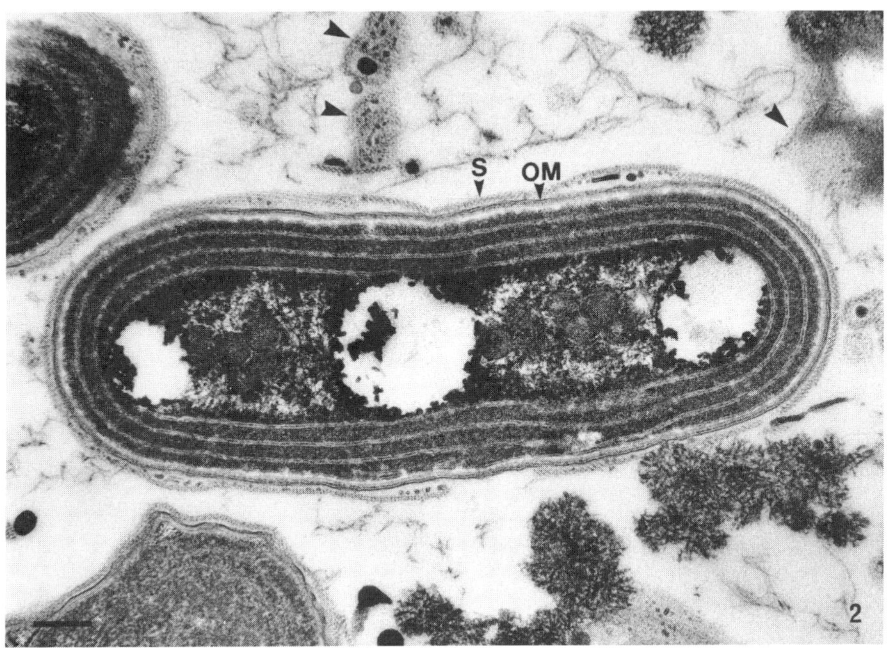

Figure 2. Longitudinal thin-section of a *Synechococcus* cell. The S-layer can be seen as a patterned structure encircling the cell. Interruptions in this layer are due to differential shrinkage of the S layer and the rest of the cell during processing. Numerous fragments that have been shed by the cell can be seen (arrows). The general features of a cyanobacterial cell are also apparent. S, S-layer; OM, outer membrane; T, thylakoid (photosynthetic) membrane; C, carboxysome; P, polyphosphate granule. Bar = 100 nm.

an open, lacy appearance (Fig. 4a). Such images have been subjected to computer image processing in order to show the approximate shape and arrangement of the constituent subunits (Fig. 4b; Schultze-Lam et al., in press).

When whole cells were probed with polycationized ferritin (PCF) at 0.5 mg/mL in 0.1 M cacodylate buffer, pH 7.1, no PCF was found on the cell which indicated that the surface is uncharged, positive, or that negative groups are not available for binding by the large (25 nm) PCF molecule due to steric hindrance (Schultze-Lam et al., 1992). However, even after pretreatment with 0.5% glutaraldehyde for 30 min prior to PCF labelling, which should cross-link amine groups and leave carboxyl groups free for PCF binding (Sára et al., 1989), no PCF was bound by *Synechococcus* cells. In all cases, some, but not all, fragments of sloughed-off membrane material were labelled. These were probably patches of outer membrane material with no overlying S-layer in which negatively charged groups on the lipopolysaccharide were exposed to PCF binding. When cells were treated with cytochrome c (also positive at neutral pH but only 3.7 x 2.5 x 2.5 nm in size; Takano and Dickerson, 1981), negatively stained, and subjected to computer image processing, it appeared that the cytochrome c bound preferentially in the larger pores of the S-layer (unpublished data). These results indicate that the intact array probably presents an uncharged surface to the external environment but that negatively charged sites exist within the large pores of the S-layer and on the LPS molecules in the outer membrane, exposed when patches of outer membrane are released from the cell.

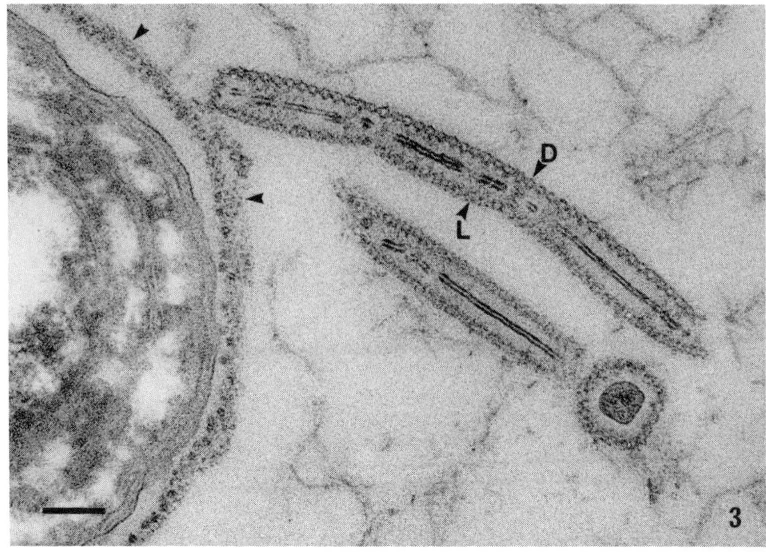

Figure 3. Thin-sectioned S-layer material showing details of the pattern. The diamonds (D) are revealed when the S-layer is cut parallel to an axis of symmetry. Parallel lines (L) are seen when a cut is made perpendicular to a line of symmetry. The dotted pattern (arrows) resulting from a tangential cut can be seen on the nearby cell. Bar = 50 nm.

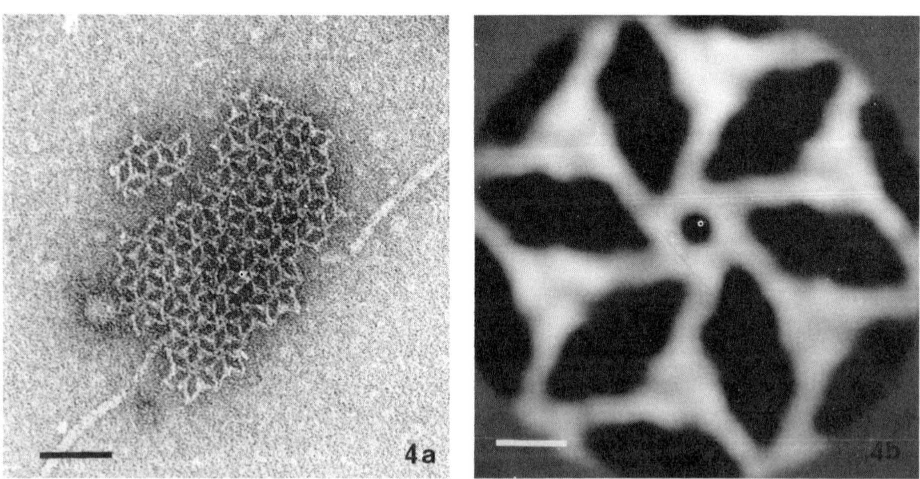

Figure 4a. S-layer fragment that has been negatively stained with 2% (w/v) aqueous uranyl acetate. The hexagonal pattern can be clearly seen and appears to be made up of numerous pinwheels (outline). Bar = 50 nm.

Figure 4b. Computer enhancement of an image similar to the one shown in Fig. 4a. The image is of a single morphological unit and details of the morphology of individual subunits can be seen. Bar = 10 nm.

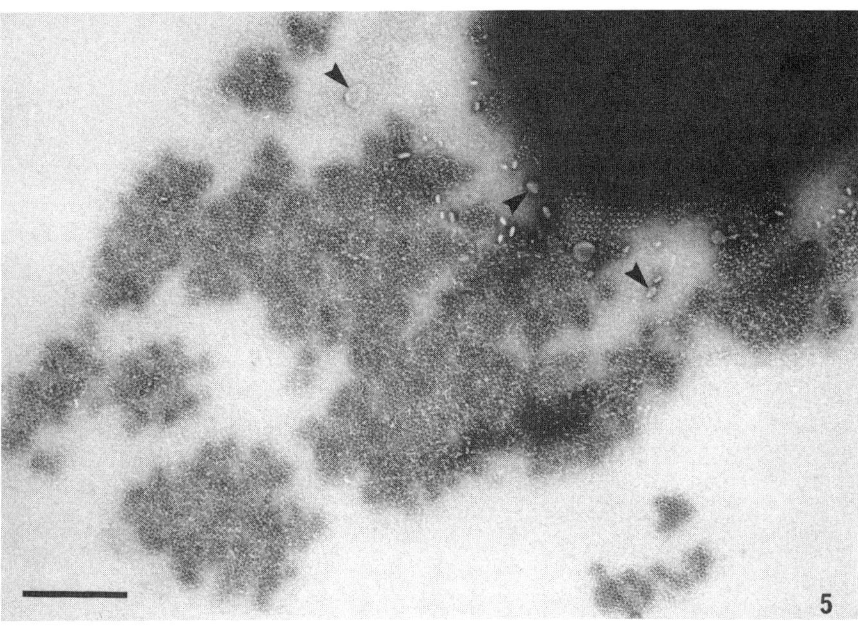

Figure 5. Suspension of *Synechococcus* cells in water and heating at 60°C for one hour leads to the release of S-layer accompanied by outer membrane material. Occasionally a sample will have very little underlying outer membrane material as in this case. Arrows indicate small outer membrane vesicles. Bar = 200 nm.

ISOLATION OF S-LAYER AND PRELIMINARY BIOCHEMICAL STUDIES

Attempts to preferentially remove the S-layer from whole cells and solubilize its constituent molecules showed that it is remarkably resistant to chemical degradation. The S-layer bound tenaciously to the underlying outer membrane, possibly by strong chemical linkage to lipopolysaccharide (LPS) molecules or outer membrane proteins. All the agents tested, which included a wide range of detergents (anionic, cationic, and nonionic), salts, denaturing agents, reducing and oxidizing agents and combinations of these at various pH and temperatures, failed to preferentially extract S-layer protein from underlying surface structures (as assessed by silver-stained SDS-PAGE gels) or cause noticeable disruption in the array pattern (as judged by electron microscopy of negatively stained whole mounts). However small patches of S-layer, often with accompanying outer membrane material, were released from cells by heating in 60°C water for one hour, and these were suitable for ultrastructural and electrophoretic analyses (Fig. 5). Bands could be seen in SDS-PAGE gels although the intensity of the stained bands tended to be low. Two bands were seen, one at M_r = 109 kDa and the other at M_r = 104 kDa. We suspect that the upper band represents a proportion of glycosylated molecules or ones which have LPS tightly associated with them. Glycosylation is difficult to verify due to the presence of contaminating LPS molecules, even with protein that has been electroeluted from SDS-PAGE gels. This strong association with LPS and the S-layer's resistance to release by strong chemical perturbants raises the possibilty of the protein being covalently bonded to an outer membrane constituent such as the LPS. The equally strong resistance to disruption of the array pattern itself also suggests the presence of very strong, possibly covalent subunit-subunit bonds.

LABORATORY BIOGEOCHEMICAL SIMULATIONS

When cells were grown in lake water to observe mineral formation, a significant (but not unexpected) observation was made. Cells in the early stages of mineral formation clearly showed the S-layer pattern in unstained whole mounts where the only contrast comes from the minerals present (Fig. 6a). The pattern became obscured as mineralization progressed (Fig. 6b). The proteinaceous subunits of the S-layer were encrusted by small micro-crystals of gypsum as verified by energy dispersive x-ray spectroscopy (EDS) and selected area electron diffraction (SAED) (Schultze-Lam et al., in press). The appearance of the unstained layer was similar to negatively-stained samples in that the pores appeared dark, indicating that they were clogged with electron-dense material (Fig. 7a). When these images were subjected to computer processing, subtle changes in the shape of the large pores could be seen. They appeared to be filling-in from the edges due to encrustation of the surrounding protein by the gypsum grains (Fig. 7b). The blurred appearance of the computer image results from an inherent loss of symmetry due to the presence of the irregularly shaped mineral aggregates.

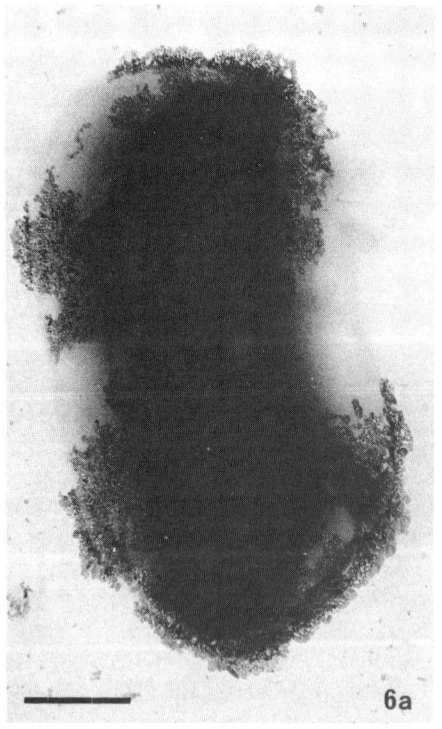

Figure 6a. Unstained whole mount of a *Synechococcus* cell in lake water. Mineralization has begun to occur on the cell surface but the S-layer pattern can still be seen in places (arrows). Bar = 200 nm.
Figure 6b. On this cell, mineral formation has progressed further until the encroaching gypsum crystals have almost obscured the S-layer pattern. Bar = 200 nm.

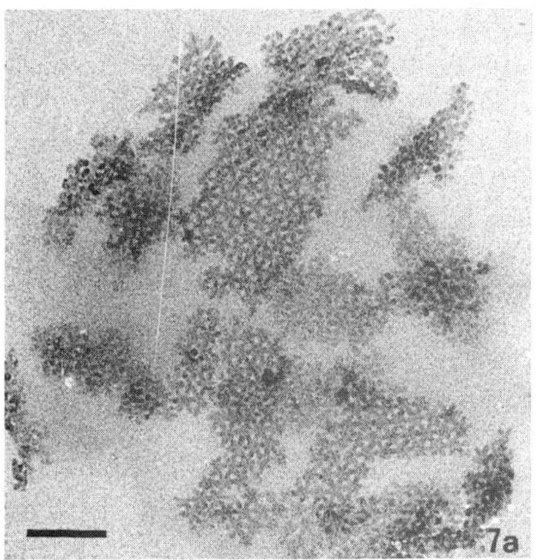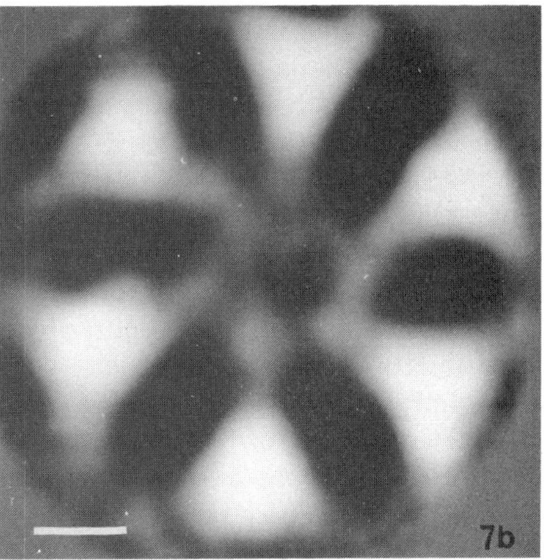

Figure 7a. Unstained whole mount of an S-layer fragment on which gypsum formation has begun. The only contrast in the image comes from the mineral grains present. The hexagonal S-layer pattern is clearly visible. Bar = 50 nm.

Figure 7b. Computer enhancement of an image similar to the one shown in Fig. 7a. The large diamond-shaped pores have become smaller and altered in morphology as the protein becomes encrusted with gypsum grains. The image appears blurred due to the presence of irregularly shaped crystals on the S-layer. Bar = 10 nm.

Mineral formation appears to begin in the pores of the S-layer, which must possess negatively charged sites to which Ca^{2+} can bind, initiating the mineralization process. The S-layer acts as a template for mineral formation by providing regularly spaced, chemically identical sites for initial binding events. By providing nucleation sites for mineral formation the template lowers the thermodynamic energy barrier, aiding and enhancing crystal formation. Eventually the pores of the S-layer are filled with fine-grain minerals. The growth is 3-dimensional and as the mineral grains increase in size the S-layer pattern is obscured. This encrustation can hardly be welcome to the cells since it is a seeming impediment to important physiological processes such as growth and division, nutrient uptake (mainly inorganic ions for *Synechococcus*) and photosynthesis. Cells shed minerals by shedding S-layer, together with some outer membrane material, on which the mineral grains continue to grow after their release. Shedding is commonly observed in our cultures (see Fig. 2) and cells in in situ lake water samples are generally found with great quantities of organic material which resembles the S-layer/outer membrane patches seen in our cultures. Thus the S-layer of *Synechococcus* GL24 seems to serve as a protective layer which intercedes between the external milieu and the delicate, physiologically important outer membrane beneath. By providing convenient sites for mineral formation, thereby actually encouraging this process on the S-layer, the cell is able to control where mineral formation will occur.

The ability of an S-layer to act as a template for mineral formation is a novel function that until now was not suggested for these structures. It also adds a new perspective to concepts of authigenic mineral formation by bacteria. It has been our experience that bacteria observed in fresh samples from both marine and freshwater environments often possess S-layers and are invariably associated with minerals of some type. Perhaps the presence of S-layers enhances the formation of minerals while at the same time protecting the bacterium from them. While mineral formation by bacteria that do not possess S-layers certainly does occur (Beveridge and Fyfe, 1985; Ferris et al., 1987a,b), the periodic nature and regularly-spaced nucleation sites provided by paracrystalline structures may allow a greater efficiency of mineral formation. We can envision that possession of S-layers represents an evolutionary adaptation to the threat of mineral encrustation by promoting mineral formation on an expendable and easily shed portion of the cell surface rather than at sites where it may interfere with the cell's metabolic activities (such as on the outer membrane of gram-negative or the cell wall of gram-positive bacteria).

The majority of S-layers are hexagonal lattices (Messner and Sleytr, 1992). Many of the most common minerals (calcite, gypsum, quartz, feldspar) also follow hexagonal symmetry schemes. It may be that over evolutionary time S-layer structure and mineral symmetries became increasingly complementary. It is clearly evident that S-layers are capable of serving as templates for mineral formation. It has been hypothesized (Cairns-Smith, 1985) that, in the 'primordial soup' that existed on Earth in its very early years, assembly and replication processes of organic molecules were catalyzed by mineral templates, which provided regularly-structured surfaces for these processes to occur. Here we have a case where minerals arise on an organic template, proving that the line between organic and inorganic structures may not be as clear as we have assumed. The Ca^{2+} that binds to the *Synechococcus* S layer is also a part of the mineral that forms on it, acting almost as a bridge between the two structures. The involvement of microorganisms, particularly cyanobacteria, in the formation of carbonate structures in depositional environments has been widely acknowledged, yet the actual mechanism of formation of the constituent grains has generally been attributed to abiogenic processes. It is widely believed that the cyanobacteria or algae merely entrap or adsorb the grains. This quite likely does occur, but we now have evidence that some may be active participants. The implications and magnitude of this association remain to be seen but we suspect that on a global scale over geological time they will prove to be significant.

ACKNOWLEDGEMENTS

This work was supported by the Natural Sciences and Engineering Research Council of Canada (NSERC) through an operating grant to T.J.B. and a PGSB award to S.S.-L. Electron microscopy was performed at the NSERC Guelph Regional STEM Facility which is partially funded by an NSERC infrastructure grant to T.J.B. The computer imaging facility at the University of Guelph Department of Molecular Biology and Genetics is funded by an NSERC equipment grant to T.J.B. and to G. Harauz, whose assistance is gratefully acknowledged.

REFERENCES

Beveridge, T.J., and Fyfe, W.S., 1985, Metal fixation by bacterial cell walls, *Earth Sci.* 22:1893.

Brunskill, G.J., 1969, Fayetteville Green Lake, New York II: precipitation and sedimentation of calcite in a meromictic lake with laminated sediments, *Limnol. Oceanogr.* 14:830.

Cairns-Smith, A.G., 1985, The first organisms, *Sci. Am.* 252:90.

Ferris, F.G., Fyfe, W.S., and Beveridge, T.J., 1987a, Manganese oxide deposition in a hot spring microbial mat, *Geomicrobiol. J.* 5:33.

Ferris, F.G., Fyfe, W.S., and Beveridge, T.J., 1987b, Bacteria as nucleation sites for authigenic minerals in a metal-contaminated lake sediment, *Chem. Geol.* 63:225.

Messner, P., and Sleytr, U.B., 1992, Crystalline bacterial cell surface layers, *Adv. Microbial Physiol.* 33: 213.

Miller, A.G., and Colman, B., 1980, Evidence for HCO_3^- transport by the blue-green alga (cyanobacterium) *Coccochloris peniocystis, Plant Physiol.* 65:397.

Sára, M., and Sleytr, U.B., 1989, Use of regularly structured bacterial cell envelope layers as matrix for the immobilization of macromolecules, *Appl. Microbiol. Biotechnol.* 30:184.

Schultze-Lam, S., Harauz, G., and Beveridge, T.J., 1992, Participation of a cyanobacterial S layer in fine-grain mineral formation, *J. Bacteriol.* (in press).

Schultze-Lam, S., Harauz, G., and Beveridge, T.J., 1992, Characterization of the S layer from the cyanobacterium *Synechococcus* GL24, Abstr. 92nd Annu. Meet. Amer. Soc. Microbiol.

Takano, T., and Dickerson, R.E., 1981, Conformation change of cytochrome C, *J. Mol. Biol.* 153:213.

Thompson, J.B., Ferris, F.G., and Smith, D.A., 1990, Geomicrobiology and sedimentology of the mixolimnion and chemocline in Fayetteville Green Lake, New York, *Palaois* 5:52.

Thompson, J.B., and Ferris, F.G., 1990, Cyanobacterial precipitation of gypsum, calcite, and magnesite from natural alkaline lake water, *Geology* 18:995.

Chapter 8

ADVANCES IN S-LAYER RESEARCH OF CHROOCOCCAL CYANOBACTERIA

Jan Šmarda

Department of Biology, Faculty of Medicine
Masaryk University
Brno, Czechoslovakia

Jiří Komrska

Department of Physics
Faculty of Mechanical Engineering
Technical University
Brno, Czechoslovakia

INTRODUCTION

Cyanobacteria are among the most ancient residents of the Earth and were the first oxygenic organisms to provide oxygen to the Earth's atmosphere by photosynthesis. Cyanobacteria appeared some 3.0-2.3 billion years ago, and have not evolved much since then. In today's biosphere they are ubiquitous, inhabiting the most diverse and extreme ecological niches. Many of them possess crystalline S-layers on the surface of their cell walls.

The recent taxonomic system of cyanobacteria (Anagnostidis and Komárek, 1985) divides these prokaryotes (Castenholz and Waterbury, 1989) into four orders: Chroococcales, Oscillatoriales, Nostocales and Stigonematales. S-layers have so far been noticed solely on the unicellular cyanobacteria which constitute the order Chroococcales (Šmarda, 1991) although their existence on the cell walls of filamentous species cannot be excluded.

Research into the S-layers on the cell walls of chroococcacean cyanobacteria during the last three years has been sporadic; the only publication of this period seems to be that by Šmarda (1991). Most of the present knowledge on the S-layers of cyanobacteria was published during the years 1977- 1988 and was briefly reviewed at the 2nd International Workshop on Crystalline Bacterial Cell Surface Layers in Vienna (Šmarda, 1988). The present chapter will review the progress in the field of S-layers of cyanobacteria since the last workshop.

Advances in Bacterial Paracrystalline Surface Layers
Edited by T.J. Beveridge and S.F. Koval, Plenum Press, New York, 1993

OCCURRENCE OF S-LAYERS IN CYANOBACTERIA

Material and Methods

For electron microscopic investigations, freeze-etching and freeze-fracturing techniques were used on 11 strains obtained from the Collection of Autotrophic Organisms (CAO), Institute of Botany, Czechoslovak Academy of Sciences, Třeboň, Czechoslovakia, and from the Collection of Algae, Biological Station, E.-M.-Arndt University Greifswald, Kloster/Hiddensee, Germany.

Enriched biological materials were fixed with 2 % (v/v) glutaraldehyde in 66 mM phosphate buffer, pH 7.2, washed twice with the same buffer and resuspended in water or in a graded series of cryoprotective glycerol (5 % - 30 %, v/v). Samples were freeze-fractured and -etched in a Balzers BA 360M device. Platinum-shadowed carbon replicas of fracture planes were cleaned with a concentrated mineral acid. Replicas were photographed in a Tesla BS 500 transmission electron microscope.

Results and Discussion

Genus: *Aphanocapsa*. An S-layer possessing a hexagonal lattice was found in the strain *Aphanocapsa rivularis* Pringsheim 1947/Camb. 1404-1 (Fig. 1) and a periodic surface layer was seen in the strain *Aphanocapsa* sp. 6308 (Šmarda, 1991).

Genus: *Merismopedia*. A periodic surface layer was found in the strain of *Merismopedia* sp. Hindák 1985/2 but not in the strain *M. elegans* Hoffman/Austin 1902 (Šmarda, 1991).

Genus: *Synechocystis*. We were able to prove that a hexagonal S-layer existed in the strain *Synechocystis aquatilis* Schiewer 428 (Šmarda, 1991) in which a periodic

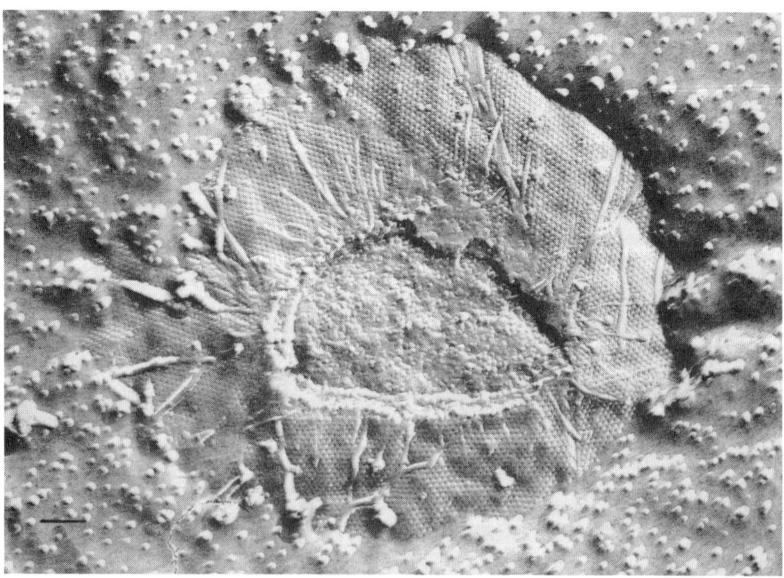

Figure 1. An S-layer in the Pringsheim 1947/Camb. 1404-1 strain of *A. rivularis* showing a hexagonal lattice. Bar = 0.1 μm.

Figure 2. Cells of the *S. aquatilis* Holubcová 1959/1 strain bear two distinct superimposed hexagonal S-layers underneath a slime sheath; the inner consists of finer subunits than the outer coarser lattice. Bar = 0.1 μm.

structure was first seen in thin sections by Schiewer and Jonas in 1977. (In fact, thiswas the very first published report of the existence of S-layers in cyanobacteria). Furthermore, two superimposed S-layers, both made up of hexagonal lattices, were discerned on the cells of *S. aquatilis*, strain Holubcová 1959/1. The inner and the outer layers consisted of fine and coarse uniform units, respectively (Fig. 2). Our finding of a hexagonal S-layer in *S. aquatilis* f. *salina* (Šmarda, 1988) was repeated in the strain Vaara 1978/CB-3 (Fig. 3). On the other hand, two strains of *S. fuscopigmentosa* (strain George 1954/Camb. 1412-4 and strain Hindák 1965/23) failed to show periodicity in their outer cell wall layer. A periodic structure, not characterized further, was found on cells of another strain of *Synechocystis* sp. Hindák 1985/3 (Šmarda, 1991).

Figure 3. S-layer in the *S. aquatilis* f. *salina* Vaara 1978/CB-3 strain showing a hexagonal lattice. Bar = 0.1 μm.

Genus: *Microcystis*. We observed a periodic cell wall surface layer in *Microcystis* sp. strain 691 (Šmarda, 1991) and a hexagonal network in *M. firma* Schiewer.

In conclusion, we found S-layers in 8 out of 11 studied strains of chroococcacean cyanobacteria (Table 1).

Table 1. Recent observations on S-layers in cyanobacteria (one strain of each taxon investigated)

Taxon	Lattice	Reference	Note
A. rivularis	p6	Šmarda 1991	Fig.1
Aphanocapsa sp.	P*	Šmarda 1991	
Merismopedia sp.	P	Šmarda 1991	
S. aquatilis	p6	Šmarda 1991	
S. aquatilis f. *salina*	p6	Šmarda 1991	Fig.3
Synechocystis sp.	P	Šmarda 1991	
M. firma	p6	this chapter	
Microcystis sp.	P	Šmarda 1991	

*P: periodic, without further characterization

We have so far observed hexagonal S-layers in five taxa of Chroococcales belonging to two subfamilies of the family Microcystaceae: in *A. rivularia*, *S. aquatilis*, *S. aquatilis* f. *salina* (Merismopedioideae), in *M. firma* and *M. incerta* (Microcystoideae).

It seems that in cyanobacteria the ability to form S-layers is not genus specific but, rather, species specific and can be regarded as an auxiliary taxomonic marker. This trait may be more useful at the level of strains as suggested for non-photosynthetic eubacteria. Strains may be found which consistently lack the ability to form S-layers (Messner and Sleytr, 1992; Sleytr et al., 1988). So far we have not detected the loss of S-layers during laboratory cultivation; presumably, S-layers are a constant, gene-encoded, structural marker for those cyanobacteria which possess them. Undoubtedly, further research will bring additional evidence of S-layer-forming strains in chroococcacean species, particularly in the family Microcystaceae.

In the five taxa studied, we then compared basic physical parameters of the S-layers in order to detect any correlations or differences among them. It was of interest to determine whether or not strains of a given taxon displayed structurally similar S-layers.

DENSITY OF S-LAYER SUBUNITS PER SURFACE AREA IN CYANOBACTERIA

Material and Methods

Suitable cut-outs from S-layer electron micrographs of *A. rivularis*, *S. aquatilis*, *S. aquatilis* f. *salina*, *M. firma* and *M. incerta* cells were obtained after enlargment of 200000x. In these, squares of 0.01 μm^2 were randomly outlined and the hexagonally-arranged uniform subunits were counted. For each species, a set of about 12 cells was evaluated statistically; each set was characterized by its average number of subunits per 0.01 μm^2 and its standard deviation of the average. The statistical significance of differences between the average subunit densities of the selected sets was estimated by the Man-Whitney test. In *S. aquatilis* only the inner S-layer was evaluated.

Results and Discussion

Averages of identical subunits per 0.01 μm^2 of S-layers on cells of the five taxa are given in Table 2.

Table 2. Density of subunits in S-layer arrays of cyanobacteria

Taxon	Number of strains evaluated	Average no. of subunits per 0.01 μm^2	Standard deviation
A. rivularis	1	65.5	± 3.2
S. aquatilis	2	43.8	± 6.1
S. aquatilis f. *salina*	2	33.8	± 2.1
M. firma	1	38.7	± 1.2
M. incerta	1	29.3	± 3.1

Clear differences were revealed between the subunit densities in the S-layers of *A. rivularis*, *S. aquatilis* and *M. incerta*, as well as between *M. firma* and *M. incerta*. The two strains of *S. aquatilis* showed the same density. Similarly, no difference in density was noted in the two *S. aquatilis* f. *salina* strains. The differences between the array densities of *S. aquatilis* f. *salina* and the species of *Microcystis* appeared less clear-cut.

Most of the differences were so large that they could not be artefacts as a result of the preparation procedures (e.g., dehydration, freezing, drying, etc.). Statistical analysis showed highly significant differences ($P < 0.01$) in the densities of hexagonal crystalline arrays of *S. aquatilis*, *A. rivularis*, *M. incerta* and, surprisingly, also *S. aquatilis* f. *salina*. The difference in the density between *S. aquatilis* and *M. firma* was not significant.

The S-layer subunit densities showed clear diversity at the species level but, because the method of estimation was not exact enough and the number of analysed strains was limited, we were not able to establish their strain specificity. The crystalline arrays of S-layers in five taxa were analysed in more detail by optical diffraction.

OPTICAL DIFFRACTION ANALYSIS OF ELECTRON MICROGRAPHS OF S-LAYERS

Material and Methods

To get objective values of subunit spacings in S-layers, the electron micrographs were analysed by optical diffraction (Horne and Markham, 1973). The optical diffractometer consisted of a 4 m long OSK-2 optical bench containing a He-Ne laser. The diffraction patterns were taken from areas involving several hundreds of subunits.

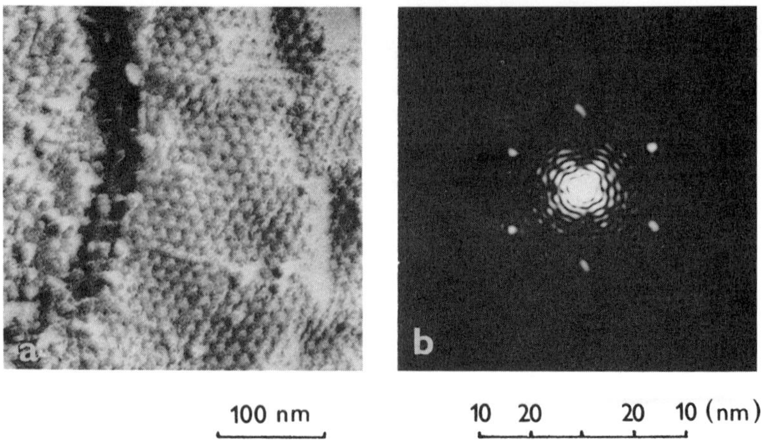

100 nm   10 20   20 10 (nm)

Figure 4. (a) S-layer in *A. rivularis* and its corresponding optical diffraction pattern (b) taken from a 160 nm diameter circle of S-layer.

The bench was set up so that the reciprocal spacings on the S-layer image produced a transform which did not exceed the 60 x 60mm film used to record them.

The five taxa of cyanobacteria described in the previous section were analysed and three images of S-layers on separate cells of each taxon were investigated.

Results and Discussion

Although only low-order reflections were produced in the optical transforms of freeze-etching images, it was apparent that each S-layer was a hexagonal network (Figs. 4 and 5).

In some cases the hexagonal array was deformed by up to 10 angular degrees; this was possibly due to the preparation procedures. Spacings of subunits in the investigated S-layers are listed in Table 3.

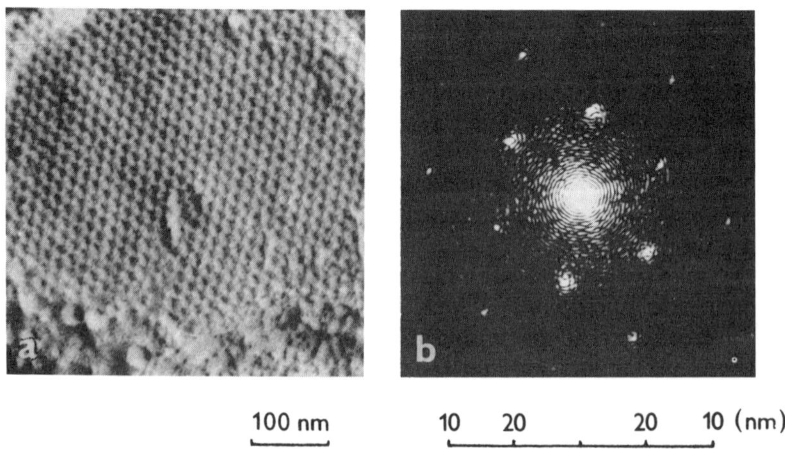

100 nm   10 20   20 10 (nm)

Figure 5. (a) S-layer in *S. aquatilis* and its corresponding optical diffraction pattern (b) taken from a 400 nm diameter circle of S-layer.

Table 3. Range of S-layer subunit spacings in the five taxa of cyanobacteria

Taxon	Range of spacings (nm)
A. rivularis	11.3 - 12.6
S. aquatilis	14.5 - 15.3
S. aquatilis f. *salina*	15.2 - 17.3
M. firma	14.3 - 16.1
M. incerta	17.0 - 18.0

The ranges of spacings showed marked differences among the five taxa. The most obvious difference was between *A. rivularis* (characterized by a low center-to-center spacing) and *M. incerta* (with a high spacing value). All spacing values were in agreement with the subunit density data reported in the previous section. No difference in subunit spacing was found between strains of the same species but considerable differences were seen between various species of one genus.

For all the strains of *S. aquatilis* and *Synechocystis* sp. so far analysed (Lounatmaa et al., 1980; Karlsson et al., 1983; Šmarda, 1991) subunit spacings in the range of 14.4 nm to 15.5 nm appear to be typical. Surprisingly, the spacing in *S. aquatilis* f. *salina* was distinctly larger. For *M. incerta*, the direct measurement in highly enlarged electron micrographs showed a spacing of 19.3 nm (Šmarda, 1991) which was slightly higher than the value found by diffraction analysis.

SUMMARY

1. Since 1988, hexagonal S-layers have been found in the following taxa of cyanobacteria: family Microcystaceae; *A. rivularis*, *S. aquatilis* and *S. aquatilis* f. *salina* of the subfamily Merismopedioideae, and *M. firma* and *M. incerta* of the subfamily Microcystoideae. In addition, periodic arrays of undistinguishable lattice were noticed in *Aphanocapsa* sp., *Merismopedia* sp., *Synechocystis* sp. and *Microcystis* sp.
2. In one strain of *S. aquatilis*, two superimposed hexagonal S-layers were found.
3. All four strains of *S. aquatilis* formed S-layers, whereas some of *M. firma* and *M. incerta* were unable to.
4. Each of the five taxa investigated showed a specific density of subunits in hexagonal patterns: in *A. rivularis*, 66 subunits per 0.01 μm^2 of the S-layer were found as compared to a mere 29 protomers in *M. incerta*. The difference in subunit density between *S. aquatilis* and *M. firma* was not significant.
5. Optical diffraction revealed distinct subunit spacings for each of the taxa examined with the exception of *S. aquatilis* and *M. firma* whose spacings were equal. The lowest spacing value was found for *A. rivularis* (11.3 nm - 12.6 nm) and the highest in *M. incerta* (17.0 nm - 18.0 nm).

ACKNOWLEDGEMENTS

The authors wish to thank V. Trávníčková for skilled technical assistance in freeze-etching procedures and to H. Koukalová for statistical analyses.

REFERENCES

Anagnostidis, K., and Komárek, J., 1985, Modern approach to the classification system of cyanophytes. 1 - Introduction, *Arch. Hydrobiol. Suppl./Algolog. Studies* 38/39:291.

Castenholz, R.W., and Waterbury, J.B., 1989, Cyanobacteria, *in*: "Bergey's Manual of Systematic Bacteriology", Vol. 3, J.T. Staley, M.P. Bryant, N. Pfennig, and S.G. Holt, eds., Williams and Wilkins, Baltimore.

Horne, R.W., and Markham, R., 1973, Application of optical diffraction and image reconstruction techniques to electron micrographs, *in*: "Electron Diffraction and Optical Diffraction Techniques", A.M. Glauert, ed., North-Holland Publishing Co., Amsterdam.

Karlsson, B., Vaara, T., Lounatmaa, K., and Gyllenberg, H., 1983, Three-dimensional structure of the regularly constructed surface layer from *Synechocystis* sp. strain CLII, *J. Bacteriol.* 156:1338.

Lounatmaa, K., Vaara, T., Österlund, K., and Vaara, M., 1980, Ultrastructure of the cell wall of a *Synechocystis* strain, *Can. J. Microbiol.* 26:204.

Messner, P., and Sleytr, U.B., 1992, Crystalline bacterial cell-surface layers, *Adv. Microbial Physiol.* 33:213.

Schiewer, U., and Jonas, L., 1977, Die Wirkung unterschiedlicher NaCl-Konzentrationen auf die Ultrastruktur von Blau-Algen II. *Synechocystis aquatilis. Arch. Protistenk.* 119:146.

Sleytr, U.B., Messner, O., and Pum, D., 1988, Analysis of crystalline bacterial surface layers by freeze-etching, metal shadowing, negative staining and ultrathin sectioning, *in*: "Methods in Microbiology", Vol. 20, F. Mayer, ed., Academic Press Ltd., New York.

Šmarda, J., 1988, S-layers in Cyanobacteria, *in*: "Crystalline Bacterial Cell Surface Layers", U.B. Sleytr, P. Messner, D. Pum, and M. Sára, eds., Springer-Verlag, Berlin.

Šmarda, J., 1991, S-layer of chroococccal cell walls, *Arch. Hydrobiol. Suppl./Algolog. Studies*, 64:41.

Chapter 9

PREDATION ON BACTERIA POSSESSING S-LAYERS

Susan F. Koval

Department of Microbiology and Immunology
University of Western Ontario
London, Ontario, Canada

INTRODUCTION

Bacteria in many habitats - terrestrial or aquatic (marine or freshwater) -are likely to find themselves in the company of potential predators. *Bdellovibrio* spp. are gram-negative eubacteria that are predacious upon other gram-negative eubacteria. Free-living phagotrophic protozoa are an important part of the microbial food cycle because they graze (feed) upon smaller protozoa and bacteria. It is important to ask if paracrystalline protein surface arrays (S-layers) serve as a protective barrier against predation by *Bdellovibrio* or protozoa. Specific descriptions of bacterial S-layers often postulate functions but these are rarely tested. Studies involving predation provide an excellent opportunity to experimentally assay a specific function. The abundance of S-layers in nature indicates that they must fulfil vital selective functions.

This chapter reviews the results of studies on predation by *Bdellovibrio bacteriovorus* (Koval and Hynes, 1991) and presents some recent observations on grazing by the ciliated protozoan *Tetrahymena thermophila*.

BACTERIAL PREY

Bacteria with S-layers that have been well characterized both structurally and biochemically were chosen as potential prey for *B. bacteriovorus* and *T. thermophila* (Table 1). S-layer variants of some of these organisms were also available. The S-layers of *Aeromonas salmonicida*, *Campylobacter fetus*, *Caulobacter crescentus* and *Synechococcus* are described in other chapters in this book. Descriptions of the other S-layers in Table 1 are given as follows: *Aquaspirillum serpens* VHA (Koval and Murray, 1981), *A. serpens* MW5 (Kist and Murray, 1984), *A. sinuosum* (Smith and Murray, 1990), *Bacillus brevis* (Yamada et al., 1981).

Advances in Bacterial Paracrystalline Surface Layers
Edited by T.J. Beveridge and S.F. Koval, Plenum Press, New York, 1993

Table 1. Bacterial Prey

Organism	No. of S-layers	Symmetry	Source
Aquaspirillum serpens VHA	1	p6	R.G.E. Murray
Aquaspirillum serpens VHL	-		
Aquaspirillum serpens MW5	2	p6	R.G.E. Murray
Aquaspirillum sinuosum	2	p4, p6	R.G.E. Murray
Aeromonas salmonicida A449	1	p4	W.W. Kay
Aeromonas salmonicida A449-3	-		
Bacillus brevis 47	2	p6	R.G.E. Murray
Campylobacter fetus 23D	1	p6	M.J. Blaser
Campylobacter fetus 23B	-		
Caulobacter crescentus CB2NY66R	1	p6	J. Smit
Caulobacter crescentus CB2A	-		
Synechococcus GL24	1	p6	T.J. Beveridge

LIFE CYCLE OF *BDELLOVIBRIO BACTERIOVORUS*

B. bacteriovorus is an aerobic, obligate predator with a biphasic life cycle (Thomashow and Rittenberg, 1979). Single, motile, attack-phase cells encounter potential prey cells by random collision. The range of susceptible prey cells varies with the *Bdellovibrio* strain but is confined to gram-negative eubacteria. The chemical nature of prey-cell identification is unknown, but is not as restrictive as that for bacteriophage attachment, so host range is broad (Gray and Ruby, 1991). The irreversible attachment allows penetration through the outer membrane and peptidoglycan sacculus by mechanical "drilling" and enzymatic attack (Tudor et al., 1990). The growth phase is in the periplasmic space of the prey, and utilizes nutrients within the prey cell protoplasm (Fig. 1). Within the first 30 minutes two important events occur: the conversion of the prey cell into a stable spherical structure termed a bdelloplast and the destruction of the prey's ability to generate energy. During the subsequent 2-3 hours, the bdellovibrio grows into a long, aseptate filament, which subsequently divides into individual, motile cells. Production of a lytic enzyme by the progeny allows their release from the bdelloplast into the environment.

PREDATION BY *BDELLOVIBRIO BACTERIOVORUS* ON BACTERIA WITH S-LAYERS

When eubacteria are first isolated from nature, those possessing S-layers are completely covered by the protein array. S-layers on gram-negative eubacteria would mask many outer membrane components and receptors, including potential attachment sites for bdellovibrios. We examined this barrier function by studying the interaction of the gram-negative bacteria listed in Table 1 with different strains of *B. bacteriovorus*. We found that not all the S-layer⁻ prey cells were susceptible to the same predator strain. For example, *A. serpens* was susceptible to *B. bacteriovorus* strains 6-5-S and 109J; *A. salmonicida* was susceptible to strain 109J, but not strain 6-5-S; *C. fetus* was susceptible to strains 6-5-S and 109J. *C. crescentus* was not susceptible to either strain 6-5-S or 109J, but was successfully attacked by a new isolate of a *Bdellovibrio*-like bacterium strain JSS (Koval and Hynes, 1991).

    B. bacteriovorus can multiply in the presence of viable, but nonproliferating prey cells. Thus, studies on predator-prey interactions can be conducted using prey cells suspended in an appropriate buffer (Koval and Hynes, 1991). Alternatively, dilute culture medium such as 1/10 strength yeast extract-peptone medium (YP/10) can be used (Stolp, 1981). We chose a buffer system or dilute medium supplemented with calcium and/or magnesium that provided optimal conditions for maintenance of S-layers and predation by *Bdellovibrio*. Predation was monitored by phase contrast microscopy, decrease in turbidity, and an increase in plaque-forming units (PFU) /ml. Control cultures of prey cells alone were included in experiments.

    Two-membered culture systems in HEPES buffer (N-2-hydroxyethylpiperazine-N'-2-ethane-sulfonic acid) plus 2 mM $CaCl_2$, pH 7.8 (HEPES plus Ca^{2+}) at 30° were used to assess the predation of *Aquaspirillum* species. At 48 hours, the decrease in turbidity of the two-membered cultures containing *A. serpens* VHA, *A. serpens* MW5, and *A. sinuosum* as prey cells was not significantly different from that of prey cells alone in buffer. No bdelloplasts were seen by light or electron microscopy, and there was no increase in PFU/ml. Therefore, these species were not attacked by *B. bacteriovorus*. An isogenic S-layer⁻ variant of *A. serpens* VHA, strain VHL, was

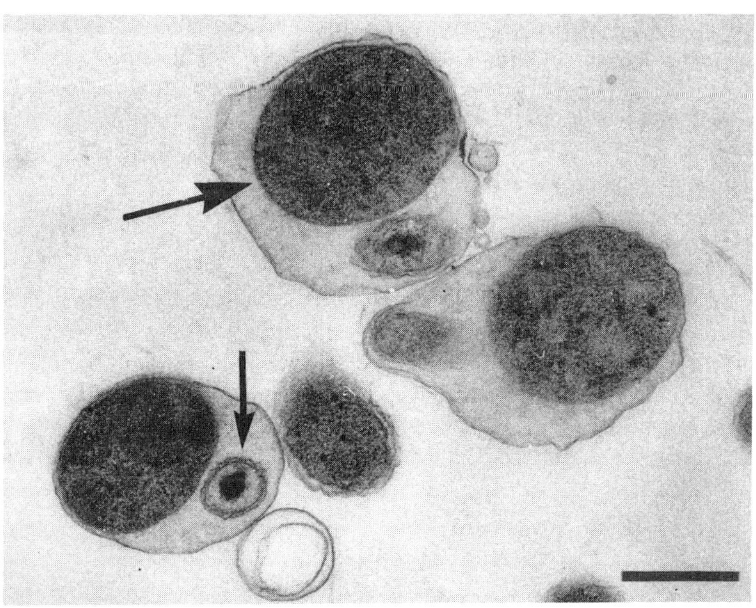

Figure 1. Thin section of bdelloplasts of *Campylobacter fetus* 23B. Large arrow, protoplast of *C. fetus*. Small arrow, *Bdellovibrio bacteriovorus* 109J. Bar = 500 nm.

rapidly attacked by *B. bacteriovorus* and produced a 10^2 increase in PFU/ml. No isogenic S-layer variants of *A. serpens* MW5 or *A. sinuosum* were available.

Two-membered culture systems in Tris buffer plus Ca^{2+} and Mg^{2+} were used to assay the predation of *A. salmonicida* A449 and A449-3 (Koval and Hynes, 1991). *A. salmonicida* cells were not stable in HEPES plus Ca^{2+}, and there was a significant decrease in turbidity of all cultures during the incubation period. This predation was performed at 22°C, because at temperatures above 24°C *A. salmonicida* cells spontaneously lose their S-layer (Ishiguro et al., 1981). *A. salmonicida* A449-3 (S-layer) was slowly attacked by *B. bacteriovorus*, and showed a 10^2 increase in PFU/ml at 48 h. *A. salmonicida* A449 (S-layer+) was not attacked.

The predation of *C. crescentus* by the *Bdellovibrio*-like bacterium strain JSS, was studied in YP/10 plus Ca^{2+} (Koval and Hynes, 1991). The S-layer strain of *C. crescentus*, strain CB2A, was stable in HEPES and Tris buffers plus Ca^{2+}, but not strain CB2NY66R (S-layer+). *Caulobacter* is often found in oligotrophic environments and readily grows in dilute media in vitro. Thus YP/10 allowed some initial growth of both strains of *C. crescentus*. In the presence of strain JSS, *C. crescentus* CB2A was slowly lysed, with a final increase of 10^2 PFU/ml. Over the same time period there was no predation on *C. crescentus* CB2NY66R.

Studies with *C. fetus* strains 23D and 23B have shown that the S-layer also protects these cells from predation by *B. bacteriovorus* (S.F. Koval and M.J. Blaser, unpublished).

Bacteria lacking an S-layer are susceptible to predation by *B. bacteriovorus*, and this means that organisms in nature must be under strong selection pressure to maintain a surface that is entirely covered by one (or more) S-layer(s). Studies with two strains of *A. serpens* that were not fully covered by an S-layer showed that *B. bacteriovorus* eventually found the exposed outer membrane surface and attached.

Gram-negative bacteria included in these studies as prey cells are all members of the *Proteobacteria* that are found in freshwater environments. They include two pathogens (*A. salmonicida* and *C. fetus*), an oligotroph (*C. crescentus*), and spirilla commonly found in freshwater. Therefore, it can be concluded that this protective role is a general function for S-layers in nature. *Bdellovibrio* spp. are widely distributed in most aquatic and terrestrial habitats (Varon and Shilo, 1980; Ruby, 1992) and the survival of gram-negative bacteria would be aided by the presence of an S-layer.

FEEDING MECHANISMS OF PROTOZOA

If bacteria in nature manage to avoid an encounter with bdellovibrios, they may in turn be ingested by protozoa. Bacteria can be a major food source for protozoa and it is therefore important to determine if S-layers protect a bacterium from predation by protozoa. The study of this biological function of S-layers is more complex than the studies with *B. bacteriovorus*, because it concerns a group of eucaryotic microorganisms that display an enormous variation in size, are structurally complex and exhibit a diversity in feeding mechanisms (Fenchel, 1987). It is necessary to briefly consider these feeding mechanisms before we consider the effect of S-layers.

Free-living protozoa that ingest particulate organic matter are called "phagotrophic" or "bacterivorus". Phagocytosis is the process whereby a food particle is enclosed in a membrane-covered vacuole in which digestion takes place. This may occur at one specialized site (cytosome), as is usual in protists with pellicles (e.g. ciliates), or, food vacuoles may be formed over much or all of the body surface, as in the naked protists of many flagellate groups.

Protozoa have developed various mechanisms allowing dilute food particles in the environment to be concentrated prior to phagocytosis. The variety of these adaptations contributes to the structural diversity of protozoan forms. Studies on protozoan bacterivory have concentrated on two main groups, the ciliated protozoa and the flagellated protozoa. Different groups of these protozoa are specialized for grazing on attached microorganisms, or for direct interception of passing cells (raptorial feeders). Raptorial feeders can feed on relatively large particles, can select between prey types, and may even specialize on one prey species (Fenchel, 1987). Other groups of these protozoa can also feed on freely suspended, relatively small food particles by a filter feeding mechanism. The ciliates exhibit the greatest diversity and specialization for filter feeding. Two essential features of filter feeders are the presence of a filter and a means of propelling water through it. Flagella or cilia are used to create the water flow through the filtering structures. The filter may be a size-limiting mechanical sieve or may be composed of sticky bars that aid adhesion of particulate matter.

Filter feeders have a constant feeding rate and lack the ability to discriminate between different kinds of particles except for their mechanical properties such as size and shape (Fenchel, 1980a). *Paramecium* and *Tetrahymena* are two well-studied ciliates that will ingest inert particulate matter as readily as nutritionally useful particles. This feature has been utilized to estimate in situ protozoan grazing on bacteria, as a part of studies on microbial food webs in aquatic ecosystems. The uptake of latex beads (Fenchel 1980a), fluorescent latex microspheres (Sanders, 1988), or fluorescently-labelled bacteria (FLB) (Scherr et al., 1987) have proved very useful for these studies. The FLB were not toxic to the protozoa assayed and could be metabolized to support rapid growth of both ciliates and flagellates.

Most protozoa have a much lower surface-to-volume ratio than bacteria and so require a higher food concentration for survival. This requirement for nutrients affects the ecology of protozoa. In the open waters of oceans and lakes, bacterial numbers are too low to sustain ciliates (Fenchel, 1980b). In marine environments, flagellated protozoa are most often responsible for bacterial mortality (Barcina et al., 1992). Ciliates are found in systems with a higher load of organic matter (sewage treatment plants, polluted waters, etc.) and in benthic sediments.

PREDATION BY *TETRAHYMENA THERMOPHILA* ON BACTERIA WITH S-LAYERS

All of the bacteria listed in Table 1 are freshwater species and ciliates were used for the initial studies with protozoa. The ciliate *T. thermophila* was chosen since it can be grown axenically in vitro and the rate of formation of food vacuoles and the process of digestion have been well studied (in *T. pyriformis*) (Nilsson, 1987). Bacterivory was estimated by use of FLB, prepared by a modification of the technique of Scherr et al. (1987), because their heat treatment at 60°C for 2 h was not compatible with the maintenance and structural integrity of the S-layers of bacteria listed in Table 1. The phototrophic cyanobacterium, *Synechococcus*, was autofluorescent. Both gram-negative and gram-positive bacteria could be used in these experiments, unlike the predation studies with *B. bacteriovorus*, because protozoa cannot discriminate between these two cell types.

Initial studies have shown that *T. thermophila* efficiently ingested S-layer⁺ cells of *A. serpens, B. brevis, C. fetus, C. crescentus,* and *Synechococcus* (S.F. Koval and D.H. Lynn, unpublished). *A. salmonicida* was not included in the study because of the instability of the S-layer above 22°C. All ingested bacteria were densely packed and brightly fluorescent within the food vacuoles. *A. serpens* VHA was the largest bacterium used as a prey cell, and *T. thermophila* could pack four or five of these cells into a food vacuole. Comparison of *T. thermophila* uptake of FLB of S-layer$^+$ bacteria and an S-layer$^-$ control, *Enterobacter aerogenes*, showed that the ciliate did not discriminate between S-layer$^-$ and S-layer$^+$ bacteria. These results agree with unpublished observations (Fig.2) by F.G. Ferris and T.J. Beveridge on the ingestion by *Tetrahymena* of the S-layer$^+$ gram-positive bacterium *Sporosarcina ureae* (Beveridge, 1979). We can conclude that S-layers do not protect bacteria from ingestion by ciliates.

Do S-layers protect bacteria from digestion inside food vacuoles? S-layers vary in their susceptibility to proteolysis and chemical denaturants (Koval and Murray, 1985; Messner and Sleytr, 1992) and this may affect digestibility. Digestion is best monitored by thin-section electron microscopy and has been well documented for *Tetrahymena* (Nilsson, 1987). Our studies have shown that 75 minutes after ingestion, *C. crescentus* CB2NY66R was readily digested inside the food vacuoles (Fig. 3) but that *Synechococcus* was more resistant to digestion (S.F. Koval and D.H. Lynn, unpublished). These results agree well with the S-layer sensitivity of *C. crescentus* (Chapter 18) and the resistance of *Synechococcus* (Chapter 7) to chemical degradation. Further studies are needed to extend these results to other bacteria.

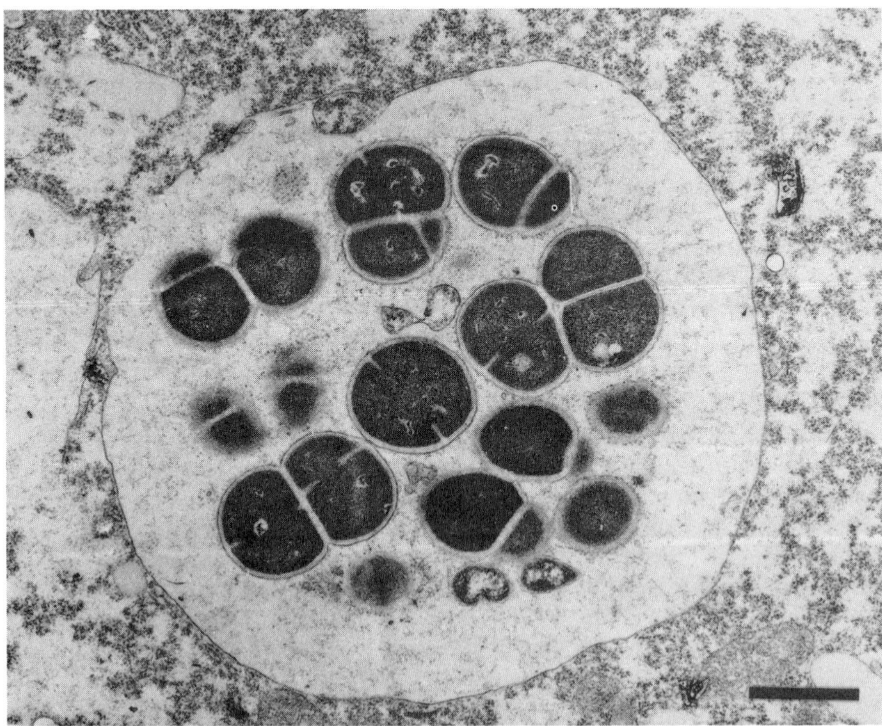

Figure 2. Thin section of *Sporosarcina ureae* inside a food vacuole of *Tetrahymena*. Bar = 500 nm. Micrograph courtesy of F.G. Ferris and T.J. Beveridge.

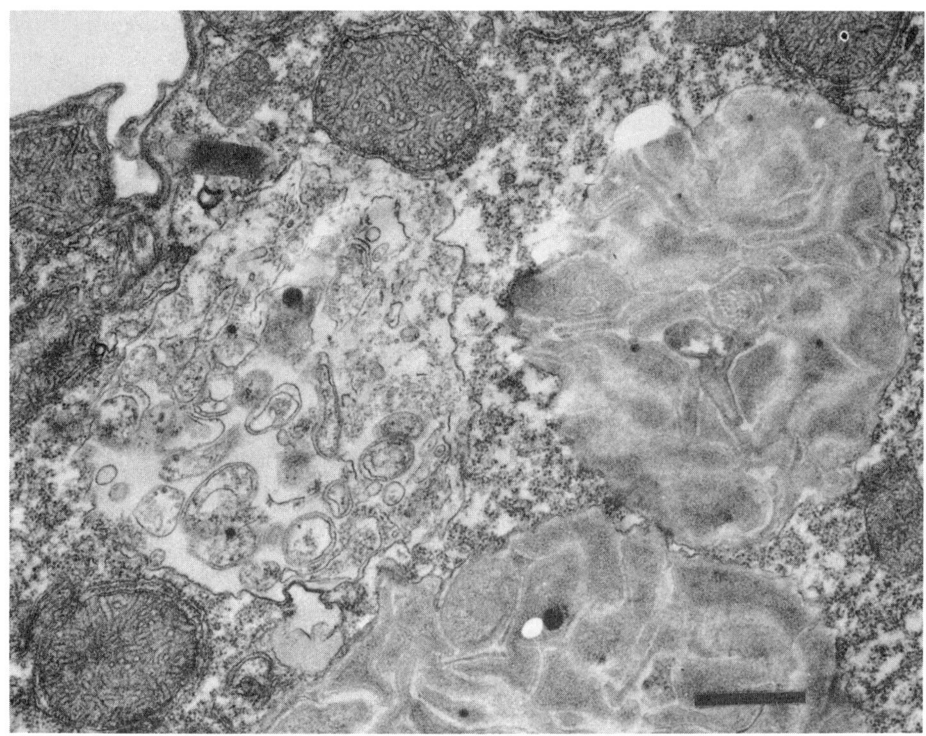

Figure 3. Thin section of digested *Caulobacter crescentus* CB2NY66R inside food vacuoles of *Tetrahymena thermophila*. Bar = 500 nm.

CONCLUSIONS

It has been conclusively demonstrated that S-layers protect bacteria from predation by bdellovibrios. The filter-feeding ciliated protozoa can efficiently ingest bacteria with S-layers, although not all ingested prey may be digested equally as well. These studies need to be extended to filter-feeding flagellates, to appreciate in situ protozoan grazing in marine environments. These predatory studies, utilizing an ecological approach and considering natural habitats, have enhanced our knowledge of the biological functions of S-layers.

ACKNOWLEDGEMENTS

The assistance of D.H. Lynn, University of Guelph, on the studies with protozoa is greatly appreciated. This research has been supported by a grant from the Medical Research Council of Canada.

REFERENCES

Barcina, I., Ayo, B., Unanue, M., Egea, L., and Iriberri, J., 1992, Comparison of rates of flagellate bacterivory and bacterial production in a marine coastal system, Appl. Environ. Microbiol. 58:3850.

Beveridge, T.J., 1979, Surface arrays on the wall of *Sporosarcina ureae*, J. Bacteriol. 139:1039.

Fenchel, T., 1980a, Suspension feeding in ciliated protozoa: functional response and particle size selection, Microb. Ecol. 6:1.

Fenchel, T., 1980b, Suspension feeding in ciliated protozoa: feeding rates and their ecological significance, Microb. Ecol. 6:13.

Fenchel, T., 1987, Ecology of Protozoa: the biology of free-living phagotrophic protists, Madison/Springer-Verlag, Berlin.

Gray, K.M., and Ruby, E.G., 1991, Intercellular signalling in the *Bdellovibrio* developmental life cycle, in : "Microbial Cell-Cell Interactions", M. Dworkin, ed., pp. 333-366, American Society for Microbiology, Washington.

Ishiguro, E.E., Kay, W.W., Ainsworth, T., Chamberlain, J.B., Austen, R.A., Buckley, J.T., and Trust, T.J., 1981, Loss of virulence during culture of *Aeromonas salmonicida* at high temperature, J. Bacteriol. 148:333.

Kist, M.L., and Murray, R.G.E., 1984, Components of the regular surface array of *Aquaspirillum serpens* MW5 and their assembly *in vitro*, J. Bacteriol. 157:599.

Koval, S.F., and Murray, R.G.E., 1981, Cell wall proteins of *Aquaspirillum serpens*, J Bacteriol. 146:1083.

Koval, S.F., and Murray, R.G.E., 1984, The isolation of surface array proteins from bacteria, Can. J. Biochem. Cell Biol. 62:1181.

Koval, S.F., and Hynes, S.H., 1991, Effect of paracrystalline protein surface layers on predation by *Bdellovibrio bacteriovorus*, J. Bacteriol. 173:2244.

Messner, P., and Sleytr, U.B., 1992, Crystalline bacterial cell-surface layers, Adv. Microbial Physiol. 33:213.

Nilsson, J.R., 1987, Structural aspects of digestion of *Escherichia coli* in *Tetrahymena*, J. Protozool. 34:1.

Ruby, E.G., 1992, The genus *Bdellovibrio*, in: The Prokaryotes", 2nd Edition, vol.4, A. Balows, H.G. Trüper, M. Dworkin, W. Harder, and K.-H. Schleifer, eds., pp. 3400-3415, Springer-Verlag, New York.

Sanders, R.W., 1988, Feeding by *Cyclidium* sp. (Ciliophora, Scuticociliatida) on particles of different sizes and surface properties, Bull. Marine Sci. 43:446.

Scherr, B.F., Scherr, E.B., and Fallon, R.D., 1987, Use of monodispersed fluorescently labelled bacteria to estimate in situ protozoan bacterivory, Appl. Environ. Microbiol. 53:958.

Smith, S.H., and Murray, R.G.E., 1990, The structure and associations of the double S-layer on the cell wall of *Aquaspirillum sinuosum*, Can. J. Microbiol. 36:327.

Stolp, H., 1981, The genus *Bdellovibrio*, in: "The Prokaryotes", 1st Edition, vol. 1, M.P. Starr, H. Stolp, H.G. Trüper, A. Balows, and H.G. Schlegel, eds., pp. 618-629, Springer-Verlag, New York.

Thomashow, M.F., and Rittenberg, S.C., 1979, The intraperiplasmic growth cycle - the life style of the bdellovibrios, in: "Developmental Biology of Prokaryotes", J.H. Parish, ed., pp. 115-138, University of California Press, Berkeley.

Tudor, J.J., McCann, M.P., and Acrich, I.A., 1990, A new model for the penetration of prey cells by bdellovibrios, J. Bacteriol. 172:2421.

Varon, M., and Shilo, M., 1980, Ecology of the aquatic bdellovibrios, in: "Advances in Aquatic Microbiology", vol. 2, M.P. Droop, and H.W. Jannasch, eds., pp. 1-48, Academic Press Inc., New York.

Yamada, H., Tsukagoshi, N., and Udaka, S., 1981, Morphological alterations of cell wall concomitant with protein release in a protein-producing bacterium, *Bacillus brevis* 47, J. Bacteriol. 148:322.

IV. CHEMISTRY AND MOLECULAR BIOLOGY OF S-LAYERS

CHAPTER 10

GLYCOPROTEIN NATURE OF SELECT BACTERIAL S-LAYERS

Paul Messner, Judith Schuster - Kolbe, Christina Schäffer,
and Uwe B. Sleytr

Center for Ultrastructure Research and the Ludwig Boltzmann
Institute for Molecular Nanotechnology
University for Agriculture, Vienna, Austria

Rudolf Christian

Scientific Software Company
Vienna, Austria

PROCARYOTIC GLYCOPROTEINS - AN OVERVIEW

Nowadays it is established knowledge that procaryotes are able to synthesize true glycoproteins. The first proven reports on bacterial glycoproteins reach back into the mid - 1970s when Mescher and Strominger (1976) published a partial structure of the glycan chains from the surface layer glycoprotein of the halophilic archaeobacterium *Halobacterium (halobium) salinarium*. According to the recently introduced taxonomic terminology we use in this article the terms archaea and bacteria instead of archaeobacteria and eubacteria (Woese et al., 1990). With the rapid ongoing research on archaea it soon became obvious that most of these procaryotes are covered by glycosylated surface layer (S-layer) glycoproteins (for reviews see Kandler, 1982; Kandler and König, 1985; König and Stetter, 1986; Sumper, 1987; König, 1988a; Lechner and Wieland, 1989). On the other hand, it was widely believed that bacteria (eubacteria) were not able to glycosylate proteins even though Sleytr and Thorne (1976) had provided early evidence for the occurrence of bacterial glycoproteins. For some time it was even considered that the surface glycosylation was a taxonomic criterion for the discrimination between archaea and bacteria (Mescher, 1981; Kandler, 1982).

Crystalline surface layers (S-layers) have been characterized as the outermost cell envelope component of many archaea and bacteria (for reviews see Sleytr, 1978; Sleytr and Messner, 1983; Smit, 1987; König, 1988b; Koval, 1988; Sleytr et al., 1988; Messner and Sleytr, 1992). The finding that some of them consist not only of pure proteins but true glycoproteins has made these structures an interesting model for studying glycosylation processes in relatively simple procaryotic systems. Glycosylated

S-layers with varying carbohydrate contents were detected in a number of strains from the Bacillaceae family such as *Bacillus sphaericus* (Word et al., 1983; Lewis et al., 1987), *B. stearothermophilus* (Küpcü et al., 1984; Messner and Sleytr, 1988a), and *Desulfotomaculum nigrificans* (Sleytr et al., 1986). Obviously, this protein modificationis of great importance because there exist only a few scattered reports of glycosylated S-layers in other gram-positive and gram-negative bacteria such as *Aquaspirillum serpens* (Buckmire and Murray, 1973), *Myxococcus xanthus* (Maeba, 1986), *Deinococcus radiodurans* (Peters et al., 1987) and *Acetogenium kivui* (Peters et al., 1989, 1992). The complete structures of all S-layer glycoproteins characterized so far were summarized recently (Messner and Sleytr, 1991). Besides S-layers, there are a few other procaryotic glycoproteins known, including exoenzymes and other structural components of the bacterial cell (for a compilation see Table 1).

Table 1. Procaryotic glycoproteins

S-layer glycoproteins

- of Archaea (for reviews see Kandler, 1982; Kandler and König, 1985; König, 1988, Lechner and Wieland, 1989, Messner and Sleytr, 1991; Mengele and Sumper, 1992)
- of Bacteria (for reviews see Messner and Sleytr, 1991; Messner et al., 1992a)

Other bacterial glycoproteins

- Muramoylhydrolase of *Enterococcus hirae* (formerly known as *Streptococcus faecium*) (Kawamura and Shockman, 1983)
- Flagella of *Halobacterium halobium* (Wieland et al., 1985)
- Larvicidal spore protein crystals of *B. thuringiensis* (Muthukumar and Nickerson, 1987)
- Cellulosome of *Clostridium thermocellum* (Gerwig et al., 1989, 1991)
- Exoenzymes (e.g. cellulase and xylanase) of *C. stercorarium* (W.L. Staudenbauer, pers. commun.)

COMPARATIVE STUDIES ON THE S-LAYER GLYCOPROTEINS OF STRAINS OF *CLOSTRIDIUM THERMOHYDROSULFURICUM*

In our studies on the distribution of S-layer glycoproteins among S-layer-carrying thermophilic strains of *Clostridium thermohydrosulfuricum* we have screened several organisms of the culture collection of the Österreichisches Zuckerforschungs-Institut, Fuchsenbigl. The following strains were chosen for the comparative studies: *C. thermohydrosulfuricum* L110-69 (DSM 568), L111-69, L77-66, L92-71, and S102-70. Hollaus and coworkers performed DNA hybridization experiments to the reference strain *C. thermohydrosulfuricum* E100-69 and found DNA homology values between 95 to 102% (Matteuzzi et al., 1978). After continuous cultivation of the bacteria (Hollaus and Sleytr, 1972) the crystalline S-layers were characterized ultrastructurally and chemically. Freeze-fracture electron microscopy of intact cells showed hexagonal S-layer lattices (Fig.1a) in all *C. thermohydrosulfuricum* strains investigated and they exhibit lattice spacings of approximately 14 to 16 nm.

After isolation of the S-layer by standard techniques (for review see Messner and Sleytr, 1988b) the purified S-layer material was subjected to sodium dodecyl sulphate polyacrylamide gel electrophoresis (SDS-PAGE) and chemical analysis of the protein and carbohydrate content (Table 2). The carbohydrate content of the glycoproteins from the different strains varied from approximately 4 to 17%. By

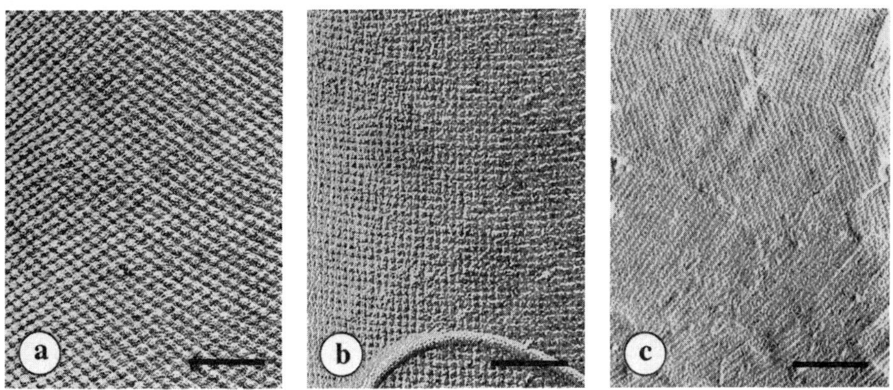

Figure 1. Electron micrographs of freeze-fractured intact cells of (a) *C. thermohydrosulfuricum* L77-66, (b) *C. thermosaccharolyticum* E207-71, and (c) *Lactobacillus plantarum* 41021/252 showing the different types of S-layer lattices. Bar = 100 nm.

SDS-PAGE high molecular weight S-layer bands were obtained in the range of approximately 135 and 152 kDa while the molecular mass of the S-layer of strain S102-70 was significantly lower. Unexpectedly, L110-69 and L111-69 showed almost identical banding patterns,and the same was observed with the strains L77-66 and L92-71 (data not shown). Carbohydrate detection was possible by either periodic acid-Schiff staining reaction on the gel, or after blotting the material on to nitrocellulose by the Boehringer DIG Glycan Detection Kit (Haselbeck and Hösel, 1990). All S-layer glycoproteins were deglycosylated by trifluoromethane sulfonic acid (TFMS; Edge et al., 1981). After deglycosylation, single sharp protein bands were visible on the gels in the range of 82 to 88 kDa (Table 2), which provided negative carbohydrate staining reactions. Notably, the decrease in molecular weight by TFMS treatment was significantly less in strain S102-70 than in the other strains. This indicates that, in intact bacteria, strains L110-69, L111-69, L77-66, and L92-71 possess rather long glycan chains in comparison to strain S102-70 (for reviews see Messner and Sleytr, 1991). The pI values of the S-layer glycoproteins were determined by isoelectric focusing. Due to the insolubility of the S-layers the procedure was performed in urea (Creighten, 1979; Görg et al., 1981). The four strains with the high molecular weight S-layers banded uniformly at pH values of approximately 5.7, while S102-70 showed a complex pattern with distinct bands between pH 5.8 to 6.0 (data not shown).

To see if the glycan moieties protected the protein portion from proteases, the purified glycoproteins were subjected to controlled proteolysis with trypsin, chymotrypsin, and *Staphylococcus aureus* V8 protease on SDS gels (Cleveland et al., 1977). Very similar results were obtained on both glycosylated and deglycosylated S-layers with all proteases (data not shown). Careful analysis of the degradation products revealed almost identical banding patterns in the low molecular mass range. No detectable bands were observed with strain S102-70. This supported the assumption that only short oligosaccharides are present in the S-layer of this bacterium. The long glycan portions of the S-layers of all other strains remained quite unaffected during proteolysis. In addition to the low molecular weight cleavage products we observed a single, slightly broadened glycopeptide band in all strains with an apparent molecular mass (M_r) of approximately 110 kDa (L110-69, L111-69) and 140 kDa (L77-66, L92-71). Obviously, the electrophoretic mobility of these glycopeptides is severely impaired by the long carbohydrate moieties.

Table 2. Characterization of purified *C. thermohydrosulfuricum* S-layer glycoproteins

| | Strain designation | | | | |
S-layer parameters	L110-69	L111-69	L77-66	L92-71	S102-70
Lattice type[a] and spacing (nm)	H, 13.9	H, 13.9	H, 14.3	H, 14.3	H,16.3
Protein content (% of dry weight)	85	83	82	87	94
Carbohydrate content[b] (% of dry weight)	14	16	17	13	4
Molecular mass of the S-layer subunit (kDa)	135	135	152	152/100	94
Molecular mass of the deglycosylated S-layer subunit (kDa)	82	82	82	82/100	88

[a]H, hexagonal.
[b]Neutral sugars and amino sugars.

The S-layer glycoproteins of all strains were subjected to exhaustive degradation with pronase. The goal of this experiment was to isolate small glycopeptide(s) which should only contain the amino acid to which the carbohydrate is bound. Purification of these fractions was performed by gel filtration on different Bio-Gel columns and cation exchange chromatography. To help determine the molecular size of these glycopeptides, the mixture of degradation products of strain L110-69 was analyzed by matrix-assisted laser desorption mass spectrometry. It revealed a broad peak with a molecular mass of ca. 10 kDa (unpublished results). As an example we describe the purification of the glycopeptide mixture of *C. thermohydrosulfuricum* L92-71 (Fig. 2). With all other strains, except S102-70, similar results could be obtained. Reversed-phase HPLC (RP-HPLC) on a C_{18} column indicated four different glycopeptides (Fig. 2a). The resulting mixtures were further purified by chromatofocusing using the Pharmacia Polybuffer system. The pH values applied ranged from approximately 8.5 to 4.0. Two separate peaks, designated pool I and II, were collected by this method (Fig. 2b). Chromatography of pool I by RP-HPLC allowed isolation of a homogenous glycopeptide (F2) (Fig. 2c). The second fraction represented a mixture of the three remaining glycopeptides F1, F3 and F4 (Fig. 2d). This mixture could eventually be separated by preparative RP-HPLC.

The purified fractions F2 of all strains, except S102-70, and the material of S102-70, which was eluted from the Bio-Gel P-4 column yielding GPIII (Messner et al., 1992a), were used for the determination of carbohydrate and amino acid compositions. Strains L110-69 and L111-69 have identical carbohydrate composition as have strains L77-66 and L92-71. The oligosaccharide of S102-70 is completely different (Table 3). Notably, tyrosine was the only amino acid found in all these fractions (Table 3). This observation, together with the assumption of there being rather short oligosaccharide chains in the S-layer glycoprotein of strain S102-70, prompted us to investigate the protein-carbohydrate linkage region of GPIII in greater detail (Messner et al., 1992a). As a result we found, for the first time in bacteria, a novel *O*-glycosidic linkage between β-D-glucose as the linkage sugar and tyrosine as the linkage amino acid of the polypeptide chain. In the meantime, tyrosine was also described as a constituent the S-layer glycoprotein of *A. kivui* (Peters et al., 1992).

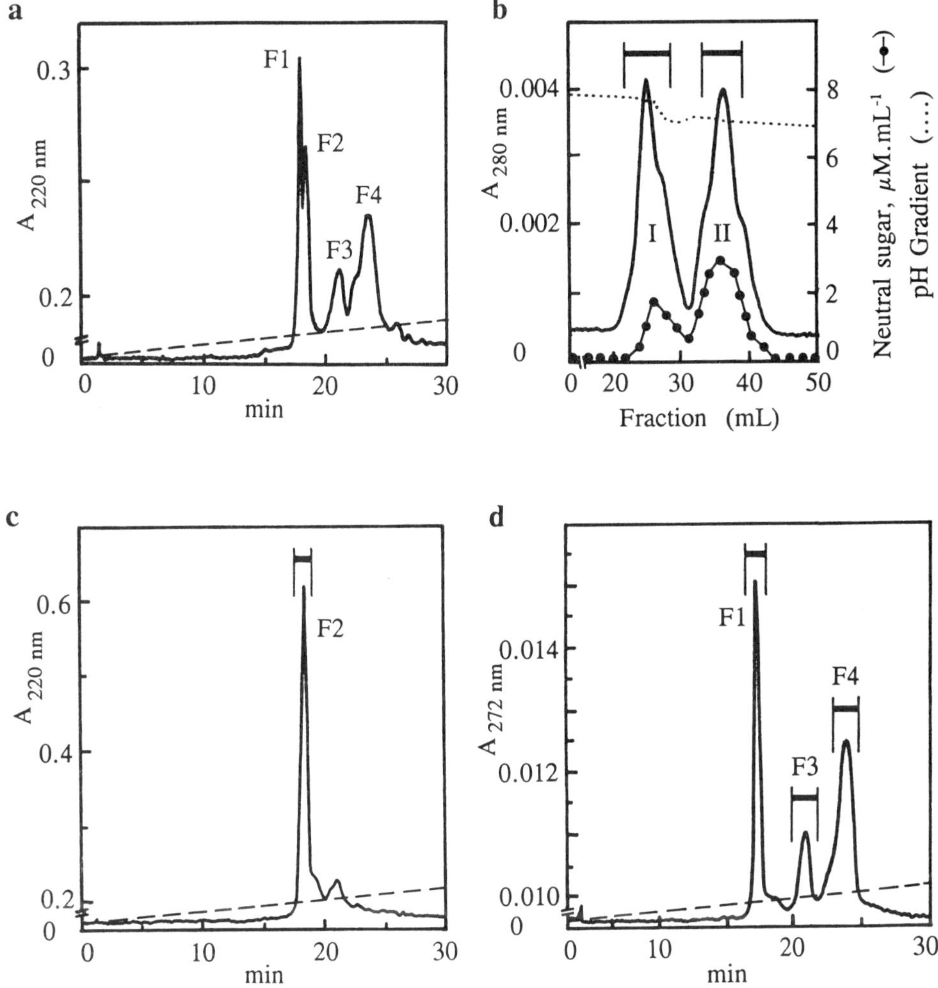

Figure 2. Separation of the glycopeptides of *C. thermohydrosulfuricum* L92-71 as obtained by pronase digestion. (a) After ion exchange chromatography a mixture of four glycopeptides F1 - F4 is obtained and is monitored by reversed phase high performance liquid chromatography (RP-HPLC; C_{18} column, acetonitrile gradient). (b) Chromatofocusing (Pharmacia Polybuffer system) in the pH range 8.5 - 4.0 separated the mixture into two pools I and II. Chromatography again by RP-HPLC of (c) pool I and (d) pool II. The bars indicate the pooled fractions.

The structural elucidation of the glycan portions of purified glycopeptides by chemical degradation methods, methylation analysis, and straightforward one and two dimensional ^{1}H and ^{13}C nuclear magnetic resonance (NMR) measurements was performed on precharacterized S-layer glycoproteins. The repeating unit of *C. thermohydrosulfuricum* L111-69 consists of linear disaccharide repeats with the structure →4)-α-D-Man*p*-(1→3)-α-L-Rha*p*-(1→ (Christian et al., 1988). Strain L77-66 has a branched tetrasaccharide repeating unit with the structure →3)-α-D-Gal*p*NAc-(1→3)[α-D-Glc*p*NAc-(1→2)-β-D-Man*p*-(1→4)]α-D-Gal*p*NAc-(1→ (Altman et al., 1992).

Table 3. Carbohydrate and amino acid composition of selected glycopeptide fractions of *C. thermohydrosulfuricum* strains

Organisms	Molar ratio						
	Man	Rha	Gal	Glc	GalNAc	GlcNAc	Tyr
L110-69 (fraction F2)	~30	~30	?	-	-	-	1
L111-69 (fraction F2)	~30	~30	?	-	-	-	1
L77-66 (fraction F2)	~25	-	-	-	~50	~25	1
L92-71 (fraction F2)	~25	-	-	-	~50	~25	1
S102-70 (fraction GPIII)	1	2	2	1	-	-	1

A summary of our comparative studies on S-layer glycoproteins of different *C. thermohydrosulfuricum* strains shows that even closely related strains can display a remarkable heterogeneity in their glycan structures. Both long chains, composed of up to 50 repeats and short chains, consisting of only a few sugar residues, were found. Further, linear and branched structures can occur. In addition to this it is very likely that tyrosine is the linkage amino acid in all S-layer glycoproteins of *C. thermohydrosulfuricum* strains.

COMPARATIVE STUDIES ON THE S-LAYER GLYCOPROTEINS OF STRAINS OF *C. THERMOSACCHAROLYTICUM*

Sleytr and Thorne (1976) analyzed the S-layer glycoproteins of *C. thermohydrosulfuricum* L111-69 and *C. thermosaccharolyticum* D120-70 and reported differences in their carbohydrate compositions. Analysis of the surface properties of both strains also revealed significant differences (Sára et al., 1988) which, at least partially, can be attributed to differences in the carbohydrate chains. After elucidating the structure of the glycan chain of *C. thermohydrosulfuricum* L111-69 (Christian et al., 1988) we determined the sugar composition from the S-layer glycoprotein of *C. thermosaccharolyticum* D120-70 (Altman et al., 1990). In contrast to the S-layers of *C. thermohydrosulfuricum* strains analyzed so far (for review see Messner and Sleytr, 1991), the S-layer subunits of *C. thermosaccharolyticum* D120-70 possess two major glycans. One is composed of a branched hexasaccharide unit with the structure →3)[β-D-Glc*p*-(1→6)]β-D-Man*p*-(1→4)-α-L-Rha*p*-(1→3)-α-D-Glc*p*-(1→4)[α-D-Gal*p*-(1→2)]α-L-Rha*p*-(1→. The other chain has mainly amino sugars as constituents of a trisaccharide repeating unit with the structure →4)-β-D-Glc*p*NAc-(1→3)[(α-D-Gal*p*)$_{0.5}$-(1→4)]β-D-Man*p*NAc-(1→ (Altman et al., 1990). On this basis we screened *C. thermosaccharolyticum* strains from the collection of the Österreichisches Zuckerforschungs-Institut. The reference strain used for the DNA hybridization experiment was *C. thermosaccharolyticum* ATCC 7956 (Matteuzzi et al., 1978). The strains of interest showed DNA homology values between 85 to 98%. Five strains were chosen as potential candidates for the S-layer glycoprotein analysis. All strains were grown in continuous culture under anaerobic conditions according to Sleytr and Thorne (1976). Thin sections of intact cells showed the typical cell envelope profile of gram-positive bacteria with a relatively thin peptidoglycan layer characteristic of

Table 4. Characterization of purified *C. thermosaccharolyticum* S-layer glycoproteins

S-layer parameters	Strain designation				
	E205-71	E207-71	S201-71	S208-71	T9-3R
Lattice type[a]	S	S	S	S	H
Protein content (% of dry weight)	91	91	95	95	93
Carbohydrate content[b] (% of dry weight)	16	16	13	10	12
Molecular mass of the S-layer subunit (kDa)	100-185	100-210	90-100	110-130	100-135
Molecular mass of the deglycosylated S-layer subunit (kDa)	83	83	88	110	97

[a]S, square; H, hexagonal.
[b]Neutral sugars and amino sugars.

most thermophilic S-layer carrying Bacillaceae. Previous analyses (Hollaus and Sleytr, 1972; Sleytr and Glauert, 1976) have shown that in contrast to *C. thermohydrosulfuricum*, all investigated strains of *C. thermosaccharolyticum* possess square S-layer lattices (Fig. 1b). *C. thermosaccharolyticum* T9-3R which was formerly known as *C. tartarivorum* (Matteuzzi et al., 1978) is an exception since it possesses a hexagonally arranged S-layer lattice. The center-to-center spacings of all lattices were in the range of approximately 11 nm. The available morphological data together with the chemical analyses are summarized in Table 4. Preliminary results of the carbohydrate analyses have revealed that, in general, there exist similarities with strain D120-70, although considerable differences in the amount of carbohydrates are found. The pronounced microheterogeneity of the S-layer glycoproteins of *C. thermosaccharolyticum* was of particular interest to us. As an example, on using SDS-PAGE strain E207-71 showed 14 detectable bands in the range of 103 to 213 kDa (Fig. 3). Even after deglycosylation with TFMS (Edge et al., 1981) several bands in the range of 82 to 84 kDa were present.

Glycopeptides were prepared by exhaustive degradation with pronase according to standard procedures. Small contaminating peptides and amino acids were removed by gel filtration over a Bio-Gel P-4 column. The glycopeptide fractions of the strains eluted mainly in the void volume of this P-4 column, but with S201-71, S208-71 and T9-3R carbohydrate-containing fractions were obtained at lower molecular weights. From the data available it appears that the S-layer glycoproteins of *C. thermosaccharolyticum* strains are considerably more heterogeneous than those of *C. thermohydrosulfuricum* strains. As a consequence, we modified the isolation and purification procedure for obtaining defined glycopeptide fractions from glycopeptide mixture from strain E207-71. After ion exchange chromatography on Dowex 50W-X8, the glycopeptides were chromatofocused between pH values of 9.2 to 5.7. Three major fractions eluting from the column at pH values between 8.3 and 7.0 were obtained which contained neutral hexoses as well as amino sugars (data not shown). These fractions are now being analyzed in detail.

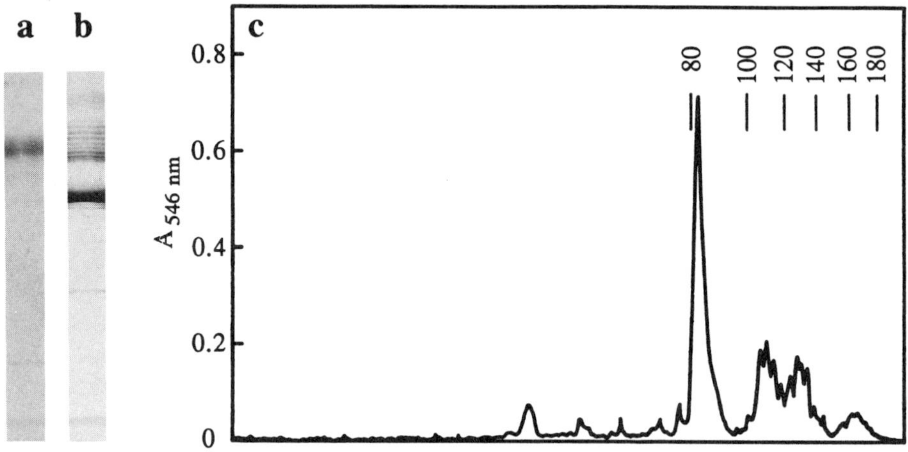

Figure 3. SDS-PAGE of (a) *C. thermohydrosulfuricum* L110-69 and (b) *C. thermosaccharolyticum* strain E207-71. On the densitometer plot (c) the microheterogeneity of the S-layer glycoprotein of *C. thermosaccharolyticum* strain E207-71 is better displayed. M_rs are expressed in kDa.

From our data the following conclusions can be drawn. *C. thermosaccharolyticum* strains, when compared to those of *C. thermohydrosulfuricum*, possess structurally more diverse S-layer glycoproteins. Isoelectric focusing has shown that most glycopeptides have rather basic pI values. At the moment there is no indication that tyrosine, as in *C. thermohydrosulfuricum* strains, serves as the common linkage amino acid in S-layer glycoproteins of *C. thermosaccharolyticum*.

ANALYSIS OF THE S-LAYER GLYCOPROTEIN OF *L. BUCHNERI*

Recent reports in the literature have indicated that some of the bacteria used in cultured milk production form polymers which are not only extracellular polysaccharides but are true glycoproteins (Macura and Townsley, 1984; Garcia-Garibay and Marshall, 1991). These observations encouraged us to extend our systematic survey on the occurrence of S-layer glycoproteins to the genus *Lactobacillus*.

Different lactobacilli from the culture collection of the Institut für Milchforschung und Bakteriologie, Universität für Bodenkultur, Wien, Austria, have been screened and two strains out of approximately 40 gave a clear positive carbohydrate staining reaction. One strain was *L. buchneri* 41021/251, the other strain was *L. plantarum* 41021/252. Electron microscopy of freeze-etched preparations showed that both strains were covered by an oblique S-layer lattice. *L. buchneri* was chosen for the glycoprotein analysis. After continuous culture and isolation of the S-layer by extraction with chaotropic agents, this material was subjected to pronase degradation. Gel permeation chromatography, ion exchange chromatography, and RP-HPLC were used to isolate a glycopeptide fraction. Together the chemical analyses, sequencing experiments, and ^{1}H and ^{13}C NMR measurements suggested that this glycopeptide contained homooligosaccharides of glucose residues linked by an *O*-glycosidic linkage via serine residues to a dekapeptide. Detailed information on this S-layer glycoprotein are presented in this book by Möschl et al.

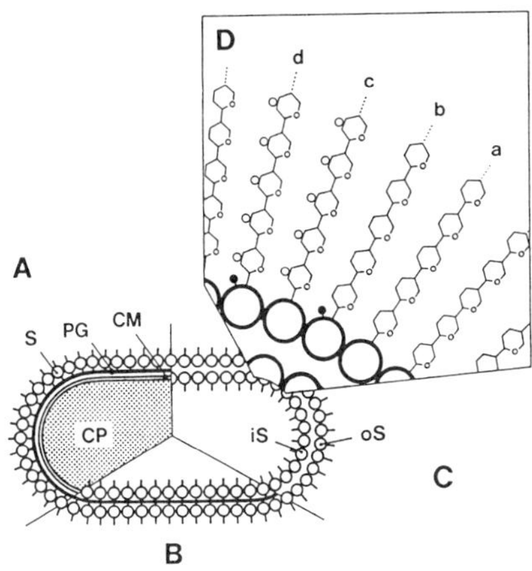

Figure 4. Preparation scheme for glutaraldehyde-fixed S-layer conjugates. (A) Intact bacterial cell; S, S-layer; PG, peptidoglycan; CM cytoplasmic membrane; CP, cytoplasm. (B) After removing the cell content and the cytoplasmic membrane an additional S-layer assembles on the inside of the peptidoglycan layer. (C) Double S-layer fragments are obtained by digesting the peptidoglycan layer with lysozyme. The two S-layers are arranged in a mirror-symmetric fashion. Functional groups of the protein and carbohydrate moieties, exposed on the inner (iS) and outer (oS) S-layer surface are available for binding ligands. (D) Schematic illustration of the possibilities for covalent binding of haptens. (a) Non-activated glutaraldehyde-fixed S-layer glycoprotein. The ligands (● , ○) can be linked either to the protein moiety (b) or to the glycan chains (c) or to both of them (d). (Reprinted with permission of Springer-Verlag).

BIOTECHNOLOGICAL APPLICATIONS

Over the last decade a considerable body of knowledge on structure, assembly, chemistry, biosynthesis, pathogenicity, and permeability properties of S-layers has accumulated. Our basic research on S-layer glycoproteins has led to a broad spectrum of biotechnological and biomedical applications (for review see Messner and Sleytr, 1992; Sleytr et al., 1993).

One major research activity was devoted to the application potential of S-layer glycoproteins as carriers for artificial haptens and antigens. Recently it has been demonstrated that S-layer fragments or self-assembly products can be used for the well-defined covalent attachment of ligands (Sleytr et al., 1987). With S-layer glycoproteins both the protein or carbohydrate portion can be used for the covalent linkage of molecules by specific immobilization reactions (Fig. 4) (Messner et al., 1991; Malcolm et al., 1993; see also Malcolm et al., this book). We used either glutaraldehyde-fixed S-layers (Messner et al., 1992b) for the immobilization of small non-immunogenic oligosaccharides (Smith et al., 1993) or unfixed S-layers for the binding of capsular polysaccharides or hydrolytically cleaved oligosaccharides of different *Streptococcus pneumoniae* serotypes (Malcolm et al., 1993; Malcolm et al., this book). With fixed S-layers, hapten-specific T-cell responses were obtained (Smith et al., 1993), whereas unfixed S-layers elicited good B-cell responses. Effective class switching from IgM to IgG subclasses was observed after secondary and tertiary immunization (Malcolm et al., 1993).

The aggregate nature of S-layer conjugates endows them with intrinsic adjuvant properties. Therefore, no additional extraneous adjuvant is necessary for vaccination with S-layer conjugates. All S-layers investigated were immunologically non-cross-reactive. Currently, tetanus and diphteria toxoids are widely used as vaccine carriers. Repeated vaccination with such toxoid carriers may induce a state of tolerance which could be overcome by use of immunologically distinct S-layers.

The preferred immunization route for induction of immunoprotective antibodies is intraperitoneal or subcutaneous (Malcolm et al., 1993; Smith et al., 1993). Nasal/Oral administration of S-layer conjugates elicited lower levels of protective antibody. In summary, the available data provide strong evidence that S-layers have good potential as carrier/adjuvants for elicting T and B cell responses to specific antigens.

ACKNOWLEDGMENTS

We thank A. Scheberl, S. Zayni and E. Mayerhofer for excellent technical assistance. Part of this work (i.e., that on *C. thermohydrosulfuricum, C. thermosaccharolyticum, L. buchneri* and *L. plantarum* was supported by grants from the Fonds zur Förderung der wissenschaftlichen Forschung in Österreich, project P8922-MOB, and the Österreichisches Bundesministerium für Wissenschaft und Forschung.

REFERENCES

Altman, E., Brisson, J.-R., Messner, P., and Sleytr, U.B., 1990, Chemical characterization of the regularly arranged surface layer glycoprotein of *Clostridium thermosaccharolyticum* D120-70, *Eur. J. Biochem.* 188:73.

Altman, E., Brisson, J.-R., Gagné, S.M., Kolbe, J., Messner, P., and Sleytr, U.B., 1992, Structure of the glycan chain from the surface layer glycoprotein of *Clostridium thermohydrosulfuricum* L77-66, *Biochim. Biophys. Acta* 1117:71.

Buckmire, F.L.A., and Murray, R.G.E., 1973, Studies on the cell wall of *Spirillum serpens*. II. Chemical characterization of the outer structured layer, *Can. J. Microbiol.* 19:59.

Christian, R., Messner, P., Weiner, C., Sleytr, U.B., and Schulz, G., 1988, Structure of a glycan from the surface-layer glycoprotein of *Clostridium thermohydrosulfuricum* strain L111-69, *Carbohyd. Res.* 176:160.

Cleveland, D.W., Fischer, S.G., Kirschner, M.W., and Laemmli, U.K., 1977, Peptide mapping by limited proteolysis in sodium dodecyl sulphate and analysis by gel electrophoresis, *J. Biol. Chem.* 252:1102.

Creighten, T.E., 1979, Electrophoretic analysis of the unfolding of proteins by urea, *J. Mol. Biol.* 129:235.

Edge, A.B.S., Faltynek, C.R., Hof, L., Reichert, Jr., L.E., and Weber, P., 1981, Deglycosylation of glycoproteins by trifluoromethanesulfonic acid, *Anal. Biochem.* 118:131.

Garcia-Garibay, M., and Marshall, V.M.E., 1991, Polymer production by *Lactobacillus delbrueckii* subsp. *bulgaricus, J. Appl. Bacteriol.* 70:325.

Gerwig, G.J., deWaard, P., Kamerling, J.P., Vliegenthart, J.F.G., Morgenstern, E., Lamed, R., and Bayer, E.A., 1989, Novel *O*-linked carbohydrate chains in the cellulase complex (cellulosome) of *Clostridium thermocellum, J. Biol. Chem.* 264:1027.

Gerwig, G.J., Kamerling, J.P., Vliegenthart, J.F.G., Morag (Morgenstern), E., Lamed, R., and Bayer, E.A., 1991, Primary structure of O-linked carbohydrate chains in the cellulosome of different *Clostridium thermocellum* strains, *Eur. J. Biochem.* 196:115.

Görg, A., Postel, W., Westermeier R., Gianazza, E., and Righetti, P.G., 1981, SDS-gel gradient electrophoresis, isoelectric focusing and high-resolution two-dimensional electrophoresis in horizontal, ultrathin-layer polyacrylamide gels, *in*: "Electrophoresis '81", R.C. Allen and P. Arnaud, eds., Walter de Gruyter, Berlin.

Haselbeck, A., and Hösel, W., 1990, Description and application of an immunological detection system for analyzing glycoproteins on blots, *Glycoconjugate J.* 7:63.

Hollaus, F., and Sleytr, U., 1972, On the taxonomy and fine structure of some hyperthermophilic saccharolytic clostridia, *Arch. Mikrobiol.* 86:129.

Kandler, O., 1982, Cell wall structures and their phylogenetic implications, *Zbl. Bakt. Hyg., I. Abt. Orig.* C 3:149.

Kandler, O., and König, H., 1985, Cell envelopes of archaebacteria, *in*: "The Bacteria", Vol. VIII, "Archaebacteria", C.R. Woese and R.S. Wolfe, eds., Academic Press, New York.

Kawamura, T., and Shockman, G.D., 1983, Purification and some properties of the endogenous, autolytic *N*-acetylmuramoylhydrolase of *Streptococcus faecium*, a bacterial glycoenzyme, *J. Biol. Chem.* 258:9514.

König, H., 1988a, Archaeobacterial cell envelopes, *Can. J. Microbiol.* 34:395.

König, H., 1988b, Archaeobacteria, *in*: "Biotechnology", Vol. 6b, H.-J. Rehm, ed., VCH Publishers, Weinheim.

König, H., and Stetter, K.O., 1986, Studies on archaebacterial S-layers, *System. Appl. Microbiol.* 7:300.

Koval, S.F., 1988, Paracrystalline protein surface arrays on bacteria, *Can. J. Microbiol.* 34:407.

Küpcü, Z., März, L., Messner, P., and Sleytr, U.B., 1984, Evidence for the glycoprotein nature of the crystalline cell wall surface layer of *Bacillus stearothermophilus* strain NRS 2004/3a, *FEBS Lett.* 173:185.

Lechner, J., and Wieland, F., 1989, Structure and biosynthesis of prokaryotic glycoproteins, *Annu. Rev. Biochem.* 58:173.

Lewis, L.O., Yousten, A.A., and Murray, R.G.E., 1987, Characterisation of the surface protein layers of the mosquito-pathogenic strains of *Bacillus sphaericus*, *J. Bacteriol.* 169:72.

Macura, D., and Townsley, P.M., 1984, Scandinavian ropy milk - identification and characterization of endogenous ropy lactic streptococci and their extracellular secretion, *J. Dairy Sci.* 67:735.

Maeba, P.Y., 1986, Isolation of a surface glycoprotein from *Myxococcus xanthus*, *J. Bacteriol.* 166:644.

Malcolm, A.J., Messner, P., Sleytr, U.B., Smith, R.H., and Unger, F.M., 1993, Crystalline bacterial cell surface layers as combined carrier/adjuvants for conjugate vaccines, *in*: "Immobilised Macromolecules: Application Potentials", U.B. Sleytr, P. Messner, D. Pum, and M. Sára, eds., Springer-Verlag, London.

Matteuzzi, D., Hollaus, F., and Biavati, B., 1978, Proposal of neotype for *Clostridium thermohydrosulfuricum* and merging of *Clostridium tartarivorum* with *Clostridium thermosaccharolyticum*, *Int. J. Syst. Bacteriol.* 28:528.

Mengele, R., and Sumper, M., 1992, Drastic differences in glycosylation of related S-layer glycoproteins from moderate and extreme halophiles, *J. Biol. Chem.* 267:8182.

Mescher, M.F., 1981, Glycoproteins as cell-surface structural components, *Trends Biol. Sci.* 6:97.

Mescher, M.F., and Strominger, J.L., 1976, Purification and characterization of a prokaryotic glycoprotein from the cell envelope of *Halobacterium salinarium*, *J. Biol. Chem.* 251:2005.

Messner, P., and Sleytr, U.B., 1988a, Asparaginyl-rhamnose: a novel type of protein-carbohydrate linkage in a eubacterial surface-layer glycoprotein, *FEBS Lett.* 228:317.

Messner, P., and Sleytr, U.B., 1988b, Separation and purification of S-layers from gram-positive and gram-negative bacteria, *in*: "Bacterial Cell Surface Techniques", I.C. Hancock and I.R. Poxton, eds., John Wiley and Sons, Chichester.

Messner, P., and Sleytr, U.B., 1991, Bacterial surface layer glycoproteins, *Glycobiology* 1:545.

Messner, P., and Sleytr, U.B., 1992, Crystalline bacterial cell-surface layers, *Adv. Microbial Physiol.* 33:213.

Messner, P., Küpcü, S., Sára, M., Pum, D., and Sleytr, U.B., 1991, Characterization and biotechnological application of eubacterial glycoproteins, *in*: "Protein Glycosylation: Cellular, Biotechnological and Analytical Aspects", GBF Monographs, Vol. 15, H.S. Conradt, ed., VCH Verlagsgesellschaft, Weinheim.

Messner, P., Christian,R., Kolbe, J., Schulz, G., and Sleytr, U.B., 1992a, Analysis of a novel linkage unit of *O*-linked carbohydrates from the crystalline surface layer glycoprotein of *Clostridium thermohydrosulfuricum* S102-70, *J. Bacteriol.* 174:2236.

Messner, P., Mazid, M.A., Unger, F.M., and Sleytr, U.B., 1992b, Artificial antigens. Synthetic carbohydrate haptens immobilized on crystalline bacterial surface layer glycoproteins. *Carbohydr. Res.*, 233:175.

Muthukumar, G., and Nickerson, K.W., 1987, The glycoprotein toxin of *Bacillus thuringiensis* subsp. *israelensis* indicates a lectin-like receptor in the larval mosquito gut, *Appl. Environ. Mirobiol.* 53:2650.

Peters, J., Peters, M., Lottspeich, F., Schäfer, W., and Baumeister, W., 1987, Nucleotide sequence analysis of the gene encoding the *Deinococcus radiodurans* surface protein, derived amino acid sequence and complementary protein chemical studies, *J. Bacteriol.* 169:5216.

Peters, J., Peters, M., Lottspeich, F., and Baumeister, W., 1989, S-layer protein gene of *Acetogenium kivui*: Cloning and expression in *Escherichia coli* and determination of the nucleotide sequence, *J. Bacteriol.* 171:6307.

Peters, J., Rudolf, S., Oschkinat, H., Mengele, R., Sumper, M., Kellermann, J., Lottspeich, F., and Baumeister, W., 1992, Evidence for tyrosine-linked glycosaminoglycan in a bacterial surface protein, *Biol. Chem. Hoppe-Seyler* 373:171.

Sára, M., Kalsner, I., and Sleytr, U.B., 1988, Surface properties from the S-layer of *Clostridium thermosaccharolyticum* D120-70 and *Clostridium thermohydrosulfuricum* L111-69, *Arch. Microbiol.* 149:527.

Sleytr, U.B., 1978, Regular arrays of macromolecules on bacterial cell walls: Structure, chemistry, assembly, and function, *Int. Rev. Cytol.* 53:1.

Sleytr, U.B., and Glauert, A.M., 1976, Ultrastructure of the cell walls of two closely related clostridia that possess different regular arrays of surface subunits, *J. Bacteriol.* 126:869.

Sleytr, U.B., and Messner, P., 1983, Crystalline surface layers on bacteria, *Annu. Rev. Microbiol.* 37:311.

Sleytr, U.B., and Thorne, K.J.I., 1976, Chemical characterization of the regularly arrayed surface layers of *Clostridium thermosaccharolyticum* and *Clostridium thermohydrosulfuricum, J. Bacteriol.* 126:377.

Sleytr, U.B., Sára, M., Küpcü, Z., and Messner, P., 1986, Structural and chemical characterization of S-layers of selected strains of *Bacillus stearothermophilus* and *Desulfotomaculum nigrificans, Arch. Microbiol.* 146:19.

Sleytr, U.B., Mundt, W., and Messner, P., 1987, Pharmazeutische Struktur, Eur. Patent Application 0 306 473 A1.

Sleytr, U.B., Messner, P., Pum, P., and Sára, M., eds., 1988, "Crystalline Bacterial Cell Surface Layers", Springer-Verlag, Berlin.

Sleytr, U.B., Messner, P., Pum, D., and Sára, M., 1993, Crystalline bacterial cell surface layers: General principles and application potentials, *in*: "Symposium Series, Society for Applied Bacteriology", Blackwell, Oxford, in press.

Smit, J., 1987, Protein surface layers of bacteria, *in*: "Bacterial Outer Membranes as Model Systems", M. Inouye, ed., John Wiley and Sons, New York.

Smith, R.H., Messner, P., Lamontagne, L.R., Sleytr, U.B., and Unger, F.M., 1993, Induction of T helper cell immunity to oligosaccharide antigens immobilized on crystalline bacterial surface layers (S-layers), *Vaccine*, in press.

Sumper, M., 1987, Halobacterial glycoprotein biosynthesis, *Biochim. Biophys. Acta* 906:69.

Wieland, F., Paul, G., and Sumper, M., 1985, Halobacterial flagellins are sulfated glycoproteins, *J. Biol. Chem.* 260:15180.

Woese, C.R., Kandler, O., and Wheelis, M.L., 1990, Towards a natural system of organisms: Proposal for the domains archaea, bacteria, and eucarya, *Proc. Natl. Acad. Sci. USA* 87:4576.

Word, N.S., Yousten, A.A., and Howard, L., 1983, Regularly structured and non-regularly structured surface layers of *Bacillus sphaericus, FEMS Microbiol. Lett.* 17:277.

Chapter 11

S-LAYER GLYCOPROTEINS FROM MODERATELY AND EXTREMELY HALOPHILIC ARCHAEOBACTERIA

Manfred Sumper

Lehrstuhl Biochemie I
University of Regensburg
Regensburg, Germany

INTRODUCTION

The first procaryotic glycoprotein was discovered in the S-layer of halobacteria of the genus *Halobacterium* by Mescher and Strominger (1976). Saturated sodium chloride solutions are the natural habitat of these rod-shaped and flagellated archaeobacteria. The original finding was that the halobacterial S-layer contains a single glycoprotein with an M_r of 200 kDa and a carbohydrate content of about 10 % by weight. Since procaryotes lack all the organelles engaged in eucaryotic glycoprotein biosynthesis, this discovery has stimulated further work on the structure and biosynthesis of this glycoprotein (for a review, see Sumper 1987; Lechner and Wieland, 1989). The halobacterial S-layer is very tightly joined to the plasma membrane as these cells lack a rigid sacculus as well as an outer membrane. S-layer proteins represent the outermost component of the cell envelope and, as a consequence, these proteins are in immediate contact with the environment. Therefore they are considered ideal model systems to study the adaptation of protein structures to unfavourable conditions. The extreme habitat of halobacteria imposes particularly serious problems with respect to the stabilization of the protein structure. Since other members of the halobacterial family only tolerate moderately halophilic conditions (e.g., *Haloferax volcanii* requires 2.3M NaCl for optimum growth), a comparison of the corresponding S-layer protein structures should give hints on how adaptation to high salt conditions works.

STRUCTURE OF THE S-LAYER GLYCOPROTEIN FROM *HALOBACTERIUM HALOBIUM*

More than 10 years ago, we found that cells of *H. halobium* synthesize a defined set of sulfated proteins (Wieland et al., 1980). One of these sulfated proteins had an unusually high molecular mass around 200 kDa and turned out to be identical with the S-layer glycoprotein. In addition, the remaining sulfated components in the 30 kDa range turned out to also represent glycoproteins and were identified as the halobacterial flagellins (Wieland et al., 1985; Gerl and Sumper, 1988). These facts initiated our interest to investigate the structures and the biosynthetic pathways of these novel types of glycoproteins.

We started our work with a detailed analysis of the saccharides covalently attached to the protein. The S-layer glycoprotein is extremely acidic, more than 50 sulfate residues are found to be covalently attached per glycoprotein molecule (Wieland et al., 1980). Exhaustive pronase digestion of the purified glycoprotein and subsequent chromatography on a Biogel P-10 column resulted in the separation of three glycopeptide fractions.

Work over several years gave a detailed picture of the chemical structures involved. To summarize these data, a schematic representation of the S-layer glycoprotein of *H. halobium* is given in Fig. 1.

At amino acid 2-position of the glycoprotein, a high molecular mass saccharide is attached and this saccharide consists of repeating units of a pentasaccharide. Remarkable features of this repeating unit saccharide are the occurrence of a methylated galacturonic acid and a furanosidic galactose. Both these unusual sugars are linked peripherally to the backbone of the polysaccharide that consists of a linear chain of GalNAc-GalA-GlcNAc repeats (Paul and Wieland, 1987). The overall chain length of this polysaccharide ranges between 10 and 20 repeats of the pentasaccharide unit. Each pentasaccharide bears two sulfate ester residues, one attached to the 4-position of GalNAc. The repeating unit saccharide is directly attached to the protein in a N-glycosidic bond as was proven by the isolation and

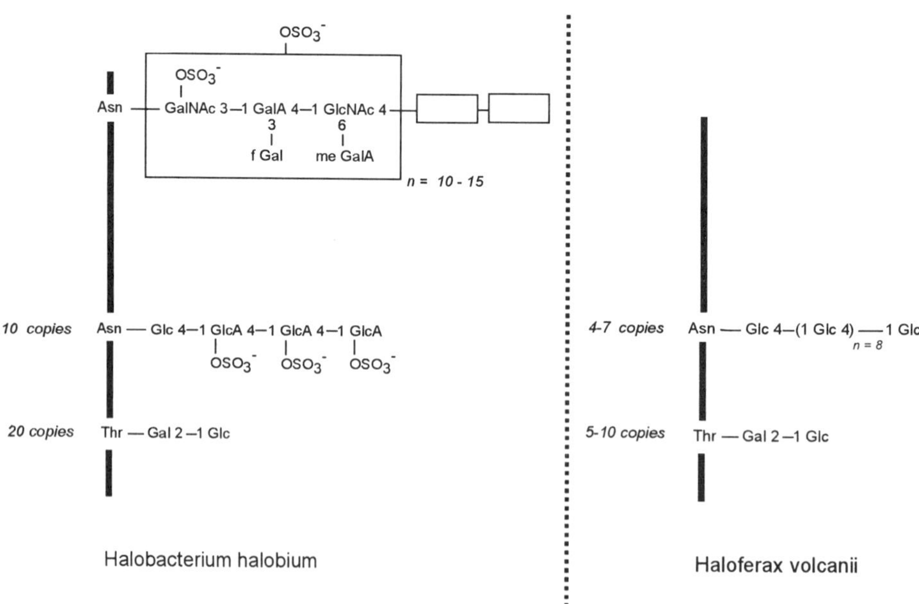

Figure 1. Structures of saccharides found in halobacterial S-layer glycoproteins. The thick solid line from which the saccharides arise represents the protein.

chemical characterization of the novel linkage unit asparaginyl-GalNAc by Paul et al. (1986). The repeating unit saccharide is present only once per glycoprotein molecule.

Another type of sulfated saccharide with a low molecular mass is found to be present in about 10 copies per protein molecule (Wieland et al., 1982; Lechner et al., 1985). These sulfated oligosaccharides are again linked in a N-glycosidic bond to the polypeptide, since asparaginylglucose could be characterized as another novel linkage unit (Wieland et al., 1983). This linkage unit is extended by a chain of two or three β-(1-4)-linked glucuronic acids. A sulfate residue in ester linkage to the 3-position of each glucuronic acid further increases the negative charge density of these saccharides.

Finally, about 15 copies of a neutral disaccharide Glc-(1-2)-Gal are attached to threonine residues in O-glycosidic linkage (Mescher and Strominger, 1976; Wieland et al., 1982). These disaccharide units occur in a highly clustered arrangement within the polypeptide chain as was concluded from amino acid sequence data. The same type of disaccharide is known to occur in collagen.

In order to establish the complete primary structure of the cell surface glycoprotein, we constructed a halobacterial genomic library in an expression vector and screened this library with antibodies raised against the S-layer glycoprotein. In addition, we determined partial amino acid sequences from the glycoprotein. These data allowed the identification and complete characterization of the gene encoding the cell surface glycoprotein (Lechner and Sumper, 1987). Some of the features of the deduced amino acid sequence are summarized as follows:

1. The mature glycoprotein consists of a polypeptide chain with 818 amino acids.
2. The open reading frame encodes a N-terminal leader peptide of 34 amino acid residues reminiscent of a typical signal peptide.
3. Whereas the complete polypeptide chain consists mainly of polar and negatively charged amino acids there is a single exception at the C-terminal end. Only three amino acid positions away from the C-terminus, a stretch of 21 amino acid residues consists of hydrophobic amino acids and this hydrophobic peptide most probably serves as a membrane anchor. Close to this membrane binding domain, a cluster of 14 threonine residues could be identified as the binding site of all the neutral disaccharides attached to the glycoprotein. Perhaps this structural element, located immediately above the membrane binding domain, serves as a spacer to the extracellular domain of the glycoprotein.

As known from eucaryotic glycoproteins, the N-glycosylation sites always match the amino acid sequence Asn-X-Ser(Thr). We find a total of 12 potential N-glycosylation ("sequon") sites throughout the polypeptide chain. As proven from amino acid sequence data of isolated peptides, halobacteria use the same type of glycosylation sites for their novel N-glycosidic linkages. The very first sequon structure is found at 2-position of the mature polypeptide chain and it is this Asn-residue which is linked to the repetitive pentasaccharide structure occurring once per glycoprotein molecule. All the other sequon structures distributed over the polypeptide appear to be linked with the sulfated oligosaccharides via the novel asparaginylglucose linkage. In this glycoprotein, we have the unique situation of two different types of N-glycosidic bonds at defined positions of the polypeptide chain. Therefore additional, and as yet unknown, recognition signals are required to denote individual glycosylation sites.

With these structural data and based on a three-dimensional reconstruction from electron micrographs of the S-layer glycoprotein of *H. volcanii*, Kessel et al. (1988) proposed a model for the halobacterial S-layer. The main result of this work was that six glycoprotein molecules associate to form an outward facing dome. A main problem, however, remained; data from two different halophiles were mixed together. This was another reason to establish the primary structure of the S-layer glycoprotein from the moderate halophile, which exhibits optimum growth at 2.3 M NaCl corresponding to a half-saturated sodium chloride solution.

STRUCTURE OF THE S-LAYER GLYCOPROTEIN FROM *H. VOLCANII*

Using the *H. halobium* gene as a heterologous probe, we were unable to detect the corresponding gene in a genomic DNA library of *H. volcanii*. This fact already indicated substantial differences in the primary structures of both S-layer glycoproteins. Therefore we generated a specific probe by PCR using the amino acid sequence information obtained from tryptic peptides for the synthesis of sense and antisense oligonucleotide primers. This specific probe enabled us to clone the *H. volcanii* glycoprotein gene (Sumper et al., 1990). The deduced amino acid sequence from the cloned gene indicated that the mature polypeptide contains 793 amino acids which is 25 residues less than found for the *H. halobium* protein.

A comparison of the amino acid sequences of the S-layer glycoproteins from *H. halobium* and *H. volcanii* shows that the proteins share 40.5 % identity (Fig. 2). Remarkably, there are regions of high homology and these are arranged in a regular pattern. Stretches of nearly complete homology are interrupted by stretches of unrelated sequences and the degree of homology drops strikingly towards the N-terminal end, i.e., towards the most extracellular domain. Possibly, the regions of highly conserved sequences indicate sites of essential protein-protein interactions. This interpretation is supported by the schematic drawing shown in Fig. 3. This drawing compares both glycoproteins with respect to the location of N-glycosylation sites and regions of high homology. With only a single exception, none of the glycosylation sites are located within the regions of highest homology, supporting the view that these stretches may indeed be involved in protein- protein interactions. As can also be seen from this drawing, the distribution of N-glycosylation sites differs substantially, between the two glycoproteins and, in addition the moderate halophilic protein exhibits only 7 instead of 12 glycosylation sites. Common to both polypeptides is the location and sequence of the putative membrane binding domain with the adjacent clusters of threonine residues.

Unexpected, however, was the structure of the N-glycosidically-bound saccharides of the *H. volcanii* glycoprotein that turned out to be completely different (Mengele and Sumper, 1992). We found only glucose and galactose as the main sugar components of the glycoprotein. We could not detect amino sugars, uronic acids, nor sulfate residues. This fact immediately implies a completely different set of N-glycosidically-bound saccharides compared to the extremely salt tolerant glycoprotein. The absence of amino sugars excludes the existence of the repeating unit saccharide. From tryptic peptides, we isolated glycopeptides covering 4 out of the 7 N-glycosylation sites and analysed the saccharide structures by permethylation as well as enzymatic degradation studies. Again, asparaginylglucose was identified as the linkage unit, but instead of being linked with a few sulfated glucuronic acids, a chain of 9 or 10 β-(1-4)-linked glucose residues follows; completely uncharged saccharides replace the sulfated oligosaccharides found in the extremely salt tolerant glycoprotein (Fig. 1). With respect to the O-glycosidically bound disaccharides, we could not detect any differences between both glycoproteins (Sumper et al., 1990).

```
               20        30        40        50        60        70
HALO.AMI     YTYSTASEDGKTEGSVASGATIFQGEEDVTFRKLDNEKEVSPATLSRTGGSDEGVPLQMP
                          ...::. .: :::  :    :.... . ..: ....: . ... ..:
VOLC.AMI     ERGNLDADSESFNKTIQSGDRVFLGEEISTDAGLGASNPLLTGTAGNSEGVSLDLSSPIP
                     10        20        30        40        50        60
               80        90       100       110       120       130
HALO.AMI     -IPEDQSTGSYDSNGPDNDEADFGVTVQSPSVTMLEVRNNADNDVTGGVL-NTQQDEFSI
              ..:.:. :.:: .:......    .::. .:..:  :. ....::::... ....... .
VOLC.AMI     QTTENQPLGTYDVDGSGSATTP-NVTLLAPRITDSEILTSSGGDVTGSAISSSDAGNLYV
                     70        80        90       100       110
              140       150       160       170       180       190
HALO.AMI     AVDYNYYAAEDLELTVEDEDGLDVTDEILAADQSGGAYEDGTGNNGPNTLRFDIDPNNVD
             ..:::: .::..:.:::: .: :.:.:.....    . ....:......    .:: .. :
VOLC.AMI     NADYNYESAEKVEVTVEDPSGTDITNEVLSGTDTFVD-DGSIGSTSSTGGGVGIDMSDQD
          120       130       140       150       160       170
              200       210       220       230       240       250
HALO.AMI     AGDYTVSVEGVEDLDFGDATESASVTISSSNKASLNLAEDEVVQGANLKYTIENSPEGNY
             ::.::. .::.::::::::::. ..:::: .. ...:....:.::....::..:. .::
VOLC.AMI     AGEYTIILEGAEDLDFGDATETMTLTISSQDEIGIELDSESVTQGTDVQYTVTNGIDGNE
            180       190       200       210       220       230
              260       270       280       290       300       310
HALO.AMI     HAVTIDSSDFRDSSSGADAAKVMRSVGDTVDTGLVVDNDSTTEIVDDYENTSISDVDYAY
             :.:..: ::.........: .:.:..::: ..:  ..:.:.:.. .. ........: ::
VOLC.AMI     HVVAMDLSDLQNDATTEQAKEVFRNIGDTSEVG--IANSSATNTSGSSTGPTVETADIAY
            240       250       260       270       280       290
              320       330       340       350       360       370
HALO.AMI     AIVEIDDGNGVGSIETQYLDDSSADIDLYPASDTEDAPDYVNSNEELTNGSALDGVSTDD
             :.::::...::.:::::::::..:...:                 .........:. : ....
VOLC.AMI     AVVEIDGASAVGGIETQYLDDSEVDLEVY--------------DAGVSATAAV-GQDATN
              300       310       320                      330       340
              380       390       400       410       420       430
HALO.AMI     DTDFDVTQGDIMLDNPTGAYVVGSEVDINGTANEGTDDVVLYARDNNDFELVTVDGEKSI
             :........:.. :..:::.::::::::::::: ...:.:..:.::..:..:....:...:
VOLC.AMI     DITLTIEEGGTTLSSPTGQYVVGSEVDINGTA-TSSDSVAIYVRDDGDWQLLEIGGDNEI
                    350       360       370       380       390       400
              440       450       460       470       480       490
HALO.AMI     EVDSDDTFEEEDITLSDGDKGGDDILGLPGTYRLGIIAKSDA-VNSSGGVKDNIDTSDFN
             .::::::::::::.::. . .:..::.:.::::.:.:. ::: :...:.:.:...::.:.
VOLC.AMI     SVDSDDTFEEEDIALSGLSGDGSSILSLTGTYRIGVIDASDADVGGDGSVDDSLTTSEFT
                    410       420       430       440       450       460
               500       510       520       530       540       550
HALO.AMI     QGVSSTSSIRVTDTELTASFETYNGQVAD-DDNQIDVEGTAPGKDNVAAIIIGSRGKVKF
               ::::..::::::: .::. :.: :::::  ... .:..:::.: ..: .:....::.:..
VOLC.AMI     SGVSSSNSIRVTDQALTGQFTTINGQVAPVETGTVDINGTASGANSVLVIFVDERGNVNY
                    470       480       490       500       510       520
                560       570       580       590            600
HALO.AMI     QSISVDSDDTFDEEDIDISELRQGSASAHILSSGRDGKFGE-------DTANSISDLEDE
             :..::::::.:.::.::.... : ::...::::: :::. .:.         .......::..
VOLC.AMI     QEVSVDSDGTYDEDDITVG-LTQGRVTAHILSVGRDSAIGDGSLPSGPSNGATLNDLTGY
                    530       540       550       560       570
              610       620       630       640       650       660
HALO.AMI     VGNYTSGPPTGDQIRDRILSNTVDDTASDDLIVTQQFRLVDGLTTIEA--TEGGEAGG--
             ... . .  .:.::.. : :.:::.::::::::::. :::... :.:.. ....::.:
VOLC.AMI     LDTLDQNNNNGEQINELIASETVDETASDDLIVTETFRLAESSTSIDSIYPDAAEAAGIN
          580       590       600       610       620       630
                       670       680       690       700       710
HALO.AMI     ------SVTVMGTTNRKADDNTITVELLQGD-ASIEINSTDEWNSDGQWSVDVPLSNVEP
                   .... :.:: :.:::::..:. ..: .:.....::::.:::: :..  .. :.
VOLC.AMI     PVATGETMVIAGSTNLKPDDNTISIEVTNEDGTSVALEDTDEWNNDGQWMVEIDTTDFET
          640       650       660       670       680       690
              720       730       740       750       760       770
HALO.AMI     GNYTVEADDGDNTDRQNVEIVEELEEPDQTTVDQPENNQTMTTTMTETTTETTTEMTTTQ
             :..:::::::::::: ::::.:   .:...::   ...:.: ::: :...:::::
VOLC.AMI     GTFTVEADDGDNTDTVNVEVV---SEREDTT--TSSDNATDTTTTTDGPTETTT-----T
          700       710       720                  730       740
              780       790       800       810
HALO.AMI     ENTTENGSEGTSDGESGGSIPGFGVGVALVAVLGAALLALRQN
             ...::...: :.........::::..:::::.::::::::..
VOLC.AMI     AEPTETTEEPTEETTTSSNTPGFGIAVALVALVGAALLALRRE
          750       760       770       780       790
```

Figure 2. Sequence alignment for the S-layer glycoproteins from *H. halobium* and *H. volcanii*.

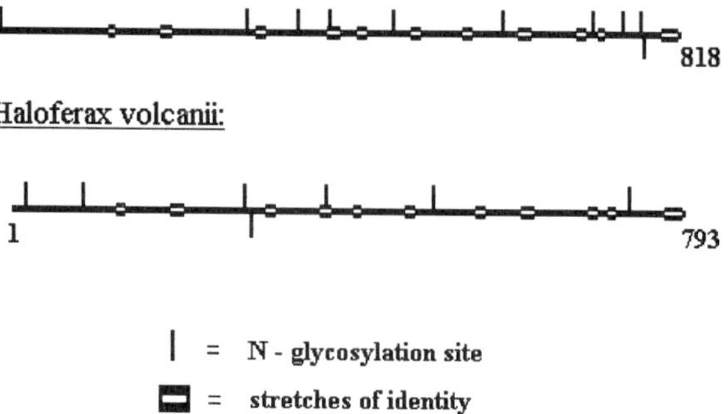

Figure 3. Schematic comparison of the S-layer glycoproteins from *H. halobium* and *H. volcanii* with respect to location of glycosylation sites and regions of high homology.

DID ACIDIC SACCHARIDES EVOLVE AS AN ADAPTATION TO HIGH SALT CONDITIONS ?

The marked difference in the glycosylation pattern of the otherwise related (40.5 % overall identity) S-layer glycoproteins drastically alters the net surface charges and, as discussed below, this probably reflects the adaptation from a moderately to an extremely halophilic environment. This drastic increase in negative charge density is illustrated in Table 1. First, unlike the S-layer proteins from non-halophilic organisms (Sleytr and Messner, 1983; Kandler and König, 1985) there is a large excess of negatively charged amino acids in the polypeptide chains of both proteins. The ratio of negatively versus positively charged amino acids is about the same for both polypeptide chains and the calculated isoelectric points are as low as 3.2 and 3.4, respectively. In the case of the moderate halophile no additional charges are introduced by the neutral saccharide structures. In sharp contrast, the saccharides of the extreme halophile contribute at least 120 additional negative charges. This is nearly a doubling of negative surface charges.

For a discussion of these observations, it is helpful to point out a few facts concerning the biosynthetic pathways of the sulfated saccharides in *H. halobium* (for a review, see Sumper, 1987). As already mentioned, two different N-glycosyl linkages are synthesized within the same polypeptide chain, and the two biosynthetic pathways differ in the type of saccharide precursors involved. A dolichol monophosphate (Lechner et al., 1985) is used as the carrier for the sulfated oligosaccharides whereas a lipid pyrophosphate (most likely a dolichol diphosphate) serves as precursor of the repeating unit saccharide. The oligosaccharides are completed and sulfated while still attached to dolichol on the cytosolic side of the cell membrane. Thereafter, they are translocated to the cell surface, where the transfer to the polypeptide chain takes place. The extracellular location of the saccharyltransferases make the halobacterial cell surface functionally equivalent to the luminal side of the endoplasmatic reticulum of the eucaryotic cell. Due to the specific differences of the biosynthetic pathways of the sulfated oligosaccharides and the repeating unit saccharide, it is possible to selectively inhibit one of these pathways. The antibiotic bacitracin is known to inhibit lipid

Table 1. Estimation of the net negative charge in the S-layer protein from *H. volcanii* and *H. halobium*, respectively

H. volcanii	*H. halobium*
Number of charges introduced by amino acid residues	
Asp 89	113
Glu 64	70
Arg 14	19
Lys 4	15
Negative charges introduced by saccharides	
no charges	about 120

pyrophosphate-dependent glycosylations. Indeed, S-layer glycoprotein isolated from halobacteria grown in the presence of bacitracin lack the repeating unit saccharide (Wieland et al., 1980). The biological consequence is quite spectacular. Bacitracin-treated bacteria are no longer rods but grow as spheres. This is good evidence for a functional role of the repeating unit saccharide in maintaining the crystalline structure of the bacterial surface layer and should be discussed in the context of possible strategies of adaptation to high salt conditions.

From a detailed study of a halophilic malate dehydrogenase, Zaccai et al. (1989) have proposed a model for the stabilization of halophilic proteins. They assume that the halophilic enzyme has a core structure very similar to the corresponding non-halophilic enzyme, but in addition has highly negatively charged inserts protruding as loops. These negatively charged loops are assumed to organize (under high salt conditions) an energetically favoured protein-water-salt network around the core structure. In other words, these loops should operate as clamps stabilizing the core structure under high salt conditions. Sulfated and uronic acid-containing saccharides as found in the S-layer glycoprotein from the extreme halophile are able to contribute an even higher negative charge density (two charges per sugar residue) than any peptide loop. Therefore, the drastic structural modifications of the saccharides found at the transition from a moderately to an extremely halophilic S-layer glycoprotein may have the important function of stabilizing the protein structure in saturated salt solutions. This is schematically illustrated in Fig. 4. It is assumed that the three dimensional structure of the negatively charged polypeptides of both S-layer proteins remains stable up to about 2 M NaCl concentrations. This is the environment of *H. volcanii* and, in this case, glycosylation is not required for stabilization. However, in saturated salt solutions, stabilization of the protein structure requires additional surface charges and these are provided by the greatly modified saccharides of *H. halobium*. This interpretation is supported by the following experimental observations. First, as mentioned above, bacitracin selectively inhibits the attachment of the repeating unit saccharide to the S-layer glycoprotein, thereby reducing the surface charge by about 60 charges per molecule. The modified S-layer protein is no longer able to maintain the integrity of

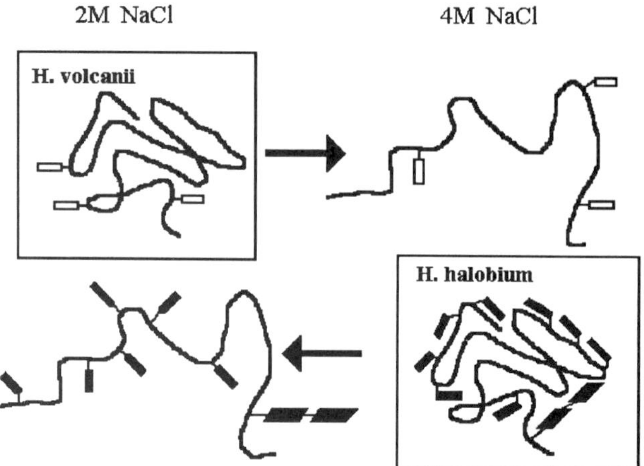

Figure 4. Glycosylation modulates the range of stability of protein structure of otherwise related polypeptide chains.

the S-layer at 4 M NaCl. On the other hand, untreated *H. halobium* cells are converted from rods to spheres if the growth medium is diluted to moderately halophilic conditions (2.3 M NaCl), again indicating a collapsing S-layer structure. The extreme, additional accumulation of negative surface charges in the *H. halobium* glycoprotein no longer supports an intact protein structure at salt concentrations that are optimal for the related *H. volcanii* polypeptide. The saccharides may modulate the requirements for the stabilization of a given protein structure.

REFERENCES

Gerl, L., and Sumper, M., 1988, Halobacterial flagellins are encoded by a multigene family, *J. Biol. Chem.* 263:13246.

Kandler, O., and König, H., 1985, Cell envelopes of archaebacteria, *in*: "The Bacteria", Woese, C.R., and Wolfe, R.S., eds., Academic Press, New York.

Kessel, M., Wildhaber, I., Cohen, S., and Baumeister, W., 1988, Three-dimensional structure of the regular surface glycoprotein layer of *Halobacterium volcanii* from the Dead Sea, *EMBO J.* 7:1549.

Lechner, J., Wieland, F., and Sumper, M., 1985, Biosynthesis of sulfated oligosaccharides N-glycosidically linked to the protein via glucose, *J. Biol. Chem.* 260:860.

Lechner, J., and Sumper, M., 1987, The primary structure of a procaryotic glycoprotein, *J. Biol. Chem.* 262:9724.

Lechner, J., and Wieland, F., 1989, Structure and biosynthesis of procaryotic glycoproteins, *Annu. Rev. Biochem.* 58:173.

Mengele, R., and Sumper, M., 1992, Drastic differences in glycosylation of related S-layer glycoproteins from moderate and extreme halophiles, *J. Biol. Chem.* 267:8182.

Mescher, M.F., and Strominger, J.L., 1976, Purification and characterization of a procaryotic glycoprotein from the cell envelope of *Halobacterium salinarium*, *J. Biol. Chem.* 251:2005.

Paul, G., and Wieland, F., 1987, Sequence of the halobacterial glycosaminoglycan, *J. Biol. Chem.* 262:9587.

Paul, G., Lottspeich, F., and Wieland, F., 1986, Asparaginyl-N-acetylgalactosamine: linkage unit of halobacterial glycosaminoglycan, *J. Biol. Chem.* 261:1020.

Sleytr, U.B., and Messner, P., 1983, Crystalline surface layers on bacteria, *Annu. Rev. Microbiol.* 37:311.

Sumper, M., 1987, Halobacterial glycoprotein biosynthesis, *Biochim. Biophys. Acta* 906:69.

Sumper, M., Berg, E., Mengele, R., and Strobel, I., 1990, Primary structure and glycosylation of the S-layer protein of *Haloferax volcanii*, *J. Bacteriol.* 172:7111.

Wieland, F., Dompert, W., Bernhardt, G., and Sumper, M., 1980, Halobacterial glycoprotein saccharides contain covalently linked sulphate, *FEBS Lett.* 120:110.

Wieland, F., Lechner, J., and Sumper, M., 1982, The cell wall glycoprotein of Halobacteria: structural, functional and biosynthetic aspects, *Zbl. Bakt. Hyg., I. Abt. Orig. C* 3:161.

Wieland, F., Heitzer, R., and Schaefer, W., 1983, Asparaginylglucose: novel type of carbohydrate linkage, *Proc. Natl. Acad. Sci. USA* 80:5470.

Wieland, F., Paul, G., and Sumper, M., 1985, Halobacterial flagellins are sulfated glycoproteins, *J. Biol. Chem.* 260:15180.

Zaccai, G., Cendrin, F., Haik, Y., Borochov, N., and Eisenberg, H., 1989, Stabilization of halophilic malate dehydrogenase, *J. Mol. Biol.* 208:491.

Chapter 12

THE UNIQUE CHEMICAL FORMATS AND BIOSYNTHETIC PATHWAYS OF METHANOGENIC SURFACES

Helmut König, Evamarie Hartmann, Günther Bröckl
and Uwe Kärcher

Applied Microbiology
University of Ulm
Ulm, Germany

INTRODUCTION

The third domain of life, the archaea (Woese et al., 1990), is divided into two main lineages: the methanogenic branch (euryarchaeota) and the branch of extreme thermophilic sulfur metabolizing bacteria (crenarchaeota). The methanogens were not only the first archaea detected by microbiologists, but they also represent the largest group within the archaeal domain. A common feature, which distinguishes the methanogens from all other procaryotes, is the production of methane (Balch et al., 1979). Sequence analysis of their 16S rRNA and other biochemical and molecular features show that methanogens are not a phylogenetically homogeneous group. They do not possess a common cell wall polymer, but rather diverse cell envelope types are found (Table 1; Kandler and König, 1985; König, 1988). All methanogens lack murein, the common eubacterial cell wall polymer.

The cell walls of the gram-positive methanogens consist of pseudomurein or methanochondroitin. An additional S-layer is present in some species. The gram-negative methanogenic bacteria have cell walls composed of single-layered crystalline protein or glycoprotein subunits which make up an S-layer. *Methanospirillum* and *Methanothrix* form long chains, which are held together within a tubular proteinaceous sheath. The individual cells are surrounded by a distinct layer of unknown chemical nature and are separated by spacer plugs.

The methanobacterial surface polymers exhibit some chemical and structural similarities with cell wall components of other eubacterial and eucaryotic organisms. Are common biosynthetic principles involved in the cell wall polymers of methanogens and other organisms? In order to answer this question we have

---

Dedicated to R.S. Wolfe on the occasion of his 70th birthday

investigated the biosynthesis of three different methanobacterial cell wall polymers: a peptidoglycan-like polymer (pseudomurein), a heteropolysaccharide (methanochondroitin) and a glycoprotein. The proposed biosynthetic pathways for the carbohydrate moieties of these polymers are summarized and common principles are discussed.

Table 1. Distribution of cell wall polymers among methanogens

Organisms	Cell wall polymer
1. Methanobacteriales	
1.1. Methanobacteriaceae	pseudomurein
1.2. Methanothermaceae	pseudomurein/S-layer[a]
2. *Methanopyrus kandleri*	pseudomurein[c]
3. Methanomicrobiales	
3.1. *Methanosarcina* spp.	methanochondroitin[c]
3.2. *Methanomicrobium mobile*	S-layer[b]
3.3. *Methanogenium cariacii*	S-layer[b]
3.4. *Methanoculleus marisnigri*	S-layer[a]
3.5. *Methanoplanus limicola*	S-layer[a]
3.6. *Methanolobus tindarius*	S-layer[a]
3.7. *Methanothrix soehngenii*	sheath[a,c]
3.8. *Methanospirillum hungatei*	sheath[a,c]
4. Methanococcales	S-layer[b]

[a]glycoprotein subunits
[b]protein subunits
[c]additional layer present which probably corresponds to an S-layer

CHARACTERISTIC FEATURES AND BIOSYNTHESIS OF PSEUDOMUREIN

The gram-positive rods or cocci of the order Methanobacteriales and of the genus *Methanopyrus* possess an electron dense cell wall sacculus which is composed of pseudomurein (Kandler and König, 1985; König, 1988; Kurr et al., 1991). The members of the genera *Methanothermus* and *Methanopyrus* have a surface layer in addition to the cell wall sacculus. The pseudomurein contains a set of three *L*-amino acids (Lys, Glu, Ala), the N-acetylated amino sugars glucosamine or galactosamine and N-acetyl-L-talosaminuronic acid, which form glycan strands consisting of alternating $\beta(1\rightarrow3)$-linked N-acetyl-*D*-glucosamine and $\beta(1\rightarrow3)$-linked N-acetyl-*L*-talosaminuronic acid residues, which are crosslinked via short peptides attached to the carboxylic group of N-acetyl-*L*-talosaminuronic acid (Fig. 1). Due to its overall chemical (König and Kandler, 1985; König, 1988) and three dimensional (Leps et al., 1984; Formanek, 1985) structure the pseudomurein has to be classified as a peptidoglycan (Sharon, 1986), although it is a basically different type of peptidoglycan and not a new variant of the eubacterial murein.

A pathway of the biosynthesis of pseudomurein was proposed on the basis of the structure of putative precursors isolated from cell extracts (Hartmann and König, 1990; Fig. 1). In cell extracts the monomers UDP-GlcNAc and UDP-GalNAc and the disaccharide UDP-GlcNAc-(3<——1)β-NAcTalNA were present. Since a monomeric derivative of talosaminuronic acid was not found it is assumed that N-acetyltalosaminuronic acid is formed during the synthesis of the disaccharide, probably by epimerization and oxidation of UDP-N-acetylgalactosamine.

The synthesis of the pentapeptide moiety is supposed to start with a UDP-activated glutamic acid residue followed by the stepwise formation of UDP-activated peptides up to a pentapeptide. UDP was directly linked to the N^α-amino group of the glutamic acid residue. Finally, the UDP-activated pentapeptide is linked to the disaccharide to give a UDP-activated disaccharide pentapeptide which is most likely transferred to undecaprenyl monophosphate resulting in an undecaprenyl pyrophosphate-activated disaccharide pentapeptide. In the last step of pseudomurein biosynthesis a terminal alanyl residue is split off and a glutamic acid residue is linked to a lysine residue of an adjacent peptide.

I.

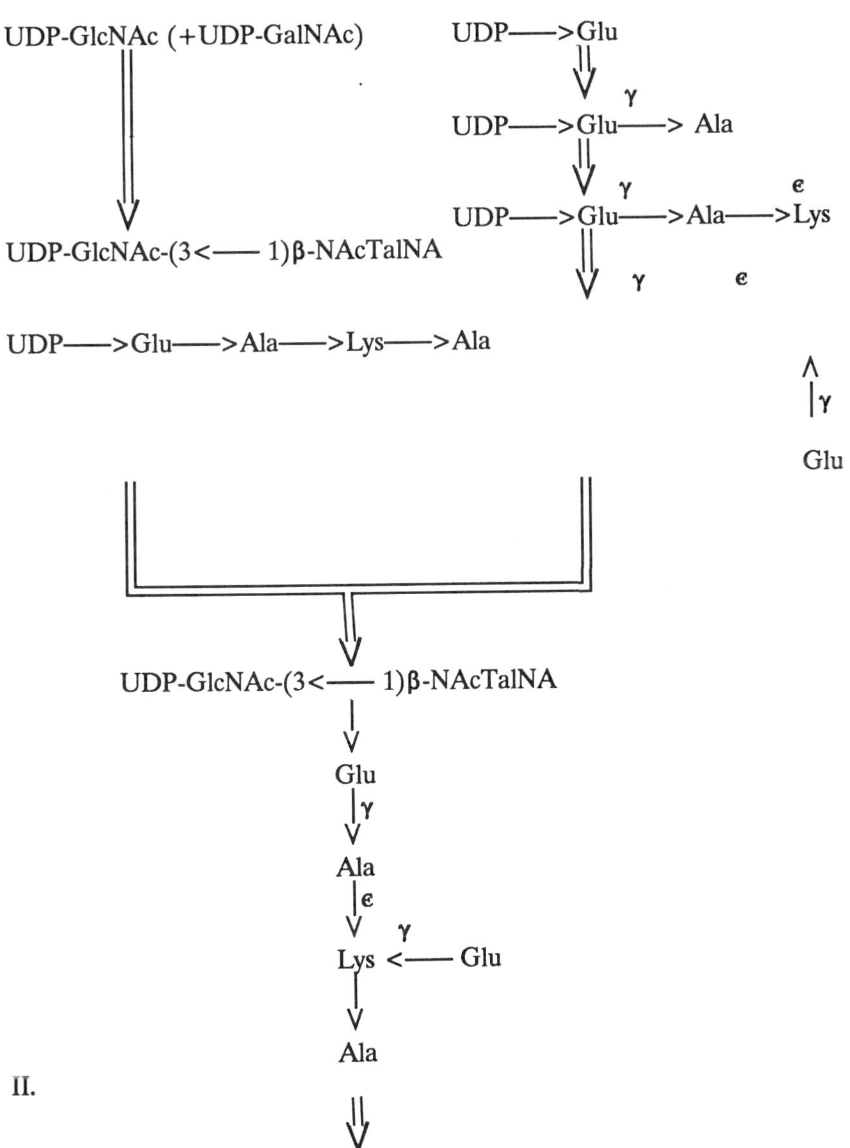

II.

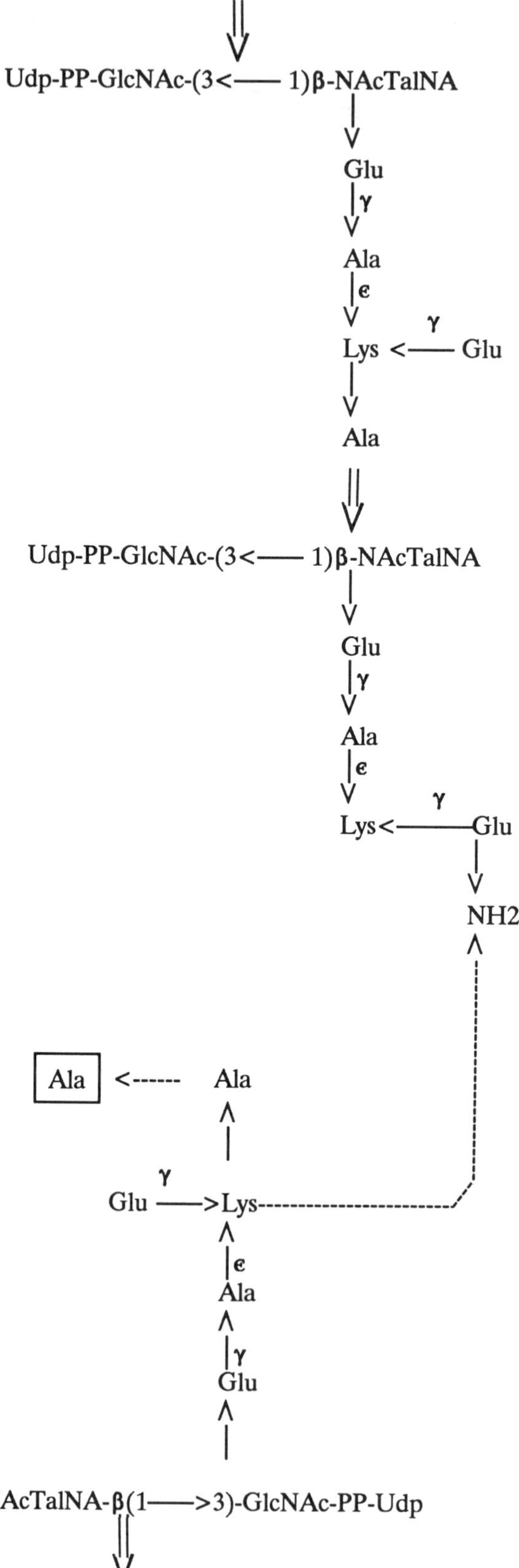

Udp-PP-GlcNAc-(3<——— 1)β-NAcTalNA
Glu
γ
Ala
ε
Lys <——— Glu
γ
Ala
III.
Udp-PP-GlcNAc-(3<——— 1)β-NAcTalNA
Glu
γ
Ala
ε
Lys<———Glu
γ
NH2
Ala
Glu ——>Lys
γ
Ala
ε
Ala
γ
Glu
NAcTalNA-β(1——>3)-GlcNAc-PP-Udp

IV.

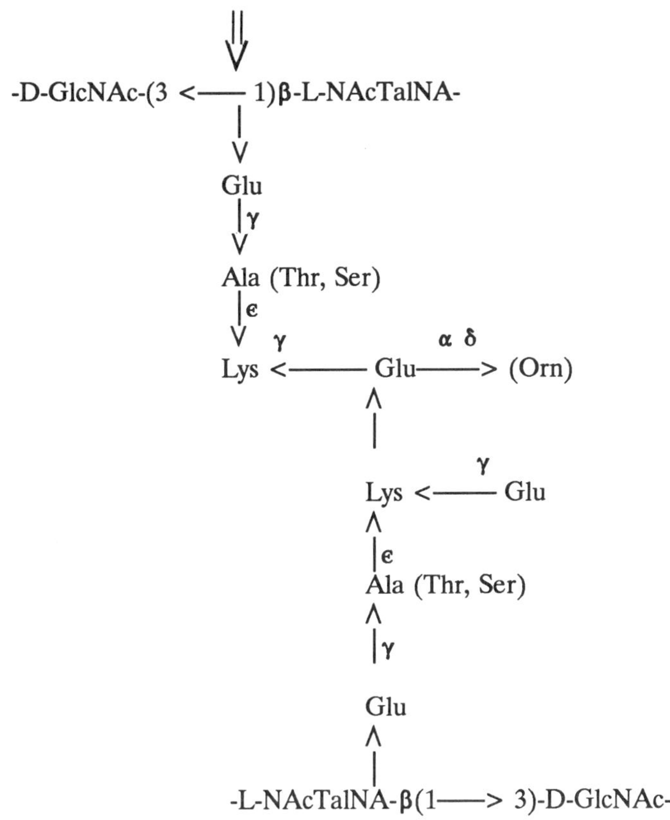

Figure 1. Proposal for the biosynthesis of the pseudomurein. I, II, III = three stages, IV = primary structure with modifications. (From Hartmann and König, 1990).

CHARACTERISTIC FEATURES AND BIOSYNTHESIS OF AN S-LAYER GLYCOPROTEIN

Methanothermus fervidus has a double-layered cell wall. The inner pseudomurein sacculus is covered by an S-layer composed of glycoprotein subunits which are hexagonally arranged. The S-layer glycoprotein could be isolated by extracting whole cells with trichloroacetic acid (5%-10%; v/v) and subsequent reversed-phase chromatography using aqueous formic acid (10%; v/v) as eluant (Nußer et al., 1988). The primary sequence of the peptide was derived from the nucleotide sequence of the corresponding gene (Bröckl et al., 1991). The mature glycoprotein consists of 593 amino acids. Compared to mesophilic S-layer glycoproteins, this hyperthermophilic glycoprotein possesses significantly higher amounts of isoleucine, asparagine and cysteine. It also has 14% more β-sheet structures. A typical leader peptide and 20 sequon structures (Asn-Xaa-Ser/Thr) as potential N-glycosylation sites are present. The isolated heterosaccharide consists of mannose, 3-O-methylmannose, 3-O-methylglucose and N-acetylgalactosamine, whereas in the glycoprotein small amounts of galactose and N-acetylglucosamine were also found. The heterosaccharides, which are linked to asparagine via N-acetylgalactosamine, consist of alternating mannose and 3-O-methylmannose/3-O-methylglucose residues (U. Kärcher and H. König, manuscript in preparation; Fig. 2).

α-D-3-O-MetMan*p*-(1→6)-α-D-3-O-MetMan*p*-[(1→2)-α-D-(Man*p*]₃-(1→3/4)-D-GalNAc

Figure 2. Proposed structure of a glycan chain from the S-layer glycoprotein of *M. fervidus*. 3-O-MeMan could be partly replaced by 3-O-MeGlc.

Based on isolated precursors (Hartmann and König, 1989; Fig. 3) the biosynthesis of the glycan chains is suggested to start with the formation of C-1-phosphate derivatives of the corresponding sugars. They are then converted to the corresponding nucleotide activated derivatives which form UDP activated oligosaccharides. Glc and 3-O-methylmannose/3-O-methylglucose are not found as monomers, but they are constituents of the oligosaccharides. The oligosaccharides are

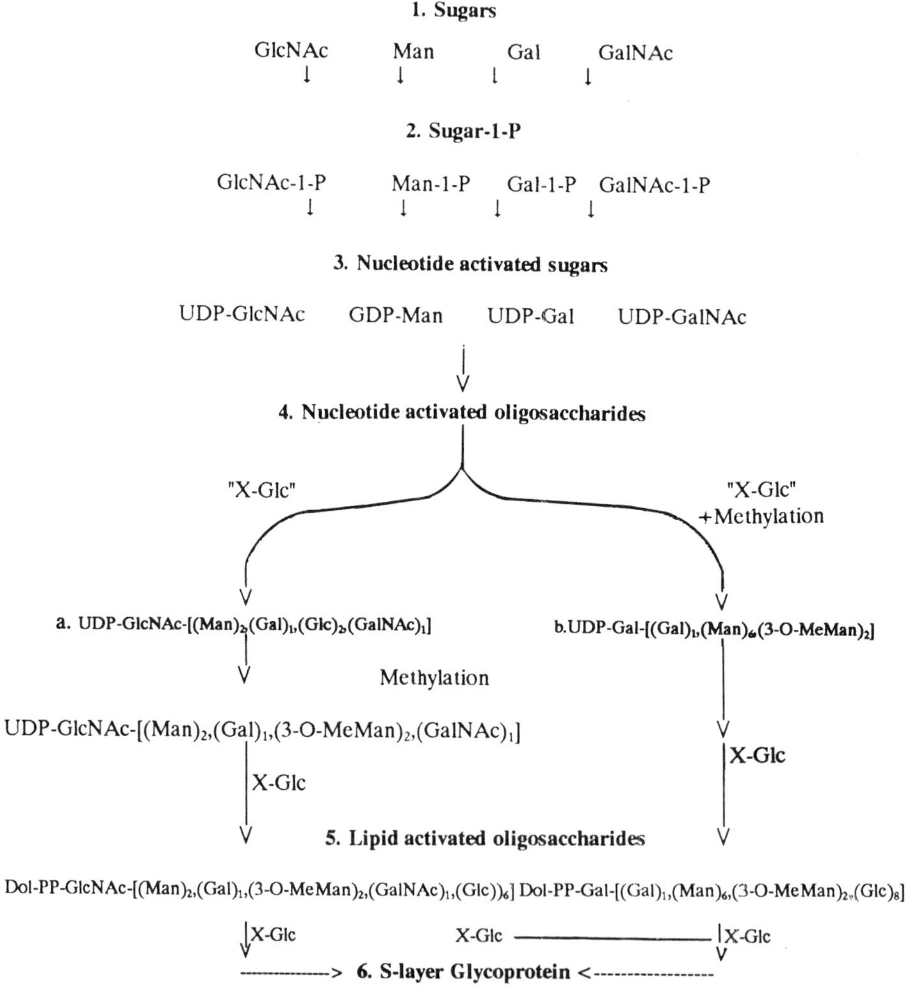

Figure 3. Proposal for the biosynthesis of the oligosaccharide chains of the S-layer glycoprotein of *M. fervidus*. 3-O-MeMan could be partly replaced by 3-O-MeGlc. (From Hartmann and König, 1989).

A) [——>)-β-D-GlcA-(1——>3)-β-D-GalNAc-(1——>4)-β-D-GalNAc-(1——>]$_n$
B) [D-GlcA-(1——>3)-GalNAc-(1——>4)-]$_{23-25}$
C) [-β-D-GlcA-(1——>3)-β-D-GalNAc-(1——>4)-β-D-GlcA-1——>]$_n$
        (4 or 6 sulfates)

Figure 4.   Comparison of the building blocks of methanochondroitin (A) of *Methanosarcina sp.*, teichuronic acid (B) of *Bacillus licheniformis* and of chondroitin sulfate (C) of animal connective tissue. (From Kreisl and Kandler, 1986; modified).

most probably transferred to C_{55}-dolichyl phosphate forming dolichyl pyrophosphate-activated oligosaccharides. The lipid activated oligosaccharides contain additional Glc residues, which are removed during further processing. A transient glycosylation has also been found in *Halobacterium salinarium* (Sumper, 1987; Lechner and Wieland, 1989).

CHARACTERISTIC FEATURES AND BIOSYNTHESIS OF METHANO-CHONDROITIN

Cells of *Methanosarcina* species are usually surrounded by a rigid, sometimes laminated, cell wall matrix consisting of methanochondroitin (Kreisl and Kandler, 1986; Fig. 4). The cell wall material is composed of *D*-galactosamine, *D*-glucuronic acid or *D*-galacturonic acid, *D*-glucose, acetic acid and trace amounts of mannose. As structural elements, chondrosine and a trimer have been isolated from partial acid hydrolysates. The trimer could be identified as being the building block of the glycan (Kreisl and Kandler, 1986; Fig. 4).

Because of its similarities to chondroitin sulfate, the shape-maintaining cell wall polymer of *Methanosarcina* has been named methanochondroitin.

Based on 4 UDP-activated and 1 undecaprenyl pyrophosphate-activated putative precursors of methanochondroitin, isolated from extracts of *M. barkeri*, a tentative pathway for methanochondroitin biosynthesis has been proposed (Hartmann and König, 1991; Fig. 5). According to this proposal, UDP-N-acetylchondrosine is formed from UDP-N-acetylgalactosamine and UDP-glucuronic acid and is followed by the transfer of N-acetylchondrosine to UDP-N-acetylgalactosamine, thus yielding a UDP-activated trisaccharide, which possesses the same structure as the suggested repeating unit of methanochondroitin. In the next step, the UDP residue of the trisaccharide is most probably replaced by undecaprenyl pyrophosphate, and the trisaccharide may then be polymerized onto methanochondroitin at the outer face of the cytoplasmic membrane.

BIOSYNTHESIS OF A EUBACTERIAL S-LAYER GLYCOPROTEIN

Cells of *Bacillus alvei* are surrounded by a murein sacculus and an additional S-layer composed of a glycoprotein. A trimer consisting of mannosamine, galactose and glucose forms the building block of the glycan chains (Fig. 6; Altman et al., 1991). The proposed scheme for the biosynthesis, derived from isolated intermediates, is depicted in Fig. 7 (E. Hartmann, H. König and P. Messner, manuscript in

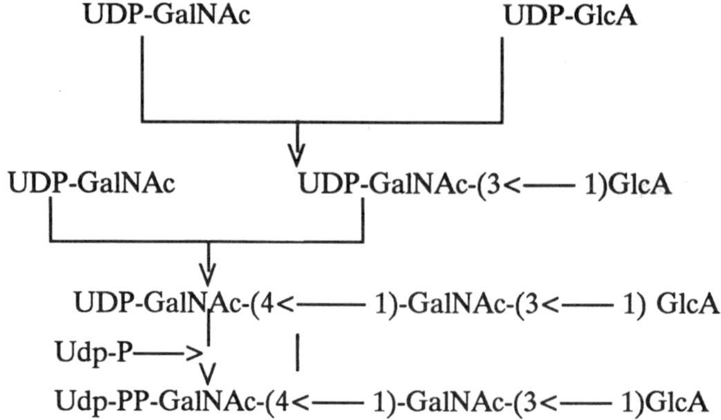

Figure 5. Proposal for the biosynthesis of methanochondroitin. The trisaccharide represents the repeating unit. (From Hartmann and König, 1991).

preparation). The biosynthesis starts with the corresponding nucleotide-activated monomers and proceeds via nucleotide- and lipid-activated oligosaccharides. Additional glucosamine residues are linked to the oligosaccharides which are split off during further processing.

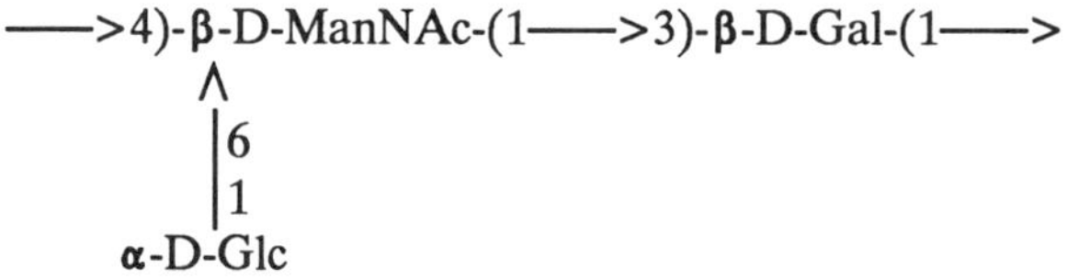

$$\longrightarrow 4)\text{-}\beta\text{-D-ManNAc-}(1\longrightarrow 3)\text{-}\beta\text{-D-Gal-}(1\longrightarrow$$

$$\wedge$$
$$\big|6$$
$$\big|1$$
$$\alpha\text{-D-Glc}$$

Figure 6. Structure of a glycan chain from the S-layer glycoprotein of *B. alvei*. (Modified from Altman et al., 1991).

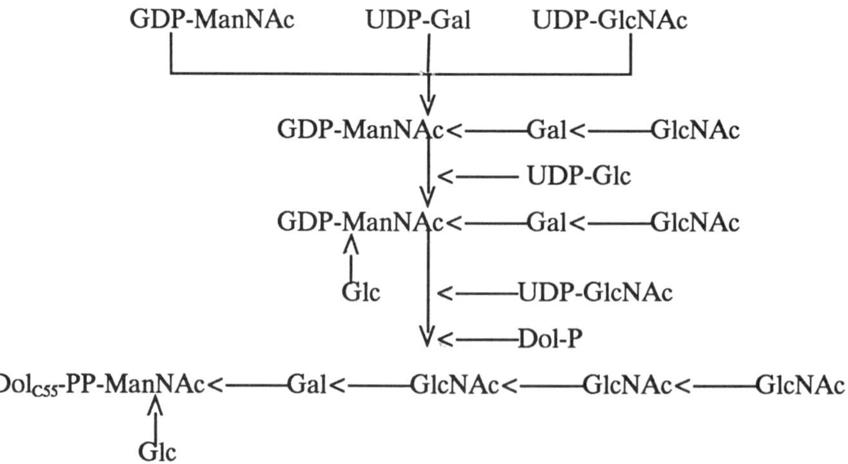

Figure 7. Proposal for the biosynthesis of the S-layer glycoprotein of *B. alvei*. (E. Hartmann, P. Messner and H. König, manuscript in preparation).

CONCLUDING REMARKS

1. Since nucleotide-activated oligosaccharides participate in the biosynthesis of three different methanobacterial cell envelope polymers such as pseudomurein, an S-layer glycoprotein and methanochondroitin, it appears that they are a typical common feature of the biosynthesis of glycan components in methanogens. Usually, oligosaccharides occuring in bacterial cell wall polymers are formed at the lipid stage (Tipper and Wright, 1979; Wright and Tipper, 1979). An exception are eubacterial glycoproteins, where nucleotide-activated oligosaccharides are also involved.

2. Undecaprenol seems to be the universal lipid carrier involved in the biosynthesis of cell wall polymer from all procaryotes, except glycoproteins, where dolichol is involved as the common lipid carrier in pro- and eucaryotes.

3. Due to the occurrence of a nucleotide-activated oligosaccharide and N^{α}-activated glutamic acid residues and peptides, it becomes clear that the route of biosynthesis for murein and pseudomurein follows different pathways. Therefore, pseudomurein could not have been derived from murein during evolution.

4. As shown for methanochondroitin, the biosynthesis of the structurally very similar eucaryotic chondroitin sulfate (Fig. 4), which is attached to protein, follows a quite different route. Here, the oligosaccharides are directly formed from their nucleotide-activated monomers (Róden 1970) at distinct serine or threonine residues of the growing polypeptide chain. Also, the teichuronic acid of *B. licheniformis*, (Fig. 4) exhibits a structure similar to methanochondroitin (Rogers et al., 1980) and is built from monomers at the lipid stage with undecaprenyl pyrophosphate as lipid carrier.

REFERENCES

Altmann, E., Brisson, J.-R., Messner, P., and Sleytr, U. B., 1991, Chemical characterization of the regularly arranged surface layer glycoprotein of *Bacillus alvei* CCM2051, *Biochem. Cell Biol.* 69:72.

Balch, W.E., Fox, G.E., Magrum, L.J., Woese,C.R., and Wolfe, R.S., 1979, Methanogens: Reevaluation of a unique biological group, *Microbiol. Rev.* 43:260.

Bröckl, G., Behr, M., Fabry, S., Hensel, R., Kaudewitz, H., Biendl, E., and König, H., 1991, Analysis and nucleotide sequence of the genes encoding the surface-layer glycoproteins of the hyperthermophilic methanogens *Methanothermus fervidus* and *Methanothermus sociabilis, Eur. J. Biochem.* 199:147.

Formanek, H., 1985, Three-dimensional models of the carbohydrate moieties of murein and pseudomurein, *Z. Naturforsch.* 40c:555.

Hartmann, E., and König, H., 1989, Uridine and dolichyl diphosphate activated oligosaccharide intermediates are involved in the biosynthesis of the surface layer glycoprotein of *Methanothermus fervidus, Arch. Microbiol.* 151:274.

Hartmann, E., and König, H., 1990, Comparison of the biosynthesis of the methanobacterial pseudomurein and the eubacterial murein, *Naturwissenschaften* 77:472.

Hartmann, E., and König, H., 1991, Nucleotide activated oligosaccharides are intermediates of the cell wall polysaccharide of *Methanosarcina barkeri, Biol. Chem. Hoppe-Seyler* 372:971.

Kandler, O. and König, H., 1985, Cell envelopes of archaebacteria, *in*: "The Bacteria. A Treatise on Structure and Function, Vol. VIII, Archaebacteria", pp. 413-457, C. R. Woese and R. S. Wolfe, eds., Academic Press, New York.

König, H., 1988, Archaeobacterial cell envelopes, *Can. J. Microbiol.* 34:395.

Kreisl, P., and Kandler, O., 1986, Chemical structure of the cell wall polymer of *Methanosarcina, Syst. Appl. Microbiol.* 7:293.

Kurr, M., Huber, R., König, H., Jannasch, H.W., Fricke, H., Trincone, A., Kristjansson, J.K., and Stetter, K.O., 1991, *Methanopyrus kandleri*, gen. nov. and sp. nov. represents a novel group of hyperthermophilic methanogens, growing at 110°C, *Arch. Microbiol.* 156:239.

Lechner, J., and Wieland, F., 1989, Structure and biosynthesis of prokaryotic glycoproteins, *Annu. Rev. Biochem.* 58:173.

Leps, B., Labischinski, H., Barnickel, G., Bradaczek, H., and Giesbrecht, P., 1984, A new proposal for the primary and secondary structure of the glycan moiety of pseudomurein, *Eur. J. Biochem.* 144:279.

Nußer, E., Hartmann, E., Allmeier, H., König, H., Paul, G. and Stetter, K.O., 1988, A glycoprotein surface layer covers the pseudomurein sacculus of the extreme thermohile *Methanothermus fervidus, in*: "Crystalline bacterial cell surface layers" U. B. Sleytr, P. Messner, D. Pum, and M. Sára, eds., pp. 21-25, Springer Verlag, Berlin.

Róden. L., 1970, Biosynthesis of acidic glycosaminoglycans (mucopolysaccharides), *in*: "Metabolic Conjugation and Metabolic Hydrolysis", W. H. Fishman, ed., Vol. 2, pp. 345-442, Academic Press, New York.

Rogers, H.J., Perkins, H.R., and Ward, J.B., 1980, "Microbial Cell Walls and Membranes", Chapman & Hall, New York.

Sharon, N., 1986, Nomenclature of glycoproteins, glycopeptides and peptidoglycans, *Eur. J. Biochem.* 159:1.

Sumper, M., 1987, Halobacterial glycoprotein biosynthesis, *Biochim. Biophys. Acta* 906:69.

Tipper, D.J. and Wright, A., 1979, Structure and biosynthesis of bacterial cell walls, *in*: "The Bacteria: A Treatise on Structure and Function", I. C. Gunsalus, J. R. Sokatch, and L. N. Ornston, eds., Vol. 7, pp. 291-426, Academic Press, New York.

Woese, C.R., Kandler, O., and Wheelis, M.L., 1990, Towards a natural system of organisms: proposal for the domains Archaea, Bacteria, and Eucarya, *Proc. Natl. Acad. Sci. USA* 87:4576.

Wright, A., and Tipper, D.J., 1979, The outer membrane of gram-negative bacteria, *in*: "The Bacteria: A Treatise on Structure and Function", I.C. Gunsalus, J. R. Sokatch, L. N. Ornston, eds., Vol. 7, pp. 427-485, Academic Press, New York.

Chapter 13

PARACRYSTALLINE LAYERS OF *METHANOSPIRILLUM HUNGATEI* GP1

Gordon Southam and Terry J. Beveridge

Department of Microbiology, College of Biological Science
University of Guelph, Guelph, Ontario, Canada

and G. Dennis Sprott

Institute of Biological Sciences
National Research Council of Canada
Ottawa, Ontario, Canada

INTRODUCTION

Methanospirillum hungatei GP1 (Patel et al., 1976) is a methanogenic archaeobacterium which possesses an unusual envelope architecture unique among bacteria because of the paracrystalline nature of its three envelope components (Zeikus and Bowen, 1975; Shaw et al., 1985; Beveridge et al., 1991; Southam and Beveridge, 1992b). The major envelope feature of *M. hungatei* is the hollow tubular sheath which surrounds, but does not completely enclose individual cells nor chains of two or more cells. The second envelope component, the multilayered end plug (spacer plug; Beveridge et al., 1987) acts in concert with the sheath to encase individual cells. Within this sheath/end plug framework, cells are bound by a proteinaceous cell wall which is the third envelope component (Southam and Beveridge, 1992b). In this study we provide evidence for the existence of a fourth periodic envelope structure located in the cell spacer region.

The sheath is composed of hoops which stack together to form the intact sheath cylinder (Sprott et al., 1986). The outer surface of the sheath (and hoop) is paracrystalline, possessing a unit cell with p2 symmetry (a= 2.8 nm, b = 5.7 nm, γ = 86°)(Stewart et al., 1985). The sheath is chemically resilient, proving stable to typical chaotropic and non-ionic detergents, denaturants and hydrolytic enzymes (Beveridge et al., 1985). This stability is attributed to disulfide bonding because of its sensitivity to the disulfide bond breaker, β-mercaptoethanol (β-ME) (Sprott et al., 1986; Southam and Beveridge, 1991; Southam and Beveridge, 1992b).

The end plugs are multilaminar, disc-shaped proteinaceous assemblies possessing p6 symmetry. Two distinct morphological assemblies have been identified (Beveridge et al., 1991).

The cell wall of *M. hungatei* has a structural periodicity that has been observed in thin sections of plasmolysed cells (Southam and Beveridge, 1992b). Beveridge et al. (1987) demonstrated that the cell wall of *M. hungatei* facilitates its cell division.

The chemical stability of the sheath contributes to its role as the shapedetermining structure for the bacterium (Beveridge et al., 1985; Sprott et al., 1986; Southam and Beveridge, 1991 and 1992b; Conway de Macario et al., 1986). However, the maintenance of cell integrity through sheath/end plug interaction is less well understood. The ability of the *M. hungatei* envelope to withstand cell turgor pressure has been attributed to a ring of amorphous material which is located between each cell pole and end plug. This material is thought to cement the end plug to the sheath (Beveridge et al., 1987) thereby preventing the pole from being deformed by its turgor pressure. We propose that the mechanism to help resist this pressure and provide structural integrity in *M. hungatei* is conferred by a newly characterized periodic envelope component which will be referred to as the "bushing assembly".

PURIFICATION OF CELL ENVELOPE COMPONENTS

The cell envelope of *M. hungatei* GP1 consists of several boundary layers (Fig. 1) which are united with one another by complex physicochemical interactions.

The Sheath

The *M. hungatei* sheath has a resilient nature which allows for its purification through a succession of harsh treatments: 0.1 N NaOH at room temperature and 1% (w/v) sodium dodecyl sulfate (SDS) at 100°C (Southam and Beveridge, 1991). Hoops were produced by chemically splitting the intact sheath along the hoop boundaries

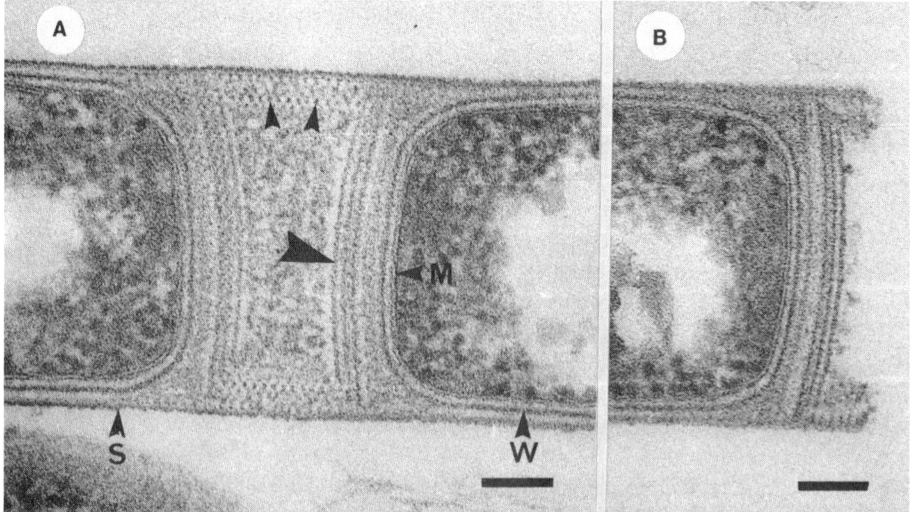

Figure 1. Thin sections of *M. hungatei* within the cell spacer region (A; from Beveridge et al., 1991) and at a filament pole (B). Note the sheath structure (S), cell wall (W), plasma membrane (M) and multilayered cell spacer (large arrow head). Also note the amorphous material between the cell wall and the spacer plug and the small particles inside the sheath within the cell spacer (small arrow heads). Bars = 100 nm.

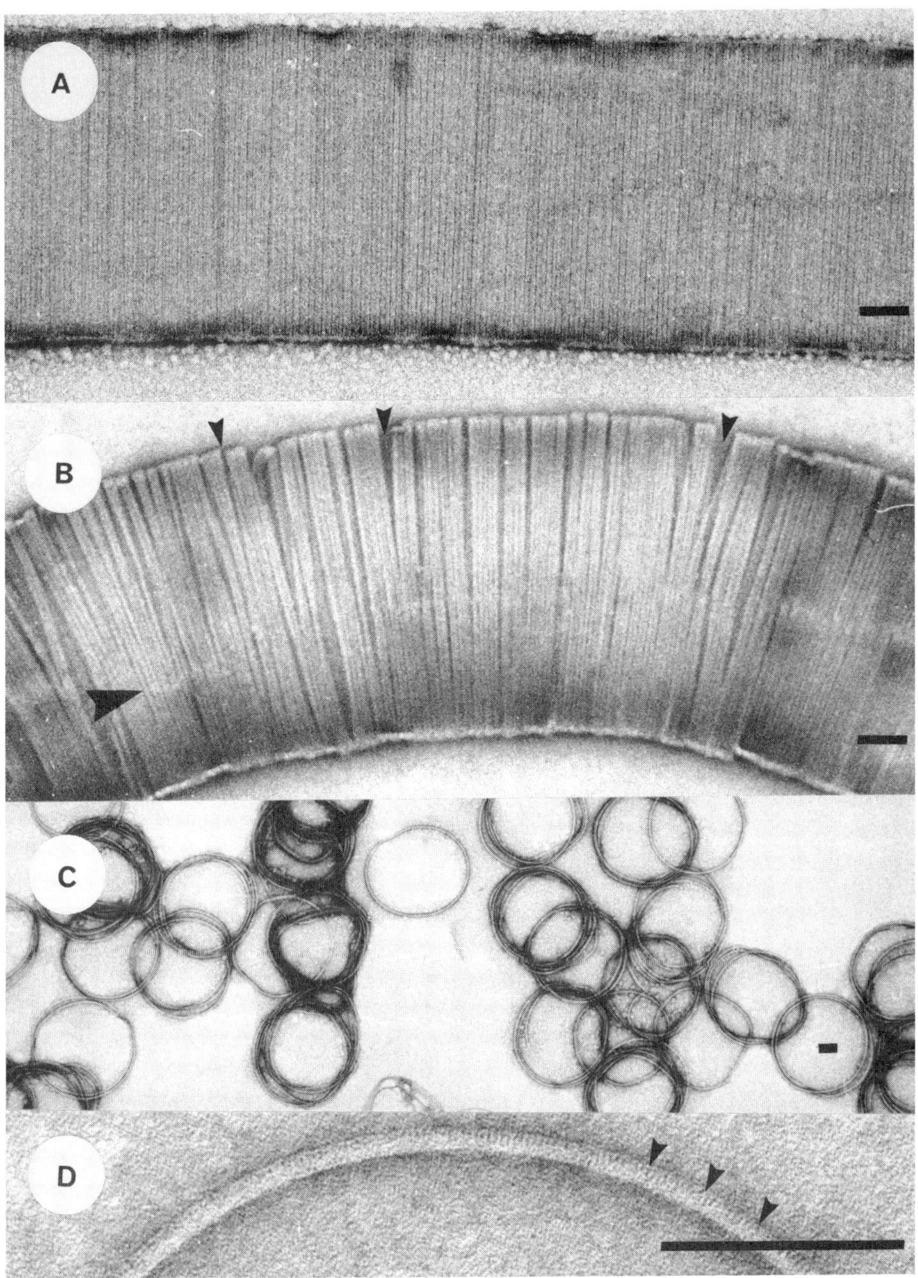

Figure 2. This is a series of negative stains (2% (w/v) uranyl acetate) of intact sheath and of sheath which has been chemically split into its constituent hoops. **A.** Electron micrograph of intact sheath. Hoop boundaries (Fig. 2B) are difficult to see due to the Moiré pattern produced by imaging through the two layers of the collapsed sheath tube. The periodicity on the surface of the sheath is a 2.8 nm repeat (Stewart et al., 1985). **B.** Sheath treated with 5% β-ME at room temperature for 2 h at 100 °C. Note the fractures along the hoop boundaries (small arrow heads) and the longitudinal fracture along the length of the sheath tube (large arrow head). **C.** Sheath treated with SDS/β-ME/EDTA at 90°C for 30 minutes which splits the sheath into individual hoops. **D.** A segment of a hoop from Fig. 2C. Note the 2.8 nm repeat along the outer circumference of the hoop (small arrow heads). Bars = 100 nm.

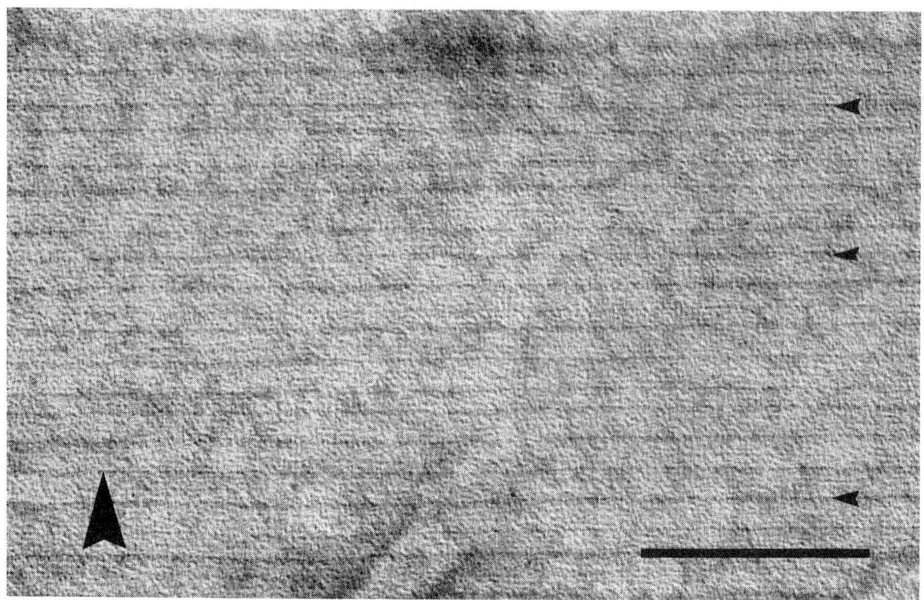

Figure 3. A negative stain (2% (w/v) uranyl acetate) of a single layered fragment of sheared sheath. The hoop boundaries are more intensely stained (small arrow heads) than the 2.8 nm periodic units within the hoops. Sheared sheath typically breaks along (near) straight lines at the hoop boundaries and longitudinally across adjoining hoops. The longitudinal orientation of the sheath is indicated by the large arrow head. Bar = 100 nm.

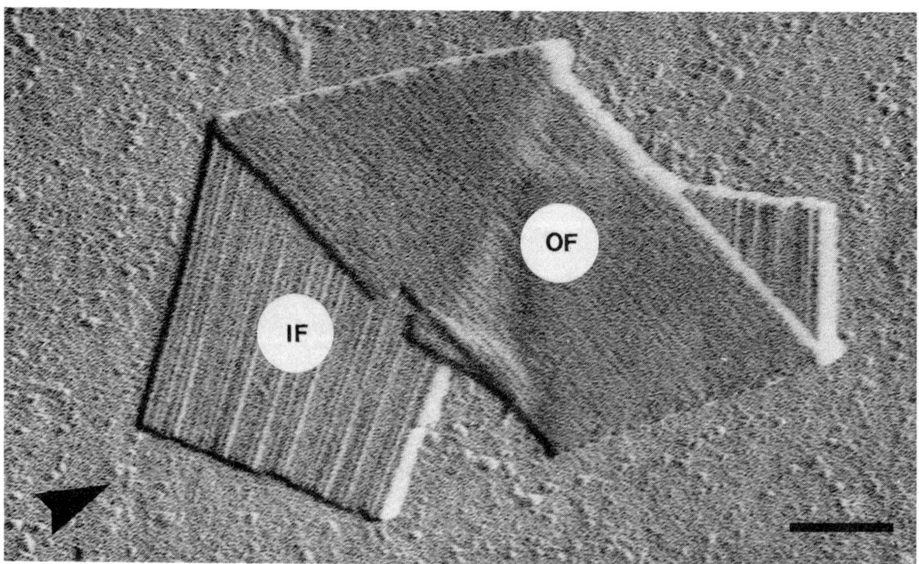

Figure 4. An electron micrograph of platinum-shadowed sheared sheath. The direction of the shadow is indicated by the arrow head. The inner face (IF) of the sheath possesses ridges, not present on the outer face (OF). These ridges correspond to the hoop boundaries. Bar = 100 nm.

with 5% (v/v) β-ME alone and with 2% (w/v) SDS and 5% (v/v) β-ME (Southam and Beveridge, 1991). Purified sheath was also reacted with 90% (w/v) phenol overnight at room temperature to separate it into phenol soluble (PS) and phenol insoluble (PI) fractions as described previously (Southam and Beveridge, 1992a).

The End plugs and the Bushing Assembly

The harshness of the sheath purification regimen would destroy the end plugs. Their isolation was, therefore, accomplished by spheroplasting and lysing the cells with 15 mM dithiothreitol (DTT) at pH 9.6 (Sprott et al., 1979). The resulting crude sheath/end plug/bushing assembly preparation was cleaned of contaminating membrane material by washing (3X) with 0.1% (w/v) SDS in DTT buffer and centrifugation (14,000 X g). The end plugs were excised from the sheath cylinder by disrupting the bushing assembly with a 1% (w/v) SDS treatment at room temperature. The sheath was pelleted by centrifugation (14,000 X g) while the end plugs remained in suspension.

The Cell Wall

For the examination of the cell wall, *M. hungatei* cells were discharged from the sheath by spheroplasting with 2% β-ME pH 7 in the presence of 0.15 M sucrose.

CHARACTERIZATION OF THE *M. HUNGATEI* ENVELOPE COMPONENTS

The Sheath

By negative stain transmission electron microscopy (TEM) analysis, the sheath takes on the rectangular shape of a collapsed sheath tube (caused by drying the specimen on the TEM grid) (see Fig. 2A). Hoop boundaries run perpendicular to the longitudinal axis of the sheath, but are difficult to see because of the Moiré pattern created by the double-layered periodic structure. Hoop boundaries are more evident in sheath which has been treated with β-ME (Fig. 2B). When used in concert with SDS, β-ME is able to chemically split the sheath into individual hoops (Fig. 2C). The 2.8 nm surface repeat of the sheath is evident along the outer circumference of these hoops (Fig. 2D). In negative stains of sheared sheath, in which single-layered fragments are evident, both the surface repeat and the hoop boundaries (darker striations) can be observed (Fig. 3). From this type of preparation, we have determined that hoops range in size from 3 to 9 times the 2.8-nm repeat (Southam and Beveridge, 1992b). In platinum-shadows of sheared sheath, the inner and outer surfaces of the sheath could be differentiated (Fig. 4). The outer surface of the sheath presents a gentle wave-form character, whereas, the inner surface of the sheath possesses ridges (0.4 nm to 0.7 nm in height; Beveridge et al., 1990) which are thought to represent the hoop boundaries.

Chemically, the sheath is composed of two groups of polypeptides. The first group is soluble by treatment of sheath with SDS/β-ME/EDTA (Fig. 5A and B, lane 2; Southam and Beveridge, 1991) and the second group can be solubilized from the bulk of the sheath structure using 90% phenol (Fig. 5A, lane 3; Southam and Beveridge, 1992a). The sheath-like structure remaining after phenol extraction (PI) (Fig. 6) consists of the SDS/β-ME/EDTA polypeptides (Southam and Beveridge, 1992a). These SDS/β-ME/EDTA polypeptides were specifically labelled with [35S] by growing *M. hungatei* in [35S]-cysteine containing medium (Southam and Beveridge,

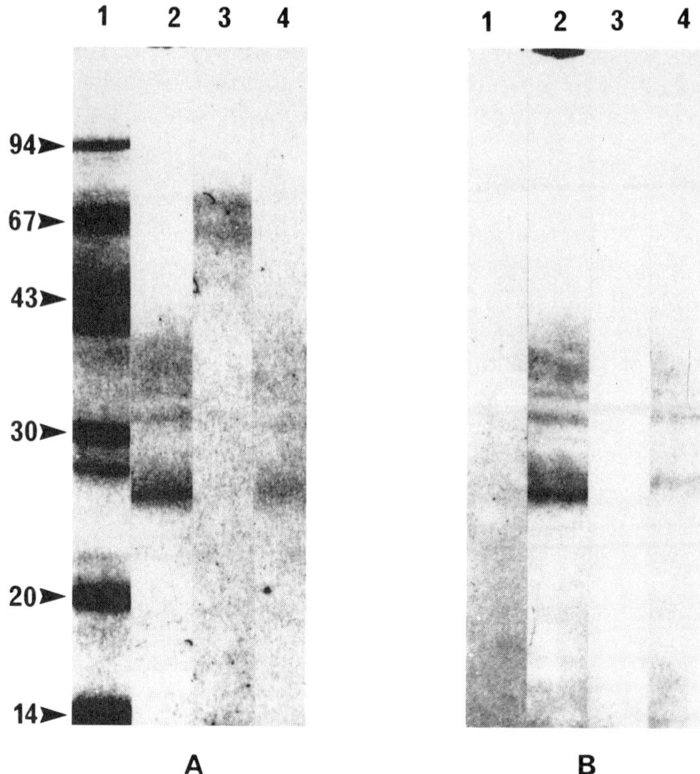

Figure 5. A. Coomassie Brilliant Blue R250 stained SDS-PAGE gel containing a Bio-Rad Low Molecular Weight Standard (lane 1), SDS/β-ME/EDTA solubilized [^{35}S] sheath (lane 2), PS fraction from [^{35}S] sheath (lane 3) and the remaining PI material after the phenol extraction (lane 4). **B.** An autoradiogram of the SDS-PAGE gel shown in panel A. (from Southam and Beveridge, 1992b).

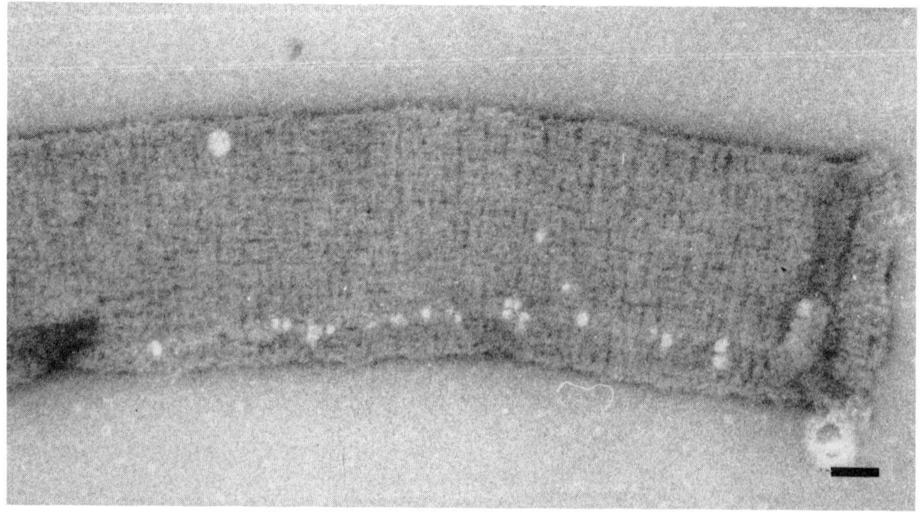

Figure 6. A negative stain of the PI sheath fraction. Note the cross-hatched pattern on the sheath which likely corresponds to the hoop boundaries and a previously unseen periodicity along the length of the sheath tube (not evident in negative stains of intact sheath; see Fig. 2A). Bar = 100 nm.

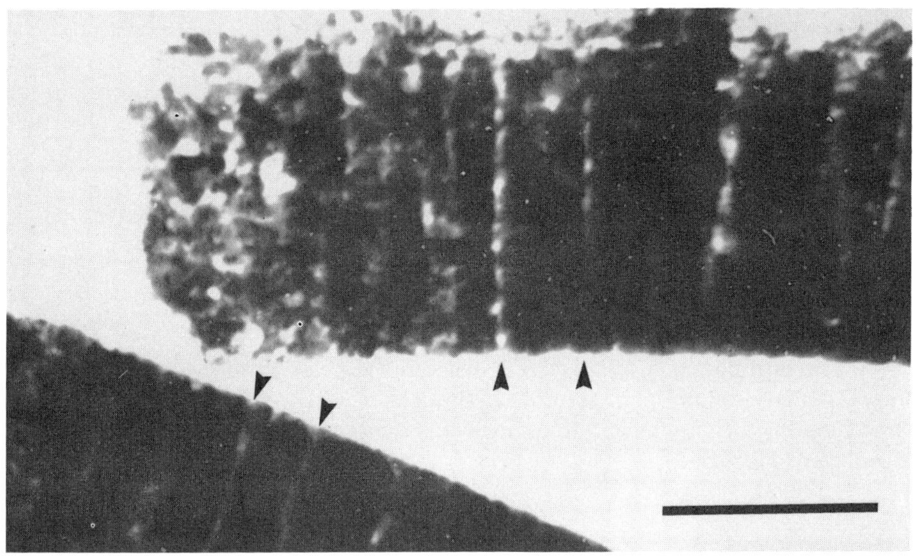

Figure 7. Thin-film TEM autoradiogram of sheath purified from a culture of *M. hungatei* grown in the presence of [^{35}S]-cysteine for 28 days. Note the presence of discrete, non-radioactive hoop-like regions (small arrow heads). Bar = 0.5 μm. Negative stains of analogous whole mounts of sheath confirmed that there was no actual (physical) separation between hoops and that the sheaths were intact (from Southam and Beveridge, 1992b).

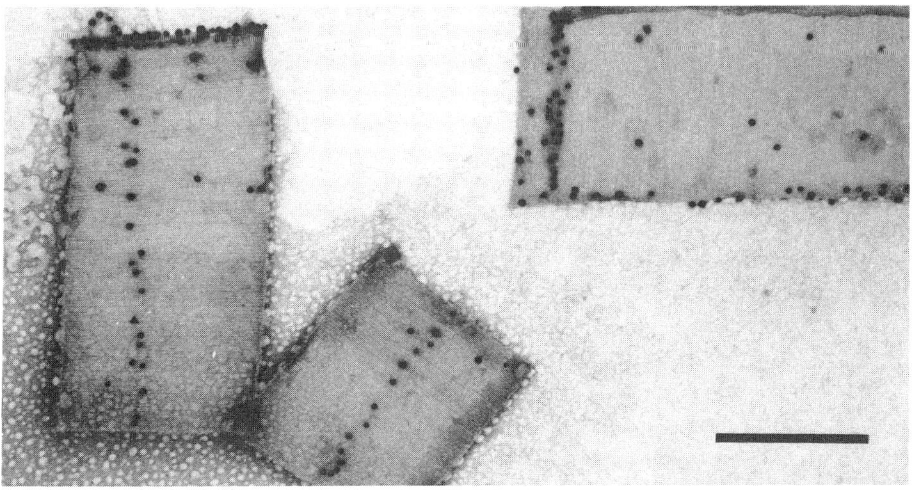

Figure 8. Negative stain of sheath probed with monoclonal antibody 1.8 (specific for the PS sheath fraction, see Southam and Beveridge, 1992b) and labelled with protein A-colloidal gold. Note the presence of colloidal gold along the length of the sheath tube at several transverse sites. This labelling pattern provides evidence for limited surface exposure of the PS proteins. Bar = 0.5 μm.

1992b). TEM autoradiography of intact radiolabelled sheath demonstrated that the [^{35}S] was distributed throughout the entire sheath (Fig. 5A and B, lanes 2 and 4; Southam and Beveridge, 1992b). TEM autoradiography also demonstrated that sheath growth is most active at the sites of cell division and that it occurs through the development of hoops (Fig. 7). Southam and Beveridge (1992a) demonstrated that the 2.8 nm surface repeat was conferred by the SDS/β-ME/EDTA soluble polypeptides so it is not surprising that the [^{35}S]-label was distributed throughout the entire sheath structure. It had also been shown that the PS polypeptides have limited surface exposure on intact sheath, occurring primarily along a longitudinal molecular fracture line in the sheath (cf. Fig. 2B and 8).

The End plug

The end plug is a multilayered structure (Fig. 1A and B) composed of disc-shaped paracrystalline (p6) assemblies (Beveridge et al., 1991). The two disc types which have been characterized to date include the particulate layer (Fig. 9A) and the holey layer (Fig. 9B).

The Cell Wall

The *M. hungatei* cells that were released from the sheath/end plug framework possessed a cell wall with p6 periodicity demonstrating that the cell wall is an S-layer (Fig. 10). Unfortunately, it is an extremely labile structure which has hindered its further characterization.

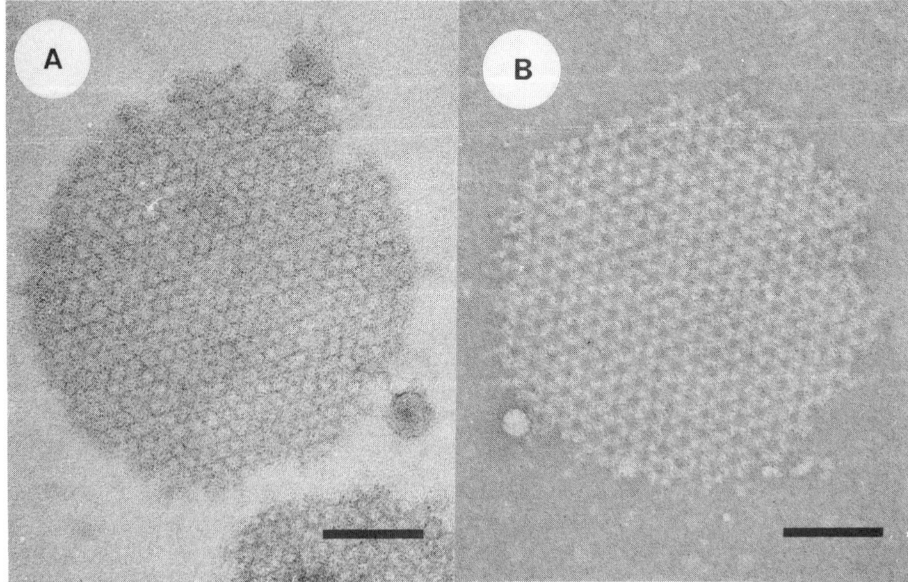

Figure 9. A. Negative stain of a particulate end plug layer which has been extruded from the DTT treated *M. hungatei* cells (from Beveridge et al., 1991). **B.** A negative stain of a holey plug layer prepared as above. Bars = 100 nm.

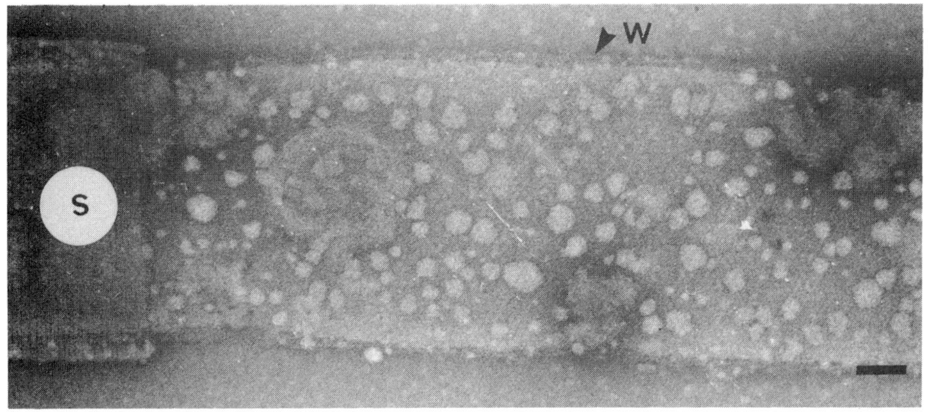

Figure 10. *M. hungatei* spheroplasted with 2% (v/v) β-ME. Note the sheath (S) and the periodicity associated with the cell wall (W). The rod shape of *M. hungatei* is maintained by the cell wall even after removal of the cell from the sheath. Bar = 0.5 μm.

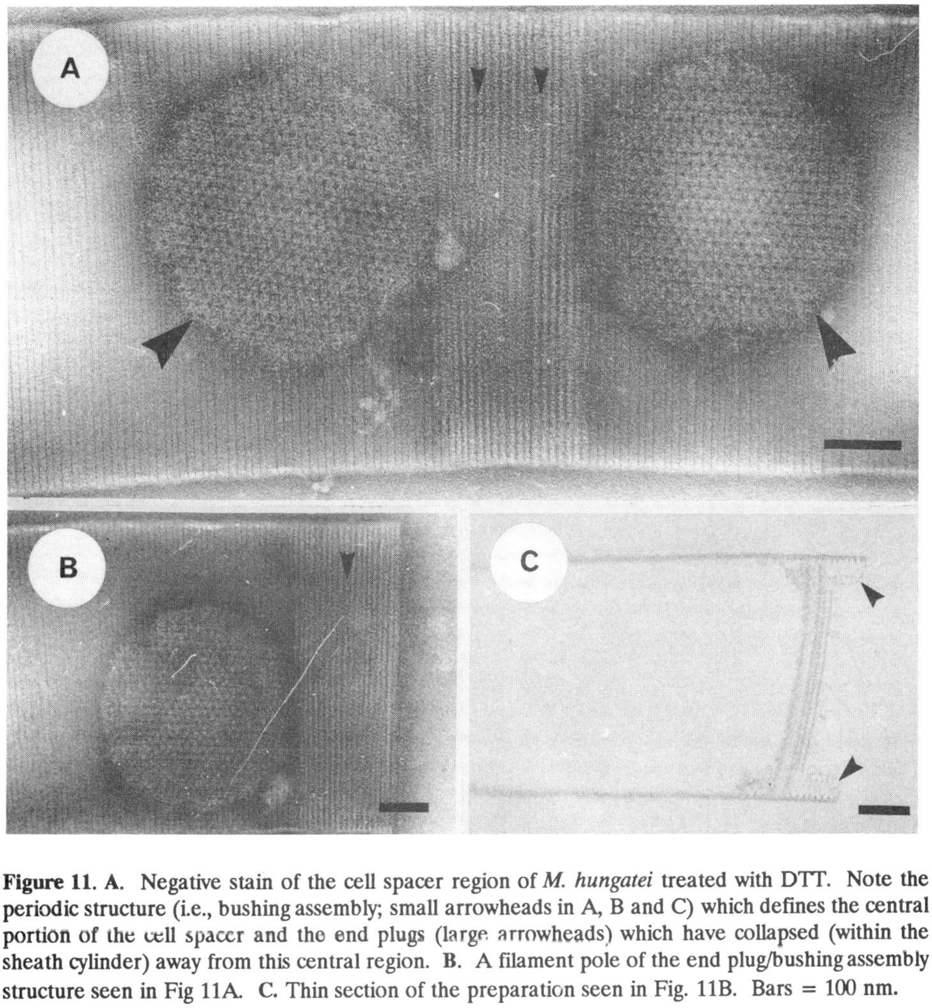

Figure 11. A. Negative stain of the cell spacer region of *M. hungatei* treated with DTT. Note the periodic structure (i.e., bushing assembly; small arrowheads in A, B and C) which defines the central portion of the cell spacer and the end plugs (large arrowheads) which have collapsed (within the sheath cylinder) away from this central region. **B.** A filament pole of the end plug/bushing assembly structure seen in Fig 11A. **C.** Thin section of the preparation seen in Fig. 11B. Bars = 100 nm.

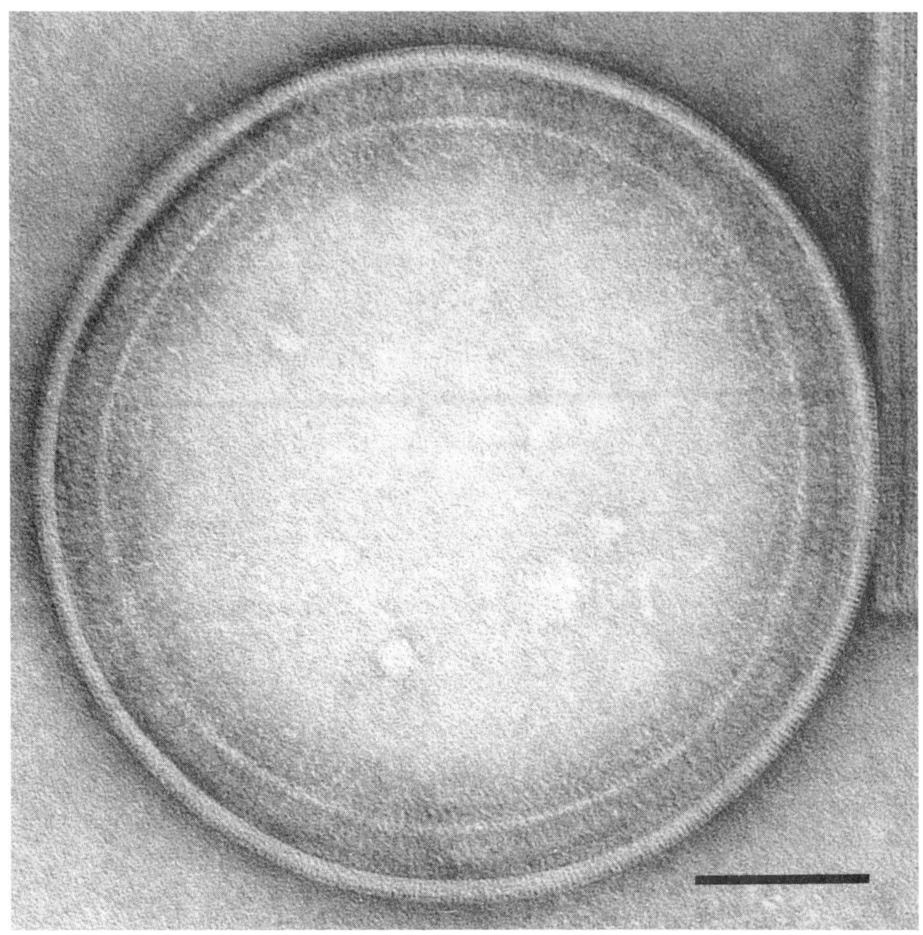

Figure 12. A negative stain of a hoop which has peeled off from the filament pole of DTT treated *M. hungatei* (see Fig. 11B) The inner disk corresponds to the basic subunit of the bushing assembly. Bar = 100 nm.

The Bushing Assembly

In the crude sheath/end-plug preparation produced by spheroplasting *M. hungatei* with DTT (Sprott et al., 1979), a novel periodic structure was observed within the cell spacer region (Fig. 11A) and at filament poles (Fig. 11B). Of particular importance, the end plugs collapsed away from this newly observed structure in negative stains (see Fig. 11A and B). This new periodic structure was named the bushing assembly because of its role in holding the end plug within the sheath cylinder. This is accomplished by reducing the diameter of the opening of the sheath within the cell spacer and at filament termini (Fig. 11C). The bushing assembly consists of a series of discs, associated with the inner surface of the sheath; one disc per hoop (Fig. 12).

DISCUSSION

Most bacteria, except the Tenericutes, possess a cell wall which serves as the shape-determining structure for the organism as well as providing protection from lysis owing to cell turgor. In the Gracilicutes and Firmicutes, the maintenance of cell integrity is conferred by a peptidoglycan-based cell wall which completely encloses individual cells. In *M. hungatei* (division: Mendosicutes), individual cells do not possess a murein sacculus but are enclosed by a proteinaceous cell wall (Fig. 10) and a complex proteinaceous sheath/end plug structure (Fig. 1A and B).

Due to the labile nature of the cell wall, it is unlikely that it plays a major role in the maintenance of cell integrity. Yet it must possess some shape-maintaining function because β-ME spheroplasted cells were rod-shaped (Fig. 10) and because it is responsible for cell division. Beveridge et al. (1987) demonstrated that cell division in *M. hungatei* occurred by septation, similar to gram-positive eubacteria. However, septation and enzymatic splitting of a septum, as described for eubacterial rods (Shockman and Barrett, 1983), cannot apply to *Methanospirillum* since they require peptidoglycan-containing cell walls. In a structural manner, the wall of *M. hungatei* is an S-layer and murein or pseudomurein is not a chemical constituent found in the cell; intuitively it must be protein or glycoprotein. Once self-assembled, opposing S-layers would be separate entities. This septation is unlike any other that has previously been described.

The sheath (Fig. 2A) is the most resilient envelope component of *M. hungatei* which suggests an important role for this structure in the maintenance of cell integrity. For the sheath to function in the physical support of *M. hungatei*, growth must occur by a mechanism which does not create openings in the sheath and make the weakened envelope subject to lysis by cell turgor.

The examination of [^{35}S]-labelled sheath by TEM autoradiography revealed that sheath growth, which is most active at sites of cell division (data not shown), occurs through the development of hoops (Fig. 7; Southam and Beveridge, 1992b). Dissolution studies with β-ME (Sprott et al., 1986; Southam and Beveridge, 1991; Southam and Beveridge, 1992b), which suggest that disulfide bonds are responsible for the maintenance of sheath integrity, have also highlighted hoop boundaries as a chemically sensitive region of intact sheath (Fig. 2B and 2C). Therefore, one aspect of sheath growth may be coordinated with the formation or hydrolysis of selected disulfide bonds as new hoop boundaries are formed.

In negative stains of single layer fragments of sheath (Fig. 3), the paracrystalline nature of the sheath is evident as is the location of hoop boundaries. These hoop boundaries consist of grooves in the sheath surface (data not shown; Southam et al., 1992). Based on structural observations, hoop boundaries are formed by an incomplete, circumferential splitting of the mid region of a large hoop (6 to 9 times the 2.8 nm repeat) into two, smaller composite hoops (i.e., this is a hoop division). This seems to be a process where hoops expand and mature, allowing cell and filament growth.

A second type of sheath maturation can be inferred from the molecular model for the 2.8-nm paracrystalline repeat proposed by Sprott et al. (1986) together with our immunogold labelling (Southam and Beveridge, 1991). In this model, the 2.8 nm repeat consists of a single layer of polypeptides on the outer surface of the sheath. However, using immuno-gold probes, these SDS/β-ME/EDTA polypeptides were localized to both the inner and outer surfaces of the sheath (Southam and Beveridge, 1991). Therefore, the sheath likely exists as a bilayer of these SDS/β-ME/EDTA polypeptides. For growth to occur, sheath polypeptides must be first incorporated at the inner surface of the sheath and second, transferred to the outer surface of the sheath where they will assume the periodic nature of the sheath's outer surface (i.e., maturation). This maturation can be visualized in platinum-shadows of folded, single layers in which different surface topographies are evident (Fig. 4). The outer surface of the sheath occurs as a gentle wave-form in contrast to the inner sheath surface which is highlighted by the presence of ridges in the circumferential or hoop orientation. By scanning tunneling microscopy, the inner surface of the sheath exists as a relatively flat plane containing ridges at hoop boundaries (Beveridge et al., 1990).

Two possibilities exist to help explain the occurrence of the 2.8 nm repeat only on the outer surface of the sheath. The 2.8 nm repeat may be explained simply by the increase in circumference at the outer surface compared to the inner surface. In this situation, a homogeneous matrix would have to be broken up into discrete subunits to accommodate an increased surface area at the outer face versus that at the inner face. Yet, the subunits would still be strongly bonded together and with the underlying matrix so that the outer face could still be a stress-bearing structure. Alternately, the formation of the 2.8 nm repeat may be produced via a stress releasing mechanism. This would allow for molecular spreading (i.e., visualization of the 2.8 nm repeat) in which the outer surface would be a reduced-stress or perhaps a non stress-bearing region.

The role of disulfide bonds in the maintenance of sheath integrity is supported by two experiments. First, labelling sheath with [^{35}S] highlighted the SDS/β-ME/EDTA polypeptides as the sulfur-containing constituents of the sheath (Fig. 5; lanes 2 and 4; Southam and Beveridge, 1992b). Second, phenol extraction did not denature the general sheath structure (Fig. 6) and did not remove any of the SDS/β-ME/EDTA polypeptides from this modified sheath (Fig. 5A and B; lane 3). This substantiates the existence of covalent bonds in the sheath, e.g., the disulfide bonds to SDS/β-ME/EDTA polypeptides.

In intact sheath, the PS polypeptides have limited surface exposure (Fig. 8) suggesting that they exist internally in native sheath. Therefore, a trilaminar arrangement of sheath polypeptides in which the PS polypeptides are sandwiched between two layers of SDS/β-ME/EDTA polypeptides has been proposed (Southam and Beveridge, 1992a). It is possible that the PS polypeptides, which are known to confer sheath rigidity, stabilize the sheath during growth and perhaps coordinate the assembly of new sheath material.

Even though the sheath is an extremely resilient cell envelope structure, it does not completely enclose each cell and its strength serves no purpose in controlling cell turgor pressure unless it is complimented by additional envelope structures which can plug the holes between individual cells within the tube made by the sheath. The mechanism to withstand cell turgor in *M. hungatei* appears to occur via the end plug/bushing assembly structure. The bushing assembly consists of a series of ring structures, smaller in diameter than the hoops, which adhere to the inner surface of the sheath (Fig. 12) inside the cell spacer and at filament poles (Fig. 11). By narrowing the diameter of the sheath cylinder, the bushing assembly prevents the end-plugs (Fig. 1A and B), from being extruded from the sheath due to cell turgor. This is demonstrated in negative stains of the sheath/end plug/bushing assembly system (Fig. 11A and B) where the bushing assembly does not allow the end plugs to cross into its region of the sheath when the sheath collapses during specimen drying prior to negative staining.

The production of spacer plugs in *M. hungatei* is coordinated in that a single spacer plug is observed prior to the assembly of the second spacer plug (Beveridge et al., 1987). The assembly of one complete spacer plug prior to the second represents a protective mechanism against damage to the cell spacer region (premature filament splitting) in which at least one of the cells is ensured survival.

Although it has been shown that the cell wall of *M. hungatei* is responsible for cell division (Beveridge et al., 1987), it is possible that the bushing assembly may contribute. The formation of the bushing assembly was at the point of cell wall ingrowth, therefore, the bushing assembly may alter the direction of cell wall expansion, towards the inside of the cell, resulting in cell division. Therefore, the function of bushing assembly may be two-fold: first to initiate cell division and second to maintain cell integrity by narrowing the diameter of the sheath cylinder at cell poles which serves to hold the end plugs in the sheath against the cell ends.

ACKNOWLEDGMENTS

This research was funded by the Natural Sciences and Engineering Research Council of Canada and the Medical Research Council of Canada.

REFERENCES

Beveridge, T.J., Harris, B.J., and Sprott, G.D., 1987, Septation and filament splitting in *Methanospirillum hungatei, Can. J. Microbiol.* 33:725.

Beveridge, T.J., Southam, G., Jericho, M.H., and Blackford, B.L., 1990, High resolution topography of the S-layer sheath of the archaebacterium *Methanospirillum hungatei* provided by scanning tunneling microscopy, *J. Bacteriol.* 172:6589.

Beveridge, T.J., Sprott, G.D., and Whippey, P., 1991, Ultrastructure, inferred porosity, and Gram-staining character of *Methanospirillum hungatei* filament termini describe a unique cell permeability for this archaeobacterium, *J. Bacteriol.* 173:130.

Beveridge, T.J., Stewart, M., Doyle, R.J., and Sprott, G.D., 1985, Unusual stability of the *Methanospirillum hungatei* sheath, *J. Bacteriol.* 162:728.

Conway de Macario, E., König, H. and Macario, A.J.L., 1986, Antigenic determinants distinctive of *Methanospirillum hungatei* and *Methanogenium cariaci* identified by monoclonal antibodies, *Arch. Microbiol.* 144:20.

Patel, G.B., Roth, L.A., van den Berg, L., and Clark, D.S., 1976, Characterization of a strain of *Methanospirillum hungatii, Can. J. Microbiol.* 22:1404.

Shaw, P. J., Hills, G.J., Henwood, J.A., Harris, J.E., and Archer, J.B., 1985, Three-dimensional architecture of the cell sheath and septa of *Methanospirillum hungatei, J. Bacteriol.* 161:750.

Shockman, G.D., and Barrett, J.F., 1983, Structure, function and assembly of cell walls of gram-positive bacteria, *Ann. Rev. Microbiol.* 37:501.

Southam, G., and Beveridge, T.J., 1991, Immunochemical analysis of the sheath of the archaeobacterium *Methanospirillum hungatei* strain GP1, *J. Bacteriol.* 173:6213.

Southam, G., and Beveridge, T.J., 1992a, Characterization of a novel, phenol soluble group of polypeptides which convey rigidity to the sheath of *Methanospirillum hungatei* strain GP1, *J. Bacteriol.* 174:935.

Southam, G., and Beveridge, T.J., 1992b, Detection of growth sites in, and protomer pools for, the sheath of *Methanospirillum hungatei* GP1 using constituent organosulfur and immunogold labelling, *J. Bacteriol.* 174:6460.

Southam, G., Firtel, M., Jericho, M.H., Xu, W., Mulhern, P.J., and Beveridge, T.J., 1992, Transmission electron microscopy, scanning tunneling microscopy and atomic force microscopy of the cell envelope layers of the archaeobacterium *Methanospirillum hungatei* GP1, *J. Bacteriol.* 175:1946.

Sprott, G.D., Beveridge, T.J., Patel, G.B., and Ferrante, G., 1986, Sheath disassembly in *Methanospirillum hungatei* strain GP1, *Can. J. Microbiol.* 32:847.

Sprott, G.D., Colvin, J.R., and McKellar, R.C., 1979, Spheroplasts of *Methanospirillum hungatii* formed upon treatment with dithiothreitol, *Can. J. Microbiol.* 25:730.

Stewart, M., Beveridge, T.J., and Sprott, G.D., 1985, Crystalline order to high resolution in the sheath of *Methanospirillum hungatei*: a cross-beta structure, *J. Mol. Biol.* 183:509.

Zeikus, J.G., and Bowen, V.G., 1975, Fine structure of *Methanospirillum hungatei*, *J. Bacteriol.* 121:373.

Chapter 14

STRUCTURAL AND FUNCTIONAL ANALYSIS OF THE S-LAYER PROTEIN FROM *BACILLUS STEAROTHERMOPHILUS*

Beatrix Kuen and Werner Lubitz

Institute of Microbiology and Genetics
University of Vienna
Vienna, Austria

Geoffrey J. Barton

Laboratory of Molecular Biophysics
University of Oxford
Oxford, England

INTRODUCTION

One feature common to many bacteria, regardless of their phylogenetic origin within the kingdoms Eucarya or Archaea, is the presence of a regularly ordered (glyco)protein border as the outermost macromolecular layer of the cell envelope. A recent list of organisms with such crystalline surface layers (S-layers) cites approximately 300 different prokaryotic species (Messner and Sleytr, 1992). However, most of the S-layers have only been described by electron microscopical or biochemical investigations and DNA sequence data of the corresponding genes are available for very few species (approximately 20). Here we report on the structural and functional properties of the S-layer of *Bacillus stearothermophilus* strain PV72 which have been deduced from computer analysis of DNA-sequence data (for sequencing details of the gene see the contribution by Kuen, Sára, Sleytr, and Lubitz in this book).

A comparison of 39 different *B. stearothermophilus* strains by SDS-PAGE and electron microscopy showed a remarkable heterogenity in the M_rs and lattice structures of the S-layer from different strains of this species (Messner et al., 1984). Strain PV72 exhibits a hexagonal type of S-layer lattice with a centre-to-centre spacing of 22.5 nm. The estimated molecular weight of the S-layer monomer from this strain determined by SDS-PAGE is about 130 kD (Sleytr et al., 1986). Investigations of various aspects of surface properties of the S-layer carrying (S$^+$) strain PV72 and the S-layer deficient (S$^-$) mutant PV72/T5 showed that cell adhesion to glass was less influenced by environmental changes in strain PV72 than it was with the mutant strain and that PV72 carries more positively charged groups on the cellsurface than the mutant PV72/T5 (Gruber and Sleytr, 1991). The sequence data of the S-layer protein from *B. stearothermophilus* are interpreted in this communication, and structural predictions and comparisons should help to understand future functional investigations of this molecule.

PRIMARY STRUCTURE OF THE S-LAYER MONOMER OF *B. STEAROTHERMOPHILUS* PV72

The mature S-layer monomer of *B. stearothermophilus* PV72 is composed of 1199 amino acids corresponding to a molecular weight of 128,041 Da. An additional 30 amino acids form a signal sequence which is not present in the mature protein. The sequence and distribution of the different amino acids of the PV72 S-layer monomer is given in the short communication by Kuen, Sára, Sleytr, and Lubitz in this book.

The five amino acids threonine (13 %), alanine (11.3 %), valine (10.9 %), lysine (9.1 %) and aspartic acid (6.9 %) make up approximately 50 % of the total amount of amino acids. The amino acid composition of the PV72 S-layer protein is different from the average amino acid compositon of globular cytoplasmic proteins probably reflecting specific requirements for the self-assembly of the surface layer. The negatively charged amino acids aspartic acid and glutamic acid are represented by 6.9 and 5.3 % of the total amino acids respectively, whereas the percent ratio of the positively charged amino acids lysine, arginine and histidine is 9.1, 1.8 and 0.4 %, respectively, resulting in 146 negative versus 135 positive charges. Thus, approximately one quarter of the amino acids of the S-layer protein from *B. stearothermophilus* are charged, suggesting the importance of ionic interactions and/or ion bridges within the molecule in binding to the underlying layer and macromolecular matrix formation. Computation of the isoelectric point (pI) based on the total number of residues and the corresponding pK values of the individual amino acids gives a pI value of 5.06 for the mature form of the S-layer protein.

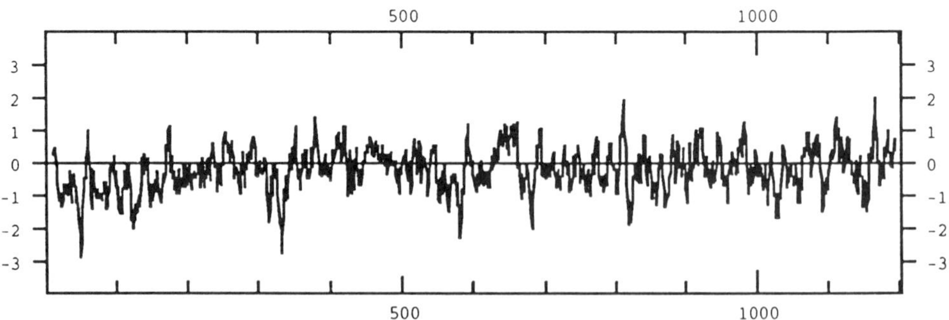

Figure 1. Hydrophobicity plot of the *B. stearothermophilus* PV72 S-layer monomer. The parameters are averaged over five amino acid residues. Positive values indicate hydrophobic regions and negative values, hydrophilic regions. Increasing values and peak size correspond to increasing degrees of hydrophobicity or hydrophilicity.

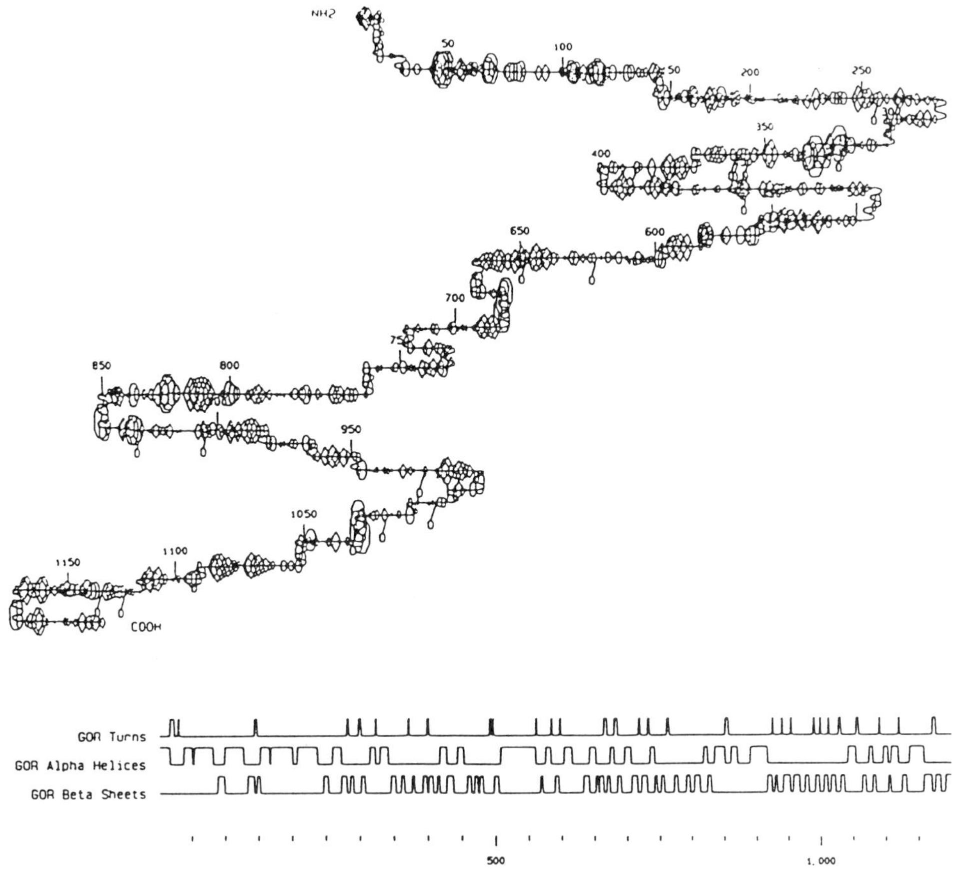

Figure 2. Structural prediction of the *B. stearothermophilus* PV72 S-layer protein. Structural features represented as zig-zag ($\sim$) lines are alpha-helix; $\sim$, beta-pleated sheet; $\propto$, beta-turn region; $\nearrow$, random coil. The parameters were averaged over five amino acid residues. Hydrophobic regions are symbolized by diamonds and hydrophilic regions by ovals. Increasing symbol size corresponds to increasing degree of hydrophobicity or hydrophilicity.

The hydrophilic and hydrophobic amino acid side chains represented in the PV72 S-layer protein do not form very extended clusters within the protein, alternating instead between hydrophilic and hydrophobic segments with a dominance of a hydrophilic area at the very N-terminal portion of the protein (Fig. 1).

SECONDARY STRUCTURE PREDICTION OF THE *B. STEAROTHERMOPHILUS* S-LAYER PROTEIN

Using the computer algorithm of Garnier et al. (1978), the secondary structure of the S-layer protein is predicted to contain alpha-helical, beta-pleated sheet and random coil areas as well as beta-turns overlayed by hydrophilic and hydrophobic regions (Fig. 2).

```
 63 KKDAYLAD LQKEYETYVFKANPKSGEARVATYIDAYNYATKL DEMRQE  110
 38 KISGYAKEAVQSLVDAGVIQGDANGNFNPLKT ISRAEAATIFTNALELE   86

111 LEAAVQAKDLEKAEQYYHKIPYEIKTRTVILDRVYGKTTRDLLRSTFKAK  160
 87 AEGDVNFKDVKADAWYYDAIAATV ENG IFEGV SATEFAPNKQLTRSE  133

161 AQELRDSLIYDITVAMKAREVQDAVKAGNLDKAKAAVDQINQYLPKVTDA  210
134 AAKILVD AFELEGEGDLSEFADASTVKPWAKSYLEI AVANGVIKGSEA  181

211 F KTELTEVAKKALDADEAALTPKVESVSAINTQNKAVELTAVPVNGTLK  259
182 NGKTNLNPNAPITRQDFAVVFSRTIENVDATPKVDK IEV VDAK TLN  227

260 LQLSAAANEDTVNVNTVRIYKVDGNIPFALNTADVSLSTDGKTITVDAST  309
228 VTLSDGTKE TVTLEKALEPNKETEVTFKIKDVEYK AKVTYVVTTATAV  275

310 PFENNTEYKVVVKGIKDKNGKEFKEDAFTFKLRNDAVVTQVFGTNVTNNT  359
276 KSVSATNLKEVVVEFDGTVDKETAEDAANYALKSGKTIKSVSLAADNKTA  325

360 SVNLAAGTFDTDDTLTVVFDKLLAPE TVNSSNVTITDVETGKRIPVIAS  408
326 TVTL TDKLNNNKADAISISNVKAGDKEINVKNVEFTAVD NK IPEVTE  372

409 TSGSTITITLKEALVTGKQYKLAINNVKTLTGYNAEAYELVFTANASAPT  458
373 VKSLG TKAVKVTLSEPVE NLSSTNF TLDG KA YFGNV VHGAGNKT  416

459 VATAPTTLGGTTLSTGSLTTNVWGKLAGGVNEAGTY YPGLQFTTTFATK  507
417 VILTPYSSSALSVGDHKLTVSGAKDFAGFVSLNSTHEFKVVEDKEAPTVT  466

508 LDESTLAD NFVLVEK ESGTVVASELKY NADAKM VT LVPKADLKEN  552
467 EATATLETVTLTFSEDIDMDTVKASNVYWKSGDSKKEASEFERIADNKYK  516

553 TIYQIKIKKGLKSDKGIELGTVNEKTY EFKTQDLTAPTVISVTSKNGDA  601
517 FVFK GSEKTLPTGK VDVYVEDIKDYSDNKIAKDTKVTVTPEIDQTRPE  564

602 GLKVTEAQEFTVK  FSENLNTFNATTVSGSTITYGQVAVVKAG ANLSA  648
565 VRKVTALDEKTIKVTFSKTVDGESAIKTGNYTVKDKDDKVVSVDKVTVDS  614

649 LTASDIIPASVEAVTGQDGTYKVK V AANQLERNQ GYKLVVFGKGA T  694
615 KDSKSVIIDLYSKVSVGENTITIKNVKDATKLNNTMLDYT GKFTRSDKE  663

695 AP VKDAANANTLATNYIYTFTTEGQDVTAPTVTK VFK GDSLKD ADA  740
664 GPDYEHVINADAKAKKVVLKFDKKMDAASLADYSNYLVKINDTLQTLSED  713

741 VTTLTNVDAGQKFTIQFSEELKTSSGSLVGGK VT VEKLTNNGW VDAG  787
714 VATLSVSNDATVVTITFAETIKGDDVVFASGKAISGSGKVNVNELQVHGV  763

788 TGTTVSVAPK TDANGKVT AAVVTLTGLDNNDKDAKLRLVVDKSSTDGI  835
764 KDTSGNVHKKFNGSENKITLSSTSTPLKLAKIDKDYDAKYTAELVDRKTV  813

836 ADVAGNVIKEKDILIRYNSWRHTVASVKAAADKD GQNASAAFPTSTA I  883
814 KVKFSTVINSAAA NAFTSESHKIDSIQVNGTSTVTVKFKDEINTNASDL  862

884 DTTKSL  LVEFNETDLAEVKPENI V VK DAAGNAVAGTVTALDGSTN  928
863 DLKVNLSKLVDIAGNESTNNTPIAIKAGINLLDSVAPVVVGEPVVDKETI  912

929 KFVFTPS QELKAGTVYSV  TIDGVRDKVGNTISKYITSFKTVSANP T  974
913 TFTFSENLTSVSIGEVLSTDFTVTRVSDNKDLAIKDYSVAIANNNQVVIT  962

975 LS SISIADG AVNVDRSKTITIEFSDSVPNPTITLKKADGTSFTNYTLV  1022
963 LSDNREVATAYKVTAKNAKLITDDNGDK KNAIADFTKTTATKVEASGTL  1011

1023 NVNNENKTYKIVFHKGVTLDEFTQYELAVSKDFQTGTDIDSKVTFITGSV  1072
1012 SLDAAKTNLNNEITKAKDAKATGTEGTAATNQI VGSKDALQVAIDVAEL  1060

1073 ATDEVKPALVGVGSWNGTSYTQDAAATRLRSVADFVAEPVALQFSEGIDL  1122
1061 VKNDTAATLQQLTD AKTDLTAAITAYNAAKVEDISSLLVAPDLVLGTT   1108

1123 TNATVTVTNITDDKTVEVISKESVDAD HDAGATKETLVINTVTPLVLDN  1171
1109 DNGTITGFVAGTGETLKVTSDSAANVEVTDPTGLAVTAKAKGEANILVQV  1158

1172 SKTYKIVVSG VKDAAGN   1188
1159 LKGDKVIKTGTVKVTVSE  1176
```

Figure 3. Comparison and alignment of the S-layer proteins of *B. stearothermophilus* PV72 and *B. sphaericus* 2362. Identical amino acid residues are indicated by an asterisk whereas homologous pairs are marked by points.

```
  1                  VLSVLSTTLVASVAASAFAAPKD            GIYIGG          HIKK    YYST        DVLFE      MTPQAKA
  1            ISGYAKEAVQSLVDAGVIQGDANGHFNPLKTI    SR  AEAATIFTNALELEAEGDVHFKDVKADAWYYDAIAATVEHGIFEGVSATEFAPHKQL
  1    ATDVATVVSQAKAQFKKAYYTYSHTVTETGEFPHINDVYAEY NKAKKRYRDAVALVNKAG      GAKKDA YLADLQKEYETYVFKAHPKSGEARVATY
  1            DSGFKKKDRSTNIPQEQFVYTRGGEHKVMKKVVNSVLASALAITVAPMAFAAEDTTT    APKMDAAMEKTVKRLEALGLVAGYGHGDFGADKTI

101 T        VASELHAMASDFN   NV VFVDYKGKGA           SIEELFTKGS    KVALG  EPLKKEDPADLYK                VVHK
101 T        RSEAAKILVDAFELEGBG  DLSEFADASTVKPWAKSYLEIAVANGVIKGSEANGKTNLNPHAPITRQDFAVVFSRTIENVDATPKVDKIEVVDAK
101    IDAYHYATKLDEMRQELEAAVQAKDLEKAEQYYHKIPYEIKTRTVILDRVYGKTTRDLLRSTFKAKAQELRDSLIYDITVAMKAREVQDAVKAGHLDKAK
101    TRA   EFATLIVRARGLEQGAKLAQFNTTYTDVRSTDWFAGFVNVASGEEIVKGFP     DKSFKPQNQVTYAEAVTMIVRALGYEPSVRGVWPHSMISKG

201 DGSSTATE DARAKV        DPTPTGDLN             VESVSANHLKEVV VTFDKAVDADTAGDKA  YYTFTAHK LAV        D
201 TLHVTLSD GTKETVTLEKALEPNKETEVTFKIKDVEYKAKVTYVVTTATAVKSVSATNLKEVV VEFDGTVDKETAEDAA  NYALKSGKTIKS       V
201 AAVDQIHQYLPKVTDAFKTELTEVAKKALDADEAALTPKVESVSAINTQNKAVELTAVPVHGTLKLQLSAAAHEDTVNVHTVRIYKVDGHIPFALNTADV
201 SELHIAKGIHNPHMQQFAATIFKMLDNALRVKLME    QIEYGTDIRLNVTDETLLTKYLKVTV RDMDWAHEKGHHSDE   LPLVTHVPAIGL      G

301 KVTVSGKTVVLTLAAKAENQASYELNVDGIKG LVKTT  KEVKP  FDNTTPT     VAAVAAIG    PKQVKVTFSEPL  SAKPSFSVHH
301 SLAADHKTATVTLHTDKLNHHKADAISISNVKA GDKEINVGNVEFTAVDNKIPE     VTEVKSLG    TKAVKVTLSEPV  ENLSSTNFTL
301 SLSTDGKTITVDASTPFENHTEYKVVVKGIKDKHGKEFKEDAFTFKLRNDAVVTQVPGTNVTNHTSVNLAAGTFDTDDTLTVVFDKLLAPETVHSSHVTI
301 SLKANEVTLNGK DADLGSNTTYKVA BGINP   NAFDGQKVQVWIKDDRENV   IVWMEGSE    DEDVVMDRVSALYLKGKAFTDDIV

401 GALAVVADNFVE GTKEVILTL GAQPTASTH TVTIVEGGADYASYKVEKVTKDFTVVAD      TTPPTVSVKKASAKQVVLEFSEDVQHVQDKNVVF
401 DGKAYPGNVVMGAGNKTVILTPYSSSALSVGDH KLTVSGAKDPAGFVSLNSTHEFKVVED    KEAPTVTEATATLETVTLTPSEDIDMDTVKASHV
401 TDVETGKRIPVIASTSGSTITITLKEALVTGKQYKLAIDNVKTLTGYNAEAYELVFTANASAPTVATAPTTLGGTTLSTGSLTTHVWGKLAGGVNEAGTY
401 KDLSKSDLDDVKIEMDGSEKSYRLTEDTKITYNFTRFNDPVDALSKIYKDNDTPGVKVVLNDHNEVAYLHIIDDQTIDKSVKGVKYGSKVISKIDADKKK

501 YH      TTKGHEGYKGTILGVDGKEVTIS      FVNP  L   PEGQFKIFVDYVVDHGTQISDLHGHKLPEQV ITGTPAADT TPPTVTKV
501 YW      KSGDSKKEASEFERIADNKYKFV     FKGSEKTL   PTGKVDVYVE     DIKDYSDNKIAKDTKVTVTPEIDQ TRPEVRKV
501 YPGLQFTTTFATKLDESTLADNFVLVEKESGTVVASELKYNADAKMVTLVPKADLKENTIYQIKIKKGLKSDKGIELGTVNEKTYEFKTQDLTAPTVISV
501 I       TNLDNSKFSDLEDQDBGKDFLVP     LDGQPAK    GDLKESDVYSVYY    ADGDKDKYLVFANRHVABGKVEKVVSRNKTDI

601 EAKT NTEIHVTFSETVHGAD  NKANFTLK GVTGNVIPLTKAEVVDAAKNIYKVVTTEPLNG GSYYLTVKGIEDASK  NKLVEYTATVAVADTVPP
601 TALD EKTIKVTPSKTVDGESAIKTGNYTVK DKDDKVVSVDKVTVDSKDSKSVIIDLYSKVSV GENTITIKNVKDATKLNNTHLDYTGKFTRSDKEGP
601 TSKNGDAGLKVTEAQEPTVKFSENLNTFNAT TVSGSTITYGQVAVVKAGANLSALTASDIIPA SVEAVTGQDGTYKVKVAANQLERHQGYKLVVPGKG
601 RLTVGGKTYKVYPDASYSENANKDVKKVNSDLDLISHIDGGFEVKLILLDPGGRVPHIETKDAIDDRKPLAIITKGATYNSS KDTYDFT  VHTQKGKT

701 NVKDLDPATPGTDAQLISPTKVKIAPTEPMDKASIENKNNYMP   NGFN LDSKV TLTATDSNTAVVVDFTNVVGFNG PKNGDA    ISV GRV
701 DYEHVINADA        KAKKVVLKFDKKMDAASLADYSNYLVKINDTLQTLSEDVATLSVSNDATVVTITFAETIKGDDVVFASGKA    ISGSGKVN
701 ATAPVKDAANANT     LAGNYIYTPTTBGGQDVTAPTVTK VFK GDSLKD ADAVTTLTNVDAGGKFTIQFSEELKTSSGSLVGGKVTVEKLTHNGWVD
701 QIVSLDQKD        IYDRYGVNYDKSNDKR QAFEKDLV  ELLQPKVVKEDSATDANQTVLLEVNFDSKGEVDKVKVLDSKLKYSEKSTWDKLA

801   LDTAG NPKTEMQTKVNLPN     SVSAPL       FD  KAEVTGKNTVKLYFKELIINAKADDFAVDNGE GYKAVNSISNDVVENKSV
801 VHELQVMGVKDTSGNVHKKFNGSENKITLSSTSTPLKLAKIDKDYDAKYTAELVDRKTVKVKFSTVINSAAANAFTSESHKIDSIQVNGTSTVTVKFKDE
801 AGTGTTVSV APKTDANGKVTAAV  VTLTGLDNNDKDAKLRLVVDKSSTDGIADVAGNVIKEKDILIRYNSWRHTVASVKAAADKDGQNASAAFPTSTA
801 DEDDDVVG   DYEVTDKTAVFK   MTGDLTPATGTK    RGELKNAGTAKFKDVAKKSDLKVW YSVDEDKGEVQAIFVVDGSGLGGDHQ

901 ITLTTGHDLFTTAAGVKVKTVGKVDAKHQYGVAV    ALTDVPADDKIGPHWLKAETVDTNHNGKIDQFKLT  FSEALYVASVQGS DFRIBGYTIAG
901 IN  TNASDLDLKVNLSKLVDIAGHESTHNTPIAIKAGINLLDSVAPVVVGEPVVDKETITPTFPSENLTSVSIGEVLSTDPTVTRVSDHKDLAIKDYSVAI
901 IDTTKSLLVEFNETDLAEVKPENIVVKDAAGNAVAGTVTALDGSTNKFVFTPSQELKA GTVYSVTIDGVRDKVGNTISKYITSFKT      VSANPTLSS
901 FGHVKQYGTASKQDTITIVTKDGDSVTEKE       YKLDGDADDLKVDQDIRRGDVISFTLNSDGEVIVDDVVEVVHHGHIDHTAS    KSATLMPED

1001 VETKGEVVTIKVT ELDIDDSDATPTVAVIGSVEDLKRNASGPFEPQKA  IDGVSAPDKEAPVVT GVEAGK TYNTAVTPD  SADKDIKTVVLKKD
1001 ANHHQVVITLSDNREVATAYKVTAKNAKLITDDNGDKKNAIADPTKTTATK VEASGTLSLDAAKTNLNNEITKAKDAKATGT  BGTAATHQIVGSKD
1001 ISIADGAVNVDRSKTITIEFSDSVPNPTIT    LKKADGTSFTNYTLVN VNHENKTYKIVFHKGVTLDEPT QYELAVSKDPQTGTDIDSKVTFITG
1001 ERQKAGIDKLVVARVDEVDGNTISLNYADGKTQKYYTKASTAFIDVYDGLEGIDGVDEGDYIVMIDSADIDGTRFDYVLVVSSD  DEIRTQHISTKAV

1101 GKELA GYALKTPISENGSYELVVTDNAGHTTTVK F   KVDIPAEDKKAPEIKTVTDDKVAVAD  APKWEAPKATATDDVDGDISDKIAVTYSSEDA
1101 ALQVAIDVAELVKGNDTAATLQQLTDAKTDLTAAITAYNAAKVEDISSLLVAPDLVLGTTDKGTITGFVAGTGETLKVTSDSAANVEVTDPTGLAVTAKAK
1101 SVATDEVKPALVGVGSWNGTSYTQDAAATRLRSVADFVAEPVALQFSEGIDLTNATVTVTNITDDKTVEVISKESVDADHDAGATKETLVINTVTPLVLD
1101 TDFLNKPTRLCTKSWRWGRSSHGTKVNTVNDEAVVD   GIVTLPADASVRNFNIAFDQEINSKDATVTVTHEDTLGHVTVSEVATDAKVLSFKTAKLD

1201 GSKVTDLASAGTHLGTAGHTVKVTYNVTD
1201 GEAHILVQVLKGDKVIKTGTVKVT  VSE
1201 HSKTYKIVVSGVKDAAGHVADTITFYIKZ
1201 TTKTYIITVKGLKDKNGKAVKDVTLYV
```

Figure 4. Multiple alignment of the S-layer proteins of *B. brevis*, *B. sphaericus* and *B. stearothermophilus*. Line 1 and line 4 amino acid sequences of the S-layer proteins of *B. brevis* 47 outer cell wall protein and HWP protein, respectively; line 2, amino acid sequence of the 125 K surface layer of *B. sphaericus* 2362; line 3, S-layer protein of *B. stearothermophilus* PV72.

The representation of alpha-helical or beta-sheet areas show that beta-sheets are slightly predominant within the structure (52 versus 45) of the S-layer protein. Both structural motifs are separated by 30 turn regions allowing compact packaging of the amino acid chain. In the absence of crystallographic data such secondary structure predictions are useful tools for modeling a possible conformation of the protein. We would like to use insertion mutagenesis to determine essential structural features of the S-layer protein which are needed for self-assembly properties, export to the cell surface and macromolecular properties such as permeability determinants, surface flexibility and antigen presentation.

SIMILARITY OF THE S-LAYER PROTEIN OF *B. STEAROTHERMOPHILUS* PV72 TO OTHER PROTEINS

Using the Smith-Waterman algorithm (Smith and Waterman, 1981) with Dayhoff's MDM76 matrix and a length-dependent gap-penalty of 8, the NBRF-PIR protein databank version 33 (42,215 sequences) has been screened for proteins similar to the S-layer protein of *B. stearothermophilus* PV72. Although the scores found for these comparisons were marginal (329-255), it seems likely that the sequences of the 125 kD S-layer protein from *B. sphaericus* 2362 (Bowditch et al., 1989), the 190 kD S-layer protein of *Rickettsia rickettsii* (Gilmore et al., 1989), and the two major cell wall proteins from *B. brevis* 47 (Tsuboi et al., 1986; Ebisu et al., 1990) are genuinely related to the query sequence reported here, as these sequences are all S-layer proteins. The alignment of the amino acid sequences of the S-layers from *B. stearothermophilus* PV72 and the most similar sequence of *B. sphaericus* 2362 is given in Fig. 3.

Using the program MULTALIGN which can compute protein alignments based on multiple alignment, flexible patterns and structure dependent gap-penalities (Barton and Sternberg 1987a,b and 1990; Barton, 1990), the S-layer sequences of the three most similar strains *B. stearothermophilus* PV72, *B. sphaericus* 2362 and *B. brevis* 47 are given in Fig. 4.

The multiple alignments provide only a rough guide to the similarity of the four S-layer sequences. In general terms it can be stated that the similarity found among the four S-layers is weak. Nevertheless, the similarities help to define areas in the S-layer molecule of strain PV72 which may be necessary for protein-protein interaction in macromolecular assembly of the S-layer monomers and intramolecular domains for the makeup of the molecular shape.

PERSPECTIVES

Genetic modifications of cloned S-layer genes and expression of recombinant sequences in suitable host cells would allow functional studies of cellular and S-layer protein requirements for S-layer formation. In addition, the potential of using recombinant S-layer proteins for the construction of molecular machines or biosensors, as carriers of relevant antigens of various pathogens for immunological applications, matrices for information storage, tailored ultra-filters, supports for Langmuir-Blodgett films can be envisaged as extensions of applications which are already possible with isolated S-layer proteins (Sleytr et al., 1989; Pum et al., 1991).

REFERENCES

Barton, G.J., 1990, Protein multiple sequence alignment and flexible pattern matching, *Meth. Enzymol.* 183:403.

Barton, G.J., and Sternberg, M.J.E., 1987a, Evaluation and improvements in the automatic alignment of protein sequences, *Prot. Engin.* 1:89.

Barton, G.J., and Sternberg, M.J.E., 1987b, A strategy for the rapid multiple alignment of protein sequences: confidence levels for tertiary structure comparisons, *J. Mol. Biol.* 198:327.

Barton, G.J., and Sternberg, M.J.E., 1990, Flexible protein sequence patterns - a sensitive method to detect weak structural similarities, *J. Mol. Biol.* 212:389.

Bowditch, R.D., Baumann, P., and Yousten, A.A., 1989, Cloning and sequencing of the gene encoding a 125-kilodalton surface-layer protein from *Bacillus sphaericus* 2362 and of a related cryptic gene, *J. Bacteriol.* 171:41788.

Ebisu, S., Tsuboi, A., Takagi, H., Naruse, Y., Yamagata, H., Tsukagoshi, N., and Udaka, S., 1990, Conserved structures of cell wall protein genes among protein-producing *Bacillus brevis* strains, *J. Bacteriol.* 172:1312.

Garnier, J., Osguthorpe, D.J., and Robson, B., 1978, Analysis of the accuracy and implications of simple methods for predicting the secondary structure of globular proteins, *J. Mol. Biol.* 120:97.

Gilmore, R.D., Joste, J.N., and McDonald, G.A., 1989, Cloning, expression and sequence analysis of the gene encoding the 120 kD surface-exposed protein of *Rickettsia rickettsii, Mol. Microbiol.* 3:1579.

Gruber, K. and Sleytr, U.B., 1991, Influence of an S-layer on surface properties of *Bacillus stearothermophilus, Arch. Microbiol.* 156:181.

Messner, P. and Sleytr, U. B., 1992, Crystalline bacterial cell-surface layers, Adv. Microbial Physiol., 33:213.

Messner, P., Hollaus, F., and Sleytr, U.B., 1984, Paracrystalline cell wall surface layers of different *Bacillus stearothermophilus* strains, *Int. J. Syst. Bacteriol.* 34:202.

Pum, D., Sára, M., Messner, P., and Sleytr, U.B., 1991, Two-dimensional (glyco)-protein crystals as pattering elements for the controlled immobilization of functional molecules, *Nanotechnol.* 2:196.

Sleytr, U.B., Sára, M., Küpcü, Z., and Messner, P., 1986, Structural and chemical characterization of S-layers of selected strains of *Bacillus stearothermophilus* and *Desulfotomaculum nigrificans, Arch. Microbiol.* 146:19.

Sleytr, U.B., Sára, M., and Pum,D., 1989, Application potentials of two dimensional protein crystals, *Microelectr. Engin.* 9:13.

Smith, T.F., and Waterman, M.S., 1981, Identification of common molecular subsequences, *J. Mol. Biol.* 147:195.

Tsuboi, A., Uchihi, R., Tabata, R., Takahashi, Y., Hashiba, H., Sasaki, T., Yamagata, H., Tsukagoshi, N., and Udaka, S., 1986, Characterization of the genes coding for two major cell wall proteins from protein-producing *Bacillus brevis* 47: complete nucleotide sequence of the outer wall protein gene, *J. Bacteriol.* 168:365.

Chapter 15

STRUCTURE - FUNCTION ASPECTS OF THE *AEROMONAS SALMONICIDA* S-LAYER

William W. Kay, Julian C. Thornton, and Raphael A. Garduño

Department of Biochemistry and Microbiology and
The Canadian Bacterial Disease Network
University of Victoria
Victoria, British Columbia, Canada

INTRODUCTION

S-layers are important, proteinaceous, supramolecular assemblies common to a large number of bacterial species. When present they can account for up to 20% of the total cellular protein (Messner and Sleytr, 1992). In spite of their ubiquity, for the most part their specific biological functions have resisted elucidation. However, it is intuitively understood that their main roles must either be protective, aggressive, structural or some combination of these.

One particularly notable example of an S-layer with a clear functional role is that of the common fin-fish pathogen *Aeromonas salmonicida*. The *A. salmonicida* S- layer, or A-layer as originally named (Udey and Fryer, 1978), is comprised of a single ~50 000 M_r protein (A-protein) organized into a 3D array with p4 symmetry (Dooley et al., 1989). It has been shown to be tethered to the outer membrane of this gram negative bacterium by specific interactions with the O-polysaccharide (O-chain) portion of lipopolysaccharide (LPS) (Belland and Trust, 1985, Ishiguro et al., 1988). The structural gene for the A-protein, *vapA,* has been cloned, sequenced and expressed and the large mass and linker domains identified (Chu et al., 1991).

This S-layer has now been shown to harbor an unusual multiplicity of functions related either directly or indirectly to the bacterium's mode of pathogenesis (Kay and Trust, 1991). This layer has not only been demonstrated to be essential for pathogenesis (Ishiguro et al., 1981; Kay et al., 1981), but also has been assigned the the specific virulence functions of serum (complement) resistance (Munn et al., 1982), porphyrin and immunoglobulin binding (Kay et al., 1985; Phipps and Kay, 1988), as well as adherance to the extracellular matrix proteins, fibronectin and laminin (Doig et al., 1992). In addition, as discussed here, the S-layer is required for cellular invasion.

It is difficult to reconcile these functions with a relatively inflexible structure as depicted from 3D modeling of a variety of S-layers. It is the intent of this study to attempt to dispell the view of bacterial S-layers as inflexible structures by demonstrating alternative molecular arrangements of the *A. salmonicida* S-layer and to relate at least one of these to the functions exhibited.

METHODS

Bacterial Strains and Growth

Five isogenic *A. salmonicida* mutants were used in these studies: A450, a wild-type strain, A450-3 an S-layer defective mutant, A450-1 an LPS O-chain mutant that secretes its S-layer, A450-10S, a cytochrome deficient mutant, and its partial supressor mutant A450-10R (Thornton and Kay, 1991). Bacteria were grown in a variety of media containing different levels of Ca^{2+} at 20°C (Garduño et al., 1993).

Electron Microscopy, Image Processing and Computer Simulations

All specimens for electron microscopy (EM) were negatively stained with saturated ammonium molybdate. Observations were made in an EM300 Philips transmission electron microscope using an accelerating voltage of 80 kV. EM images were densitometered with an Eikonix Model 1412 camera system and image processing was performed by B.M. Phipps using the SEMPER system (Saxton et al., 1979). Image subframes were extracted, appropriately masked to isolate areas of projected S-layer and subjected to correlation averaging (Saxton and Baumeister, 1982).

A model of the A-layer p4 symmetry array was drawn with a Macintosh PC using the program MacDraft. This model was drawn using the mass distribution of the A-layer tetrameric morphological unit with a theoretical lattice constant of 14 nm, according to Dooley et al. (1989). This array was copied to produce a second identical array which was then superimposed to the first one in the four posssible register forms.

A. salmonicida-Macrophage Associations

Murine peritoneal and trout head kidney macrophages were isolated and cultured in vitro as previously described (Garduño and Kay, 1992b). Bacteria-macrophage microscopic adsorption assays were carried out also as previously described (Garduño and Kay, 1992b).

RESULTS AND DISCUSSION

A Single S-layer Structural Type

The S-layer of *A. salmonicida* normally appears as two distinct structural types, type I and type II (Stewart et al., 1986; see also the region in Fig. 2a labelled #1) in 2D mass distribution projections of negatively stained preparations. We found that type I patterns were restricted to, and predominated in, darkly stained areas, whereas lighter staining regions exclusively displayed type II patterns. Since

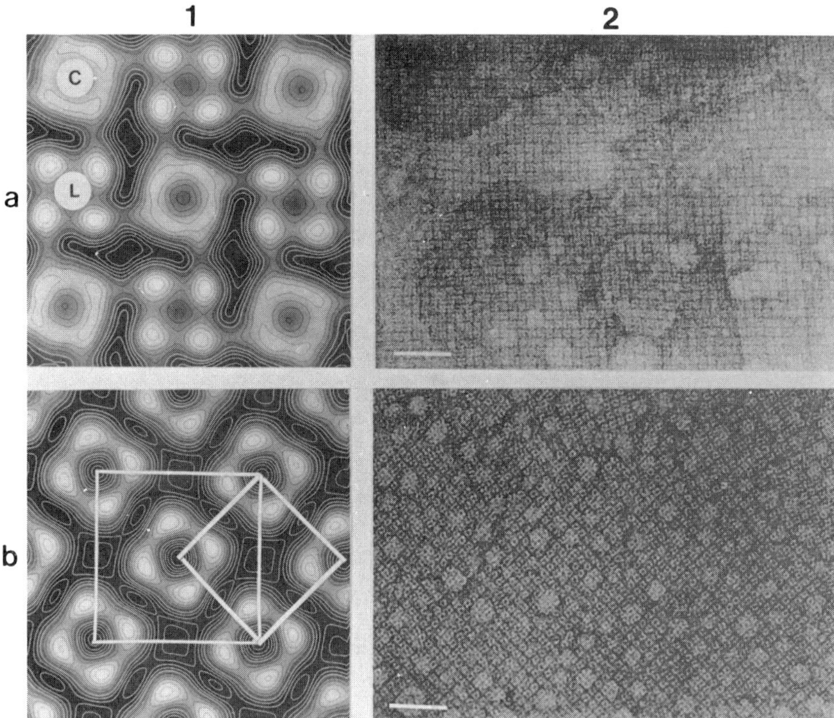

Figure 1. Comparison of the normal and the BS A-layer patterns. Correlation averages of normal (a1) and BS (b1) A-layer patterns, aligned to their corresponding micrographs (a2 and b2). The normal layer (a) shows two distinct morphological units representing the core (C) and linker (L) mass units. The lattice spacing is 0.4nm. The BS pattern (b) was obtained by growing cells in calcium-deficient medium. The overlaid boxes in 1b show the two possible unit cells. In one, all morphological units are equivalent (small box = 7.6 nm spacing) and in the other, only alternating morphological units are equivalent (large box = 10.8 nm). Both averages are presented on the same scale (22 x 22 nm). The bars represent 50nm. Correlation averaging and imaging was performed by B.M. Phipps at the Max-Plank Institute for Biochemistry, Martinsreid, Germany.

preparations of the S-layer of this organism have been shown to readily superimpose in perfect register (Dooley et al., 1989), we conducted a series of computer-simulations of various superpositions of type II patterns which faithfully reproduced type I patterns (Garduño and Kay, 1992a). This as well as other lines of evidence led to the conclusion that there normally exists only a single morphological pattern, typeII, on normally grown *A. salmonicida* cells. This observation made even more inexplicable the apparent observations of the ability of *A. salmonicida* to readily export large proteins through an S-layer of limited porosity as well as the apparent multiplicity of functions of this S-layer.

Divalent Cation-Depleted S-Layers

Divalent cation bridges were found to be involved in the integrity of the *A. salmonicida* S-layer. From a variety of differential extraction and reconstitution experiments it could readily be deduced that the S-layer monomers were held tightly to the outer membrane by non-divalent cation mediated charge-charge interactions. However, the integrity of subunit-subunit interactions were found to be profoundly affected by divalent cation-subunit interactions (Garduno et al., 1993). Two novel A-layer patterns were formed as a result of growth under severe Ca^{2+} limitation or by chelation of divalent cations on intact layers with EDTA or EGTA. Under these conditions, A-protein was sometimes released as tetrameric units, rather than in its monomeric form. The most predominant of these patterns we termed "big squares"(BS). Correlation averaging of well ordered BS patterns, such as the one shown in Fig. 1b, indicated that they were apparently composed of weakly handed, single morphological units (formed by four large domains grouped around a a major four-fold symmetry axis) with little or no mass connecting adjacent units. This is in contrast to projections through normal S-layers, which exhibit two distinct morphological units (representing the two S-layer domains) with clear connections between them (Fig. 1a). The lattice spacings of BS patterns was approximately equal to the average lattice spacings of normal S-layer divided by $\sqrt{2}$, suggesting the possibility that the core and linker A-layer domains were still present but had been altered to become similar in appearance.

This situation, in which only alternating morphological units are equivalent, is represented by the large unit cell in Fig. 1.1b (lattice spacing = 10.8 nm). The small unit cell (lattice spacing = 7.6 nm) would apply if all morphological units were in fact identical. The average shown in Fig. 1.1b was actually obtained by averaging over alternating correlation peaks, on a lattice corresponding to the large unit cell, in order to detect any minor differences between adjacent units. No differences were apparent consistent with the fact that correlation peak heights formed a continuum of values rather than partitioning into two classes. This, along with the observation that the BS patterns lie at an angle of 45° to normal S-layer patches is what would be expected if one of the two S-layer domains were disrupted while the remaining domains packed close together. A diagramatic representation of the S-layer alterations and eventual disruptions together with an image of their actual appearance in negative stain are shown in Fig. 2.

The S-layer and the bacterial association with macrophages

As one of the more important phagocytic cells, macrophages play a central role in killing of bacterial pathogens as well as in presenting important antigens to the immune system. Alternatively, ready access to the macrophage coupled to a resistance to macrophage killing would provide a potent mechanism for dissemination of the pathogen. We therefore examined the effects of the presence of the S-layer of *A. salmonicida* as well as structural alterations of it on this pathogen's ability to associate with murine and trout macrophages. To do so, we developed a procedure to culture macrophages on supported glass cover slips as well as an assay system to measure bacterial adherance and penetration (Garduño et al., 1992).

An intact S-layer was found to mediate adherence of *A. salmonicida* to macrophages even in the absence of opsonins. In contrast, unopsinized cells of an S-layer defective mutant (A⁻) with a smooth LPS layer were unable to interact with macrophages. However, this ability was recovered when the A-layer was reconstituted onto the smooth LPS surface of these A⁻ cells. Two *A. salmonicida* mutants possessing the S-layer in different disorganized states (Thornton et al., 1991) had a reduced ability to interact with macrophages (Garduño and Kay, 1992b).

In vitro experiments using A-layer coated latex beads demonstrated that the A-layer's invasin activity could be transferred to inert surfaces (Garduño and Kay,

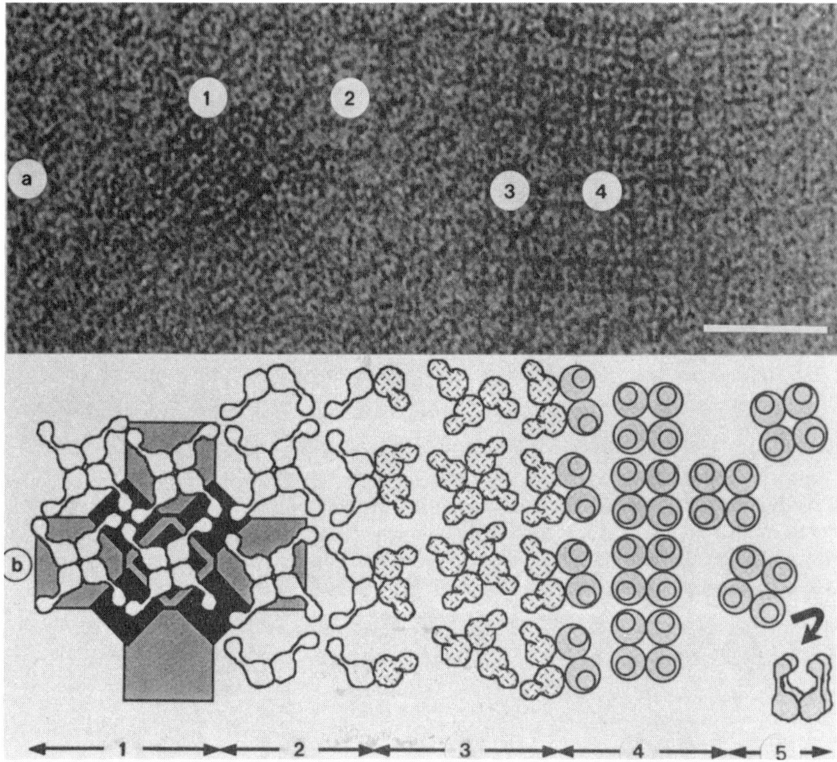

Figure 2. Hypothetical model showing the proposed structural rearrangements within the A-layer. (a) Micrograph of an A-layer fragment liberated from a cell during divalent cation removal by 0.5M EDTA. The following elements are distinguished: (1) small patch of a normal type I pattern, (2 and 3) intermediate patterns, and (4) BS pattern. Note the lattice lines of normal and altered patterns running at 45° to each other. Bar = 50 nm. (b) Hypothetical schematic representation of the changes in A-layer organization observed in (a). The normal tetragonal array (1 and filled-in squares) is disrupted by Ca^{2+} removal causing the dissociation (exaggerated in the drawing) of one of the two A-layer domains (in this case the linker domain), with the consequent formation of independent tetrameric units (2 and grey filled-in squares in 1) which upon rearrangement (3) may either be packed to form well ordered BS arrays (4) or dispersed to form free tetramers (5). One of these tetrameric units is presented in its side view (arrow in 5) to explain the appearance of "U" shaped free units seen in micrographs of disrupted layers. In the dissociation process, another pattern called "white dots" or WD may appear as an intermediate morphology between normal arrays (1) and independent units (2).

1992b). These results clearly established the role of this S-layer as an important invasin toward macrophages. Furthermore, recent experiments with fish epithelial layer cell lines have also shown that the role of this S-layer as an invasin is not restricted specifically to professional phagocytes suggesting that this S-layer's invasin activity extends beyond phagocytes to other cell types (Garduño, 1993).

When A$^+$ cells were grown under Ca^{2+}-limiting conditions (Garduño et al., 1993), S-layers were produced apparently locked into an alternative conformation, the BS pattern (Fig. 1b). These cells showed the highest levels of macrophage association even in the absence of opsonins or any other surface coating. No other qualitative or quantitative changes in the presence of other cell surface macromolecules or of

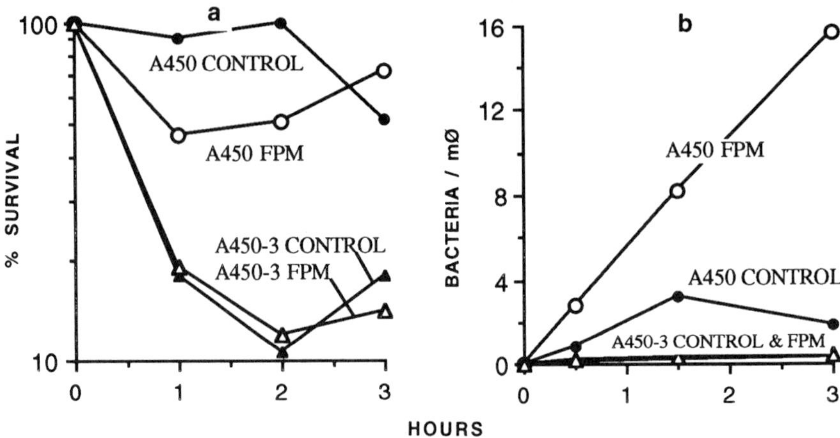

Figure 3. Functional characterization of S-layers displaying the BS pattern in *A. salmonicida* A450. Cells grown in Ca²⁺-deficient medium (FPM)(displaying the BS S-layer pattern) were compared with control cells grown in Ca²⁺-sufficient medium (TSB)(displaying normal A-layer patterns) (See Fig. 1), in order to compare the relative functional competence of BS-patterned S-layers. The S-layer negative mutant A450-3 was also included as a fine control to define S-layer independent processes. (a) Results from the serum complement resistance assay and (b) results from the macrophage association assay.

exported proteins were observed. This alternate structural arrangement made these cells slightly more succeptible to serum complement (Fig. 3a) but invoked a strikingly greater ability to associate with fish macrophages (Fig. 3b). The correlation of a preponderance of the BS alternate S-layer conformation with a higher level of macrophage association suggested that the BS conformation may be an alternate pathogenesis option of *A. salmonicida* cells presumably to facilitate intracellular spread of this particularly virulent, facultative, intracellular pathogen. We also have observed that the S-layer appears to be important in the resistance to intracellular killing by phagocytes (Garduño, 1993). Lastly, eucaryotic cell targeting by A-layer producing cells also provides a means of surviving macrophage killing by coupling the invasive ability of *A. salmonicida* with its capacity to secrete a variety of lytic toxins and enzymes. This combination results in the massive destruction of macrophages observed in vitro and in vivo.

In summary, the ability to invade and destroy host cells appears to be a new function for the *A. salmonicida* S-layer which can be added to an established littany of S-layer functions in this organism. In this regard it is hard to imagine such a plethora of S-layer functions within the confines of a relatively static structure. Thus the observation of alternative structural forms may well provide the means by which new functions are elicited.

ACKNOWLEDGMENTS

This work was supported by an operating grant to W.W.K via the Canadian Bacterial Diseases Network through the auspices of the NSERC/MRC National

Centres of Excellence . R.A.G. and J.C.T. were supported by GREAT postgraduate awards from the Science Council of British Columbia and also through the Canadian Bacterial Diseases Network. The technical assistance of M. Kuzyk and W. Elasoff as well as the critical advice of T.J. Trust is gratefully acknowledged.

REFERENCES

Belland, R.J., and Trust, T.J., 1985, Synthesis, export, and assembly of the *Aeromonas salmonicida* A-layer analyzed by transposon mutagenesis, *J. Bacteriol.* 163: 877.

Buckley, J.T., and Trust, T.J., 1981, Loss of virulence during culture of *Aeromonas salmonicida* at high temperature, *J. Bacteriol.* 148:333.

Doig, P., Emody, L.,and Trust, T.J., 1992. Binding of laminin and fibronectin by the trypsin resistant major structural domain of the crystalline virulence surface array protein of *Aeromonas salmonicida*, *J. Biol. Chem.* 267:43.

Dooley, J.S.G., Engelhardt, H., Baumeister, W., Kay, W.W., and Trust, T.J., 1989, Three dimensional structure of an open form of the surface layer from the fish pathogen *Aeromonas salmonicida*, *J. Bacteriol.* 171:190.

Garduño, R.A., 1993, Structure and function of the surface layer of the fish pathogenic bacterium *Aeromonas salmonicida*, PhD Thesis, University of Victoria, Victoria, B.C. Canada.

Garduño, R.A., and Kay, W.W., 1992a, A single structural type in the regular surface layer of *Aeromonas salmonicida*, *J. Struct. Biol.* 108:202.

Garduño, R.A., and Kay, W.W. 1992b, Interaction of the fish pathogen *Aeromonas salmonicida* with rainbow trout macrophages, *Infect. Immun.* 60:4612.

Garduño, R.A., Lee, E.J.Y., and Kay, W.W., 1992, S layer-mediated association of *Aeromonas salmonicida* with murine macrophages, *Infect. Immun.* 60:4373.

Garduño, R.A., Phipps, B.M., Baumeister, W., and Kay, W.W., 1993, Novel structural patterns in divalent cation-depleted surface layers of *Aeromonas salmonicida*, *J. Struct. Biol.* 109: in press.

Ishiguro, E.E., 1988, Thermotolerance of A-layer deficient mutants of *Aeromonas salmonicida*, *in*: "Crystallinc Bacterial Cell-Surface Layers", U.B. Sleytr, P. Messner, D. Pum, and M. Sára, eds., Springer-Verlag, Berlin.

Ishiguro, E.E., Kay, W.W., Ainsworth, T., Chamberlain, J.B., Austin, R.A., Kay, W.W., Buckley, J.T., and Trust, T.J., 1981, Loss of virulence during culture of *Aeromonas salmonicida* at high temperature, *J. Bacteriol.* 148:333.

Ishiguro, E.E., Phipps, B.M., Monette, J.P.L.,and Trust, T.J., 1980, Purification and disposition of a surface protein associated with virulence of *Aeromonas salmonicida*, *J. Bacteriol.* 147:1077.

Kay, W.W., and Trust, T.J., 1991, Form and functions of the regular surface array (S-layer) of *Aeromonas salmonicida*, *Experimentia*, 47:412.

Kay, W.W., Phipps, B.M., Ishiguro, E.E.,and Trust, T.J., 1985, Porphyrin binding by the surface array virulence protein of *Aeromonas salmonicida*, *J. Bacteriol.* 164:1332.

Messner, P., and Sleytr, U.B., 1992, Crystalline bacterial cell-surface layers, *Adv. Microbial Physiol.* 33: 213.

Munn, C.B., Ishiguro, E.E., Kay, W.W., and Trust, T.J., 1982, Role of surface components in serum resistance of virulent *Aeromonas salmonicida*, *Infect. Immun.* 36:1069.

Phipps, B.M., and Kay, W.W., 1988, Immunoglobulin binding by the surface array virulence protein of *Aeromonas salmonicida*, *J. Biol. Chem.* 236:9298.

Saxton, W.O., Pitt, T.J., and Horner, M., 1979, Digital image processing: the SEMPER system, *Ultramicrosc.* 4: 343.

Saxton, W.O., and Baumeister, W., 1982, The correlation averaging of a regularly arranged bacterial cell envelope protein, *J. Microsc.* 127:127.

Stewart, M. , Beveridge, T.J., and Trust, T.J., 1986, Two patterns in the *Aeromonas salmonicida* A-layer may reflect a structural transformation that alters permeability, *J. Bacteriol.* 166:120.

Thornton, J.C., Garduño, R.A., Newman, S.G., and Kay, W.W., 1991, Surface-disorganized, attenuated mutants of *Aeromonas salmonicida* as furunculosis live vaccines, *Microbial Pathogen.* 11:85.

Udey, L.R., and Fryer, J.L., 1978, Immunization of fish with bacterins of *Aeromonas salmonicida*, *Maricult. and Fisher. Rev.* 40:12.

Chapter 16

MOLECULAR, STRUCTURAL AND FUNCTIONAL PROPERTIES OF *AEROMONAS* S-LAYERS

Trevor J. Trust

Department of Biochemistry and Microbiology and
Canadian Bacterial Diseases Network
University of Victoria
Victoria, British Columbia, Canada

INTRODUCTION

Aeromonas are a commonly isolated group of gram-negative bacteria belonging to the proposed family Aeromonadaceae (Colwell et al., 1986). *Aeromonas* taxonomy continues to evolve and is quite complex. At least 13 DNA hybridization groups have been described in the genus and in many cases these are difficult to differentiate biochemically (Janda, 1991; Lucchini and Altwegg, 1992). For practical purposes, aeromonads are considered to fall into two major phenotypic groups, the mesophilic motile aeromonads which include the validated species *Aeromonas hydrophila*, *Aeromonas caviae* and *Aeromonas veronii* biotypes *veronii* and *sobria*, and the non-motile *Aeromonas salmonicida*. The mesophilic aeromonads are widely distributed in the environment, and have long been associated with disease of poikilothermic animals. In fish they are typically isolated from a fatal hemorrhagic septicemia. They have also been increasingly associated with a variety of diseases of homeothermic animals. In humans, at least five proposed or validated *Aeromonas* spp. have been implicated in disease (Carnahan and Joseph, 1991; Janda and Kokka, 1991; Kokka et al., 1992). They are most commonly associated with diarrheal disease, but they can also cause wound infections, and in compromised individuals can cause serious systemic infections (Janda and Duffey, 1988; Janda, 1991).

A. salmonicida is an economically significant pathogen of fish (McCarthy and Roberts, 1980; Trust, 1986). Typical strains cause furunculosis, a systemic fatal disease of salmonids, while phenotypically "atypical" strains (Belland and Trust, 1988) are isolated from a variety of diseases with different pathogenesis which are often chronic and inflammatory, and can involve surface ulceration and erythrodermatitis (Bootsma et al., 1977; Paterson et al., 1980a; Shotts et al., 1980). Atypical strains have been isolated from a variety of fishes including Atlantic salmon, European carp, goldfish, and herring. One property shared by many of the strains of mesophilic aeromonads isolated from various diseases, and by *A. salmonicida*, is the ability to produce an

S-layer (Udey and Fryer, 1978; Kay et al., 1981; Janda et al., 1987; Kokka et al., 1990; Kokka et al., 1992). In the case of *A. salmonicida* this S-layer which iscommonly known as A-layer, contributes to the ability of the organism to colonize and produce disease in fish, and is considered to be an important virulence factor (Ishiguro et al., 1981; Munn and Trust, 1983).

FUNCTION

The biological activities of *A. salmonicida* A-layer are numerous. Early studies (Ishiguro et al., 1981) showed that in typical strains, A-layer enhanced the virulence of the bacterium for salmonid fish, protecting the organism against fish defense mechanisms and promoting host colonization (Munn and Trust, 1983). A-layer contributes to protection against the bactericidal activity of both immune and non-immune serum (Munn and Trust, 1982), facilitates the association of *A. salmonicida* with macrophages, and probably influences the outcome of this *A. salmonicida*-macrophage interaction (Trust et al., 1983). A-layer also protects against the action of proteases (Chu et al., 1991), and possesses unique binding capabilities including certain porphyrins (Kay et al., 1985), and immunoglobulins (Phipps and Kay, 1988). In the case of immunoglobulins, a specific three dimensional arrangement of adjacent A-protein subunits, such as present in the tetragonal unit cell of the array, were required for binding.

Recent studies have shown that A-layer is also capable of binding a variety of extracellular matrix proteins. A-layer[+] cells bind high levels of the basement membrane components collagen IV (Trust et al., 1993) and laminin (Doig et al., 1992). Binding is specific, rapid, saturable, high affinity (collagen IV, Kd = 27 nM; laminin, Kd = 1.5 nM). It is due to A-layer because binding is inactivated by selective removal of A-layer at pH 2.2, and isogenic A-layer⁻ *A. salmonicida* mutants do not bind the matrix proteins. A-layer also specifically, rapidly and reversibly binds fibronectin via 2 classes of interactions (Doig et al., 1992). The higher affinity (Kd = 6.6 nM) interaction is similar to that of laminin, apparently involving the same binding site. Class 1 binding is of lower affinity (Kd = 220 nM) and is distinct from that of class 2 binding. This extracellular matrix protein binding by A-layer should contribute to disease in fish by allowing *A. salmonicida* to persist in the kidney in a carrier state, by promoting colonization of ulcerative lesions, by blocking the host immune response by steric masking of immunogenic epitopes, and by facilitating adhesion to host cells (e.g., macrophages via fibronectin/laminin receptors on the host cell surface). A-layer also binds vitronectin (Trust and Kay, 1992). Since vitronectin plays an important role as an inhibitor of complement-mediated cell lysis, as a regulator in the terminal phase of the coagulation system, and as a promoter of cellular adhesion, its interaction with A-layer might also influence *A. salmonicida* infections.

In contrast to the A-layer of *A. salmonicida*, the biological functions of the S-layers of the mesophilic aeromonads remain obscure. S-layer producing mesophilic *Aeromonas* normally belong to a single lipopolysaccharide (LPS) serogroup, O-11 (Sakazaki and Shimada, 1984). This serogroup is commonly associated with human infections (Janda et al., 1987; Kokka et al., 1990) and also contains strains with high virulence for fish (Mittal et al., 1980). However precise roles for S-layers in the pathogenesis of infectious disease caused by mesophilic *Aeromonas* have yet to be reported, in part because of the lack of mutants isogenic in their ability to produce S-layer. In this regard, preliminary studies in our laboratory with a single insertion transposon mutant of *A. hydrophila* TF7 which produces a truncated S-layer protein which does not assemble into a native array indicates that the S-layer does make a contribution to the ability of this strain to produce a lethal infection in fish (unpublished observations). In this case the S-layer appears to contribute to the serum resistance of the strain. The native S-layer of these mesophilic aeromonads is also resistant to protease activity, but appears not to possess the binding activities of the A-layer of *A. salmonicida*. Another function related to the pathogenic potential of S-layer producing strains of mesophilic aeromonads might be to contribute to antigenic diversity. For example, when the S-layers from isolates of *A. hydrophila* and *A. veronii* var. *sobria* belonging to serogroup O-11 were examined using polyclonal antisera prepared against the S-layer protein of *A. hydrophila* TF7, three different S-layer antigenic classes were shown among the strains examined, including those with differences in the surface-exposed epitopes of the S-layers. This antigenic diversity of the S-layer surfaces of strains belonging to a LPS serogroup which is commonly involved in the disease state suggests that this property could well provide a pathogenic advantage for this group of aeromonads.

MORPHOLOGY

The tetragonally-arranged S-layers of *Aeromonas* species are similar morphologically (Stewart et al., 1986; Murray et al., 1988; Al-Karadaghi et al., 1988; Dooley et al. 1989; Kokka et al., 1990). Three dimensional reconstructions of the *A. salmonicida* and *A. hydrophila* surface arrays have shown that the subunits constitute an array with a lattice constant of 12.0 - 12.5 nm containing a major tetragon at one four-fold axis of symmetry and a minor tetragon at the second four-fold axis of symmetry (Stewart, et al., 1986; Murray et al., 1988; Al-Karadaghi et al., 1988; Dooley et al., 1989). In the case of A-protein, the subunits contain a heavy mass domain with a linker arm to a domain of lesser mass. The major tetragonal core of A-layer is composed of the heavy mass domains of four subunits, contains a large depression in its center, and is located towards the inside of the layer (Dooley et al., 1989). The minor tetragon which is composed of the lesser mass of four subunits provides connectivity within the layer, and is raised and located at the outer surface of the layer. This structure provides the surface of the layer with a certain amount of three dimensional architecture which may contribute to the biological activities of the array.

PRIMARY STRUCTURE

The subunit molecular weight (M_r) of the *Aeromonas* S-layer proteins is in the range of $M_r = 50000 - 52000$. In the case of the mesophilic aeromonads, N-terminal amino acid sequence analysis and one dimensional endoproteinase Glu-C mapping of antigenically different S-layer proteins has shown that while the proteins are structurally related they show different primary structure (Kostrzynska et al., 1992). All mesophilic *Aeromonas* N-terminal S-layer sequences thus far determined (Kostrzynska et al., 1992; Kokka et al., 1992) are also significantly different from the conserved sequence of the *A. salmonicida* surface array protein (Chu et al., 1991). This is consistent with the lack of antigenic cross-reactivity between the S-layer proteins of these mesophilic aeromonads and the A-protein of *A. salmonicida*.

The predicted primary structure of the A-protein of *A. salmonicida* is now available (Chu et al., 1991). The species-specific structural gene (*vapA* = virulence array protein gene A) for A-protein and its upstream promoter-containing sequence was initially cloned into λ gt11 (Belland and Trust, 1987). However, this clone which appeared to express A-protein at a high level in *Escherichia coli* was unstable.

Removal of much of the upstream promoter-containing sequence allowed *vapA* to be stably cloned in the broad host range cosmid pLA2917 (Chu et al., 1991). In this case A-protein production in *E. coli* was markedly lower than that of wild type *A. salmonicida* and detection required Western blot immunoassay. A-protein was located in the cytoplasmic, inner membrane and periplasmic fractions in *E. coli*. DNA sequence analysis showed that the 1506 base pair *vapA* gene encoded a protein with a 21 residue signal peptide, and a 481 residue (M_r=50800) protein. The predicted amino acid sequence was confirmed in 31% of the mature protein sequence by N-terminal sequencing of peptides produced by protease cleavage, and by CNBr hydrolysis of purified A-protein. The sequence analysis also revealed that the original λ gt11 clone had undergone an 816 base pair deletion due to a 21 base pair direct repeat within the gene. The predicted mature A-protein contains 37.2% polar, 10.2% acidic and 8.9% basic amino acids and displays an overall negative charge. The predicted overall pI was 4.79 compared to the measured pI of 5.7 - 6.0. In terms of polar, acidic, and basic amino acid content the composition of the mature A-protein is similar to other S-layer proteins. However, with 43.7% non polar or hydrophobic amino acids, the hydrophobic content of A-protein is high for an S-layer protein. The Kyte and Doolittle (1982) hydropathic index using an interval of 9 amino acids gives an average hydrophobicity score of -0.45 (the average hydropathy for soluble proteins is -0.4), and analysis by algorithms used to predict membrane proteins classify A-protein as a peripheral rather than an integral membrane protein. Data bank searches show that A-protein is unique to *A. salmonicida*.

DOMAINS

While native *Aeromonas* S-layers are resistant to proteases, the S-layer proteins become susceptible to digestion once they have been isolated. Indeed protease digestion studies with purified *Aeromonas* S-proteins have provided biochemical evidence for two major structural domains. In the case of A-protein, treatment with trypsin results in A-protein being rapidly degraded to a major peptide of approximate M_r=38000 by SDS-PAGE, via a series of intermediates in the approximate M_r range of 48000 to 40000, and a second major peptide of approximate M_r=16700, via a M_r =1300 peptide (Chu et al., 1991). The amino-terminal larger domain is totally refractile to trypsin under non-denaturing conditions, while the overlapping 16.7 kDa carboxy-terminal peptide displays intermediate resistance to trypsin. Sedimentation analysis of the larger trypsin resistant fragment gave an apparent weight average M_r=39050 across the whole cell (unpublished observation), and based on the deduced amino acid sequence of A-protein, this would place the trypsin cleavage site after Arg370, giving an actual M_r for the tryptic fragment of 39439. The first 25 residues of the M_r=21300 peptide correspond to the deduced A-protein sequence beginning at residue 275, while the first 22 residues of the M_r=16700 peptide correspond to the deduced sequence beginning at amino acid 324. These overlapping 39.4 kDa N- and 16.7 kDa COOH-terminal polypeptides both appear to adopt compact tertiary structures because the lesser mass COOH-terminal compact peptide unit displays intermediate resistance to trypsin despite containing 14 potential trypsin cleavage sites, while the N-terminal compact peptide unit has 29 potential trypsin sites and, once formed, appears to be extremely refractile to trypsin activity.

The ability to cleave A-protein into its two major structural domains has allowed one biological activity to be assigned at the structural domain level. When both trypsin-cleavage peptides were purified and tested in binding inhibition studies, the larger mass trypsin resistant peptide was found to inhibit binding of both fibronectin and laminin to A-layer, localizing extracellular matrix protein binding function to this segment of the protein (Doig et al., 1992).

In the case of the S-layer protein of *A. hydrophila* TF7, treatment with trypsin, chymotrypsin, or endoproteinase Glu-C under non denaturing conditions results in a major peptide of approximate M_r=38000 despite very different cleavage specificities (Kostrzynska et al., 1992). This M_r=38000 protease resistant peptide maps to the N-terminal end of the linear *A. hydrophila* TF7 S-layer protein map and carries within its length the immunodominant surface exposed region of the molecule. A portion of this immunodominant region can be further mapped to an internal M_r=26000 CNBr peptide, a section of which is carried within the N-terminal M_r=38000 sequence of the protein. Under non denaturing conditions this major peptide is refractile to further cleavage by any of the enzymes tested. The peptide apparently assumes a folding which makes potential trypsin, chymotrypsin, and endoproteinase Glu-C sites inaccessible because it does not accumulate when protease digestions are performed in the presence of denaturants such as urea or SDS, and smaller M_r peptides are produced under such denaturing conditions. The lesser M_r region (12000-14000) of the S-layer subunit is readily susceptible to each of the proteases under non denaturing conditions.

Taken together these findings suggest that the two structural segments with their different protease sensitivities could reflect the two morphological domains of the *Aeromonas* S-layer proteins, the C-terminal protease sensitive peptide comprising the lesser morphological linker domain, and the N-terminal protease resistant peptide corresponding to the inner major-mass morphological core domain. This was recently confirmed in the case of *A. hydrophila* TF7 by the isolation of a Tn5 transposon insertion mutant which, instead of producing wild-type M_r=52000 S-layer protein, produced an S-layer protein of subunit M_r=38650, as determined by sedimentation analysis. Automated Edman degradation and immunochemical analysis showed that this truncated protein comprised the larger mass protease-resistant N-terminal domain of the mature *A. hydrophila* S-protein. Amino acid compositional analysis of the purified truncated protein further showed that it had an increased hydrophobic amino acid content relative to the wild-type protein (Thomas et al., 1992). Localization studies showed that the M_r=38650 truncated S-layer protein was exported via the periplasm to the cell surface but could not self-assemble into a tetragonal array, or be anchored to the cell surface. Monomers of the truncated S-protein were able to associate with each other at one end of the molecule to form a cup-shaped assembly with parallel sides when viewed from the side. The diameter of this assembly never reached a diameter greater than 6.5 nm, while the stain-filled center was approximately 3.0 nm. From the top the assembly appeared as a ring, and based on previous computer reconstructions of both the *A. hydrophila* (Murray et al., 1988; Al-Karadaghi et al., 1988) and *A. salmonicida* S-layers (Stewart et al., 1986; Dooley et al., 1989), these assemblies would consist of four copies of the truncated protein. The 6.5 nm width of the dimensions of the cup-shaped assembly was consistent with the width of the major morphological tetramer of the native *A. hydrophila* S-layer (Murray et al., 1988). Occasionally, assemblies of two major tetramers, joined by their bases were observed. Most aggregates consisted of three or four major tetramers on their sides connected by their bases forming subassembly intermediates of the S-layer.

The C-terminal segment which is missing from the truncated protein described here is clearly essential for tetragonal array assembly, appearing to contribute to the lesser morphological unit of the assembled array. This minor domain allows for correct spatial positioning of the major structural domains by contributing the molecular size and structure essential for the subunit interactions which provide connectivity within layer, and perhaps influencing the degree to which the tetramers are open. Three dimensional image reconstructions of *Aeromonas* S-layers indicate that this lesser domain of the S-layer protein is located towards the outer surface of the layer (Al-Karadaghi et al., 1988; Dooley et al., 1989). In addition, the C-terminal domain appears to participate in anchoring of the S-layer to the cell surface, presumably via interaction with LPS carbohydrate. The carbohydrate involved is probably core oligosaccharide because, while deep rough mutants cannot retain an array on the cell surface, rough mutants have an anchored array (Dooley et al., 1988). In the mutant described above, the LPS was wild-type in structure and it is the alteration in protein structure which ultimately results in the inability to anchor S-protein to the cell surface. In the case of *A. salmonicida*, the presence of O-polysaccharides chains appears to be essential for A-layer anchoring (Chart et al., 1984). These O-polysaccharide chains are homogeneous in length, as are those of the S-layer producing mesophilic aeromonads (Dooley et al., 1985). In each case the O-polysaccharides appear to be conserved antigenically.

HIGHER ORDER STRUCTURE

When *A. salmonicida* A-protein is subjected to circular dichroism (CD) analysis in 100 mM NaCl, 10 mM HEPES buffer, pH 7.4, and the secondary structure predicted by the Contin program of Provencher and Glöckner (1981) the protein was shown to contain approximately 11 - 14% α-helix, 50 - 51% β-sheet and 16 - 18% β-turns (Phipps et al., 1983; unpublished observation). At 50 - 51%, this is one of the highest β-sheet values for an S-layer protein (Messner and Sleytr, 1992). When CD analysis was performed in the presence of 0.1% SDS to model the hydrophobic environment presumed to exist in an assembled S-layer, the predicted secondary structure was dramatically different at 29% α-helix, 32% β-sheet and 0.08% β-turns. The α-helix content was essentially doubled at the expense of β-structure. Earlier studies with the S-layer protein of *A. hydrophila* had shown that the hydrophobic environment provided by 0.1% SDS also affected S-layer secondary structure (Dooley et al., 1988). However in the case of the *A. hydrophila* protein, there was an increase of β-sheet at the expense of α-helix and β-turn.

The secondary structure of the major N-terminal tryptic peptide of A-protein and the truncated M_r =38600 S-layer protein produced by the *A. hydrophila* transposon insertion mutant TF7-ST1 were also affected by 0.1% SDS, with both showing an increase of α-helix at the expense of β-structure. Taken together these findings suggest that the S-layer proteins of *Aeromonas* are capable of undergoing changes in conformation, and the proteins probably have different conformations when they are in solution and when they are assembled in the S-layer. The fact that SDS has different effects on the secondary structure of the intact S-layer proteins of *A. salmonicida* and *A. hydrophila*, but similar effects on the conformation of the major structural domains of the two proteins cannot be explained at this time. However the C-terminal domain clearly plays an important role in determining *Aeromonas* S-layer protein conformation, albeit a different role in the two species. Interestingly in the case of A-protein, CD data obtained in the presence of SDS were in close agreement with the sequence-based structural predictions suggesting that the measurements made in the presence of SDS may be more representative of the true or natural secondary structure.

A-protein does have quite a complex folding profile when assembled into a surface array. For example, monoclonal antibody (Mab) AA6 which blocks binding residues of laminin and fibronectin (Doig et al., 1992) has a surface exposed epitope (residues 401-409) which maps the C-terminal to the matrix protein binding domain (Doig et al., 1993). In contrast Mab AA1 whose surface exposed epitope (residues 333-341) maps closer to the major trypsin resistant domain than that of AA6 failed to inhibit binding. These results indicate that the blocking of laminin and fibronectin binding by AA6 result from steric interference. Presumably, when assembled into an array, the folding characteristics of A protein allow bound Mab AA6 to physically interfere with the access of the matrix proteins to their binding sequences, while the steric position of bound Mab AA1 provides no such physical impediment.

Mimeotope analysis of nonapeptides (representing the 481 residue sequence of A-protein) with a panel of 8 Mab's which bind to epitopes on the surface of A-layer, and a monospecific polyclonal antiserum have recently allowed identification of 146 residues in presumed linear epitopes exposed on the antigenically conserved surface of the A-layer (unpublished observation). Non-exposed or non-epitopic residues account for 70% of the protein. The majority of non-exposed residues reside in the N-terminal 301 residues of A-protein. Indeed, the polyclonal antiserum from which all antibodies capable of binding to linear epitopes exposed on the surface of A-layer had been removed recognized 166 possible residues, or 55% of the sequence in this region of the protein. Dispersed among these were 65 surface-exposed residues in five linear epitope clusters emphasizing the complex folding of this major structural domain of A-protein. The C-terminal 180 residues carried fewer linear epitopes but contained the major region exposed on the outer surface of A-layer, including four linear epitopes in a predominantly hydrophobic sequence as determined by Kyte-Doolittle analysis (Kyte and Doolittle, 1982). The high relative extent of surface exposure of hydrophobic residues in this C-terminal region of A-protein may explain why A-layer confers hydrophobicity to the surface of *A. salmonicida* (Trust et al., 1983; Van Alstine et al., 1986). This surface hydrophobicity appears to play a role in the biological activities of the layer including macrophage association (Trust et al., 1983) and laminin and fibronectin binding activity (Doig et al., 1992).

CONSERVATION AND STABILITY OF vapA

Polymerase chain reaction (PCR) analysis of chromosomal DNA from 28 typical and 26 atypical strains has provided convincing evidence that the *vapA* gene is conserved within the species *A. salmonicida* (Gustafson et al., 1992). The strains examined had originally been isolated between 1961 and 1988 by 22 different workers. The strains included isolates from different geographical areas, isolates from a variety of fish species, and isolates from diseases with different pathogenesis. Also included were two strains originally designated *Haemophilus piscium* but shown by several studies to belong to the species *A. salmonicida* (Paterson et al., 1980b; Belland and Trust, 1988). Fifteen of these strains had lost their ability to produce A-layer as a result of their handling since isolation, and were analyzed to provide information on the stability of the *vapA* gene. Using a primer pair to the $3'$ end of *vapA*, an appropriately sized fragment was amplified from 53 of the 54 DNAs tested. Using additional primers, six of the strains provided evidence that the *vapA* gene and its upstream flanking DNA can be subject to different deletions and rearrangements. The deletions suffered in each of these six strains was different in nature to the previously reported 795 bp internal deletion which occurred when *vapA* was cloned and expressed in *E. coli* (Chu et al., 1991). Two of the strains showing deletions were originally obtained from the American Type Culture Collection, and presumably these had received much handling over the years. All six of the strains showing deletions were incapable of producing immunologically detectable truncated A protein. The most dramatic change observed was the deletion of the DNA responsible for the entire *vapA* gene plus a length of downstream flanking DNA in the sixth strain. In this case the deletion had obviously occurred as a result of laboratory handling because the parental isolate had retained its ability to produce A-layer.

The conservation of *vapA* in *A. salmonicida* has recently been put to practical use (Gustafson et al., 1992). Despite its economic significance as a pathogen of fish, little is known about the epidemiology of disease states. In part this reflects the absence of an efficient selective medium in which to grow the bacterium and is compounded by the bacterium's slow growth characteristics which allow other organisms to overgrow it. Based on the limited data available, *A. salmonicida* is currently regarded as an obligate pathogen of fish (Popoff, 1984), and the role of environmental survival in transmission of *A. salmonicida* is poorly characterized. One important contribution to the epidemiology of *A. salmonicida* disease is thought to be the carrier state which can result in infected fish (McCarthy and Roberts, 1980). In the case of brown trout (*Salmo trutta*) up to 80% of fish in a given population are thought to be carriers of *A. salmonicida*. Carrier fish show no clinical symptoms of disease but are assumed to be capable of shedding the organism. When carrier salmonids are stressed, acute furunculosis is precipitated resulting in death. The most reliable means of detecting the *A. salmonicida* carrier state in salmonids is to stress animals by injecting them with corticosteroids and exposing them to elevated water temperatures as a mechanism of precipitating acute systemic disease (Bullock and Stuckey, 1975). Such a complex and indirect procedure is clearly unsatisfactory.

Long term control of the disease under the conditions normally employed for intensive fish culture would be facilitated by the development of a rapid, sensitive and specific method for identifying *A. salmonicida* in the fish culture environment, the identification and elimination of carrier fish, and the establishment of *A. salmonicida*-free stocks. In the case of carriers, a non-invasive method which would facilitate the detection of *A. salmonicida* being shed in fish feces or present in water samples would be highly desirable. PCR amplification of the 421 bp sequence from the $3^{\prime}$ region of *vapA* as described previously provides such a specific and sensitive method for the detection and identification of this important fish pathogen. A detection level of 5 *fg*, equivalent to approximately 1 bacterial cell was obtained using purified chromosomal DNA as template. The sensitivity of PCR detection of *A. salmonicida* directly from tissues such as kidney and spleen, and fish feces was less than 10 colony forming units (c.f.u) per mg. In tank water the detection limit was 5 c.f.u. per 500 ml. The PCR amplification method successfully and reproducibly detected *A. salmonicida* in fish showing clinical disease, in suspected carrier fish, and in tank water carrying suspected carrier fish. The assay could also be readily confirmed in equivocal cases by a simple 24 h broth outgrowth step. The *vapA*-based assay is highly reproducible, requires as little as 45 min to complete depending on the thermocyler used, and is therefore sensitive enough to be used as a non-invasive method for monitoring fish populations for the presence of carrier fish.

EXPRESSION OF *vapA*

The data summarized above indicate that *vapA* is always present in wild-type *A. salmonicida*, and that absence of A-layer on a strain reflects either an alteration in the expression of *vapA*, or deletion / rearrangement of all or part of *vapA*.

Consequently, *vapA* is normally expressed at levels which allow formation of a surface array in both typical and atypical strains. The regulation of this expression is currently being investigated. The 420 base pair sequence of DNA immediately upstream of *vapA* has been cloned, and analysis by nucleotide sequencing and polymerase chain reaction has shown that this control sequence containing DNA is conserved in wild-type *A. salmonicida* (unpublished observation). Sequencing has also revealed two promoter sequences, and primer extension analysis and Northern blot analysis has shown that *vapA* transcription in *A. salmonicida* is directed predominantly by a distal promoter P1 resulting in a 1.7 kb unit length mRNA. The 178 base leader sequence of this RNA contains at least one low free-energy, predicted stem-loop structure. A possible proximal promoter P2 is positioned at -27 bp relative to *vapA* and may direct A-protein production in *E. coli* (Chu et al., 1991).

OTHER GENES INVOLVED IN A-LAYER PRODUCTION

In addition to the structural S-protein gene, other genes and their products are obviously required for the production of a surface protein array. For example, in the case of a gram-negative bacterium like *Aeromonas*, the S-layer protein subunits have to be exported across two membranes before being assembled into an array and the array then has to be anchored to the cell surface. Little is known concerning this export pathway. However, in the case of *A. salmonicida* several Tn5 transposon mutants have been isolated which exhibit defects in S-layer protein export (Belland and Trust, 1985). Two such insertions result in periplasmic accumulation of A-protein, presumably as a result of the "knock-out" of an outer membrane translocation function. In addition to providing evidence that at least one gene product is required to translocate A-protein across the outer membrane, these mutants show that the export pathway for A-protein involves passage through the periplasmic space. The two mutations are different because in one case the LPS appears to be normal and smooth (A449-TM1), while in the other case the LPS lacks O-polysaccharides and is rough (A449-TM2). Indeed this latter mutant provides initial evidence that the ability to translocate A-protein across the outer membrane may be linked with the ability of the cells to either produce a complete LPS, or to export their LPS O-polysaccharide chains. The two Tn5 insertions map to different locations, and both insertions have been now cloned and the genes effected are being examined in greater detail (Noonan et al., 1993). In the case of mutant A449-TM2, the Tn5 insertion maps approximately 7 kb downstream of *vapA*.

A conserved *A. salmonicida* gene (*abcA*) affecting expression of *vapA* in *E. coli* has also been identified (Chu and Trust, 1993). This 924 bp gene starts 205 bp after *vapA*, and codes for a protein of deduced $M_r=34015$ containing an N-terminal P-loop and homology to a ATP-binding-cassette (ABC)-transport transport protein family. The deduced AbcA protein also contains a C-terminal leucine zipper-basic region and a central membrane-associated region. AbcA was identified using DNA-directed translation, and T7 polymerase expression, and co-purified with the cytoplasmic membrane fraction. A *lacZ* fusion containing upstream sequence and 387 bases in the 5' end of *abcA* was constructed, and the β-galactosidase activity of the fusion was similar in *E. coli* and *A. salmonicida*. The $M_r=130000$ fusion protein was purified and the 129 AbcA N-terminal residues were shown to bind ATP. Unfortunately attempts to knock out the gene in *A. salmonicida* have been unsuccessful, so the precise role of the *abcA* gene and its gene product remain to be determined.

SUMMARY

The S-layers of *Aeromonas* are excellent and exciting models for the detailed analysis of tetragonal S-layers. They are currently the best described gram-negative S-layers in terms of structure, function, biochemistry, export machinery, surface tethering and molecular genetics, and they will continue to provide valuable and exciting information on such fundamental aspects as protein-protein interactions, protein-carbohydrate interactions, protein export, gene expression and gene regulation.

ACKNOWLEDGEMENTS

These continuing studies on *Aeromonas* S-layers are funded by grants from the Natural Sciences and Engineering Research Council of Canada and the Canadian Bacterial Disease Network.

REFERENCES

Al-Karadaghi, S., Wang, D. N., and Hovmöller, S., 1988, Three-dimensional structure of the crystalline surface layer from *Aeromonas hydrophila*, *J. Ultrastruct. Mol. Struct. Res.* 101:92.

Belland, R. J., and Trust, T. J., 1985, Synthesis, export, and assembly of the A-layer of *Aeromonas salmonicida* analysed by transposon mutagenesis, *J. Bacteriol.* 163:877.

Belland, R. J., and Trust, T. J., 1987, Cloning of the gene for the surface protein array of *Aeromonas salmonicida* and evidence linking loss of expression with genetic deletion, *J. Bacteriol.* 169:4086.

Belland, R. J., and Trust, T. J., 1988, DNA:DNA reassociation analysis of *Aeromonas salmonicida*, *J. Bacteriol.* 170:499.

Bootsma, R., Fijan, N., and Blommaert, J., 1977, Isolation and preliminary identification of the causative agent of carp erythrodermatitis, *Vet. Arch.* 47: 291.

Bullock, G. L., and Stuckey, H. M., 1975, *Aeromonas salmonicida*: detection of asymptomatically infected trout, *Prog. Fish Cult.* 37:237.

Carnahan, A. M., and Joseph, S. W., 1991, *Aeromonas* update - new species and global distribution, *Experientia* 47:402.

Chart, H., Shaw, D. H., Ishiguro, E. E., and Trust, T. J., 1984, Structural and immunochemical homogeneity of *Aeromonas salmonicida* lipopolysaccharide, *J. Bacteriol.* 158:16.

Chu, S., Cavaignac, S., Feutrier, J., Phipps, B. M., Kostrzynska, M., Kay, W. W., and Trust, T. J., 1991, Structure of the tetragonal surface virulence array protein and gene of *Aeromonas salmonicida*, *J. Biol. Chem.* 266:15258.

Chu, S., and Trust, T. J., 1993, Identification and characterization of an *Aeromonas salmonicida* gene which affects A-protein expression in *Escherichia coli*, *in*: "Advances in bacterial paracrystalline surface layers", T.J. Beveridge, and S.F. Koval, eds., Plenum Publishing, New York.

Colwell, R. R., MacDonell, M. T., and Ley, J. D., 1986, Proposal to recognise the family Aeromonadaceae fam. nov., *Int. J. Syst. Bact.* 36:73.

Doig, P., Emödy, L., and Trust, T. J., 1992, Binding of laminin and fibronectin by the trypsin-resistant major structural domain of the crystalline virulence surface array protein of *Aeromonas salmonicida, J. Biol. Chem.* 267:43.

Dooley, J. S. G., Lallier, R., Shaw, D. H., and Trust, T. J., 1985, Electrophoretic and immunochemical analyses of the lipopolysaccharides from various strains of *Aeromonas hydrophila, J. Bacteriol.* 164:263.

Dooley, J. S. G., McCubbin, W. D., Kay, C. M., and Trust, T. J., 1988, Isolation and biochemical characterization of the S-layer protein from a pathogenic strain of *Aeromonas hydrophila, J. Bacteriol.* 170:2631.

Dooley, J. S. G., Engelhardt, H., Baumeister, W., Kay, W. W., and Trust, T. J., 1989, Three-dimensional structure of the surface layer from the fish pathogen *Aeromonas salmonicida, J. Bacteriol.* 171:190.

Gustafson, C. E., Thomas, C. J., and Trust, T. J., 1992, Detection of *Aeromonas salmonicida* from fish using polymerase chain reaction amplification of the virulence surface array protein gene, *Appl. Environ. Microbiol.* 58:3816.

Ishiguro, E. E., Kay, W. W., Ainsworth, T., Chamberlain, J. B., Buckley, J. T., and Trust, T. J., 1981, Loss of virulence during culture of *Aeromonas salmonicida* at high temperature, *J. Bacteriol.* 148:333.

Janda, J. M., 1991, Recent advances in the study of the taxonomy, pathogenicity, and infectious syndromes associated with the genus *Aeromonas, Clin. Microbiol. Rev.* 4:397.

Janda, J. M., and Duffey, P. S., 1988, Mesophilic aeromonads in human disease: current taxonomy, laboratory identification, and infectious disease spectrum, *Rev. Infect. Dis.* 10:980.

Janda, J. M., and Kokka, R. P., 1991, The pathogenicity of *Aeromonas* strains relative to genospecies and phenospecies identification, *FEMS Microbiol. Lett.* 90:29.

Janda, J. M., Oshiro, L. S., Abbott, S. L., and Duffey, P. S., 1987, Virulence markers of mesophilic aeromonads: association of the autoagglutination phenomenon with mouse pathogenicity and the presence of a peripheral cell-associated layer, *Infect. Immun.* 55: 3070.

Kay, W. W., Buckley, J. T., Ishiguro, E. E., Phipps, B. M., Monette, J. P. L., and Trust, T. J., 1981, Purification and disposition of a surface protein associated with virulence of *Aeromonas salmonicida, J. Bacteriol.* 147:1077.

Kay, W. W., Phipps, B. M., Ishiguro, E. E., and Trust, T. J., 1985, Porphyrin binding by the surface array virulence protein of *Aeromonas salmonicida, J. Bacteriol.* 164:1332.

Kokka, R. P., Vedros, N. A., and Janda, J. M., 1990, Electrophoretic analysis of the surface components of autoagglutinating surface array protein positive and surface array protein negative *Aeromonas hydrophila* and *Aeromonas sobria, J. Clin. Microbiol.* 28:2240.

Kokka, R. P., Vedros, N. A., and Janda, J. M., 1992, Immunochemical analysis and possible biological role of an *Aeromonas hydrophila* surface array protein in septicaemia, *J. Gen. Microbiol.* 138:1229.

Kokka, R. P., Lindquist, D., Abbott, S. L., and Janda, J. M., 1992, Structural and pathogenic properties of *Aeromonas schubertii, Infect. Immun.* 60:2075.

Kostrzynska, M., Dooley, J. S. G., Shimojo, T., Sakata, T., and Trust, T. J., 1992, Antigenic diversity of the S layer proteins from pathogenic strains of *Aeromonas hydrophila* and *Aeromonas veronii* biotype *sobria, J. Bacteriol.* 174:40.

Kyte, J., and Doolittle, R. F., 1982, A simple method for displaying the hydropathic character of a protein, *J. Mol. Biol.* 157:105.

Lucchini, G. M., and Altwegg, M., 1992, rRNA gene restriction patterns as taxonomic tools for the genus *Aeromonas*, *Int. J. Syst. Bacteriol.* 42:384.

McCarthy, D. H., and Roberts, R. J., 1980, Furunculosis of fish - the present state of our knowledge, *Adv. Aquatic Microbiol.* 2:293.

Messner, P., and Sleytr, U. B., 1992, Crystalline bacterial cell-surface layers, *Adv. Microbial Physiol.* 33:213.

Mittal, K. R., Lalonde, G., Leblanc, D., Olivier, G., and Lallier, R., 1980, *Aeromonas hydrophila* in rainbow trout: relation between virulence and surface characteristics, *Can. J. Microbiol.* 26:1501.

Munn, C. B., Ishiguro, E. E., Kay, W. W., and Trust, T. J., 1982, Role of surface components in serum resistance of virulent *Aeromonas salmonicida*, *Infect. Immun.* 36:1069.

Munn, C. B., and Trust, T. J., 1983, Role of additional protein layer in virulence of *Aeromonas salmonicida*, *in*: "Fish Diseases, 4th Cooperative Programme of Research on Aquaculture Session", ACUIGRUP, ed., EDITORA ATP, Madrid.

Murray, R. G. E., Dooley, J. S. G., Whippey, P. W., and Trust, T. J., 1988, Structure of an S-layer on a pathogenic strain of *Aeromonas hydrophila*, *J. Bacteriol.* 170:625.

Noonan, B., Cavaignac, S., and Trust, T. J., 1993, Localization of genes resulting in periplasmic accumulation of the A-protein in *Aeromonas salmonicida*, *in*: "Advances in Bacterial Paracrystalline Surface Layers," T. J. Beveridge and S. F. Koval, eds., Plenum Publishing, New York.

Paterson, W. D., Douey, D., and Desautels, D., 1980a, Isolation and identification of an atypical *Aeromonas salmonicida* strain causing epizootic losses among Atlantic salmon (*Salmo salar*) reared in a Nova Scotian hatchery, *Can. J. Fish. Aquat. Sci.* 12:2236.

Paterson, W. D., Douey, D., and Desautels, D., 1980b, Relationships between selected strains of typical and atypical *Aeromonas salmonicida*, *Aeromonas hydrophila*, and *Haemophilus piscium*, *Can. J. Microbiol.* 26:588.

Phipps, B. M., and Kay, W. W., 1988, Immunoglobulin binding by the regular surface array *Aeromonas salmonicida*, *J. Biol. Chem.* 263:9298.

Phipps, B. M., Trust, T. J., Ishiguro, E. E., and Kay, W. W., 1983, Purification and characterization of the cell surface virulence A- protein from *Aeromonas salmonicida*, *Biochemistry* 22:2934.

Popoff, M., 1984, Genus III, *Aeromonas* Kluyver and Van Niel 1936, *in*: "Bergey's Manual of Systematic Bacteriology", N. R. Krieg and J. G. Holt, eds., Williams and Wilkins, Baltimore.

Provencher, S. W., and Glöckner, J., 1981, Estimation of globular protein secondary structure from circular dichroism, *Biochemistry* 20:33.

Sakazaki, R., and Shimada, T., 1984, O-serogrouping for mesophilic *Aeromonas* strains, *Japan. J. Med. Sci.* 37:247.

Shotts, E. B. J., Talkington, F. D., Elliot, D. G., and McCarthy, D. H., 1980, Aetiology of an ulcerative disease in goldfish, *Carassius auratus* (L): characterization of the causative agent, *J. Fish Dis.* 3:181.

Stewart, M., Beveridge, T. J., and Trust, T. J., 1986, Two patterns in the *Aeromonas salmonicida* A-layer may reflect a structural transformation that alters permeability, *J. Mol. Biol.* 166:120.

Thomas, S., Austin, J. A., Cubbin, W. D. M., Kay, C. M., and Trust, T. J., 1992, Roles of structural domains in the formation, morphology and surface anchoring of tetragonal paracrystalline array of *Aeromonas hydrophila*: biochemical characterization of the major structural domain, *J. Mol. Biol.* 228:652.

Trust, T. J., 1986, Pathogenesis of infectious diseases of fish. *Ann. Rev. Microbiol.* 40:479.

Trust, T. J., and Kay, W. W., 1992, S-layers in bacterial pathogenesis: the tetragonal paracrystalline surface protein arrays of *Aeromonas* as models, *in:* "Molecular Recognition in Host-Parasite Interactions", T. K. Korhonen, T. Hovi, P. H. Makela, eds., Plenum Publishing, New York.

Trust, T. J., Kay, W. W., and Ishiguro, E. E., 1983, Cell surface hydrophobicity and macrophage association of *Aeromonas salmonicida*, *Curr. Microbiol.* 9:315.

Trust, T. J., Kostrzynska, M., Emödy, L., and Wadström, T., 1993, High affinity binding of the basement membrane protein collagen type-IV to the crystalline virulence surface protein array of *Aeromonas salmonicida*, *Mol. Microbiol.*, in press.

Udey, L., and Fryer, J. L., 1978, Immunization of fish with bacterins of *Aeromonas salmonicida*, *Mar. Fish. Rev.* 40:12.

Van Alstine, J. M., Trust, T. J., and Brooks, D. E., 1986, Differential partition of virulent *Aeromonas salmonicida* and attenuated derivatives possessing specific cell surface alterations in polymer aqueous-phase systems, *Appl. Environ. Microbiol.* 51:1309.

Chapter 17

BIOLOGY OF *CAMPYLOBACTER FETUS* S-LAYER PROTEINS

Martin J. Blaser

Division of Infectious Diseases and Department of
Microbiology and Immunology
Vanderbilt University School of Medicine
and Department of Veterans Affairs Medical Center
Nashville, TN, U.S.A.

INTRODUCTION

Campylobacter fetus (formerly called *Vibrio fetus*) is a microaerophilic, curved, motile, gram-negative bacterium. Originally described early in this century as causes of bovine infertility, these organisms now are known to be pathogenic for both animals and humans (Smibert, 1984). Two major subspecies are recognized that are closely related; *C. fetus* subsp. *fetus* and *C. fetus* subsp. *venerealis*. *C. fetus* subsp. *fetus* has a broad host range including sheep, cattle, horses, poultry, and reptiles; humans are occasionally infected. In sheep and cattle, following oral ingestion, there is bacteremia. Pregnant ewes and cows may abort because of the tropism of the organism for the placenta. In other infected ungulates, infection may be inapparent but eventually may lead to chronic biliary carriage and fecal excretion. Humans occasionally ingest *C. fetus* in foods, but illness mostly occurs in compromised hosts. Bacteremia is the hallmark of clinically significant human infections, which may then result in lesions in organs distant to the gastrointestinal tract (Blaser and Reller, 1981). *C. fetus* infection of humans may be underdiagnosed because of the fastidious nature of these organisms (Wang and Blaser, 1986). *C. fetus* subsp. *venerealis* infection is essentially confined to cattle. Bulls carry the organism in the penile prepuce and overlaying the deep epithelial crypts, and introduce it into cows via sexual intercourse. Infection causes endometritis, which leads to infertility. After several months, cows are usually able to clear this infection focus but chronic vaginal colonization may persist for months or years.

This chapter will focus on *C. fetus* subsp. *fetus* but there are important parallels in the biology of the two related subspecies. Researchers from the Cornell Veterinary School found that after experimental introduction of *C. fetus* subsp. *fetus* into the bovine vagina, strains recovered showed evidence of antigenic variation (Corbeil et al., 1975). The Cornell researchers also noted that *C. fetus* subsp. *fetus* strain 23D could spontaneously become agglutinated by O-antiserum and this phenomenon was associated with the loss of an S-layer (McCoy et al.,1975). These S⁻ strains could be ingested by bovine phagocytes whereas the wild type strain could not be ingested (Winter et al., 1978). Thus, there was evidence that *C. fetus* possessed an S-layer associated with resistance to phagocytosis.

STUDIES OF *C. FETUS* SERUM RESISTANCE

My introduction to this area of study was based on a patient who developed *C. fetus* bacteremia and meningitis. The immediate question was how an orally-acquired organism that transiently colonizes the gastrointestinal mucosa was able to cause bacteremia and meningitis. Since other gram-negative colonizing bacteria capable of causing bacteremia often resist the complement-mediated bactericidal activity present in normal human serum, we sought to address this question for *C. fetus*. We found that these organisms were profoundly serum-resistant and that this resistance could not be overcome by increasing incubation time, serum concentration, or addition of immune serum (Blaser et al., 1985; 1987). However, occasional strains were completely serum-sensitive (Pérez-Pérez et al., 1986). To understand the basis for serum resistance, we examined the structure of the lipopolysaccharide (LPS) molecule of *C. fetus*, since organisms with LPS molecules with long polysaccharide side chains are often serum-resistant. We found that all *C. fetus* strains had LPS molecules with long polysaccharide chains, regardless of whether the strain was serum-sensitive or resistant (Pérez-Pérez et al., 1986; Pérez-Pérez and Blaser, 1985). However, we found that there were two structural types A, and B, which conformed to the earlier description of two antigenic types (serotypes), (Pérez-Pérez et al., 1986). We had no strong clues about the basis for serum resistance until one day our typically serum-resistant strain 82-40 had become serum-sensitive. This new strain, 82-40HP (HP; high-passage), was found to have the same SDS-PAGE and LPS profile as the low-passage strain, (now called 82-40LP) except that it lacked a band migrating at about 97 kDa (Blaser et al., 1987). Re-examination of the Cornell strains showed the same phenomenon; the wild type strain 23D possessed the band and was serum resistant whereas the spontaneous mutant strain (23B) lacked the band and was serum-sensitive. Now we had two sets of strains with which to consider the significance of the band.

RELATION BETWEEN THE HIGH MOLECULAR WEIGHT PROTEINS AND THE SURFACE LAYER

Electron microscopy confirmed that possession of the band indicated presence of an S-layer; for the strains possessing the 97 kDa band the crystalline structure was hexagonal (Fujimoto et al., 1991). Examination of a wide variety of *C. fetus* strains indicated that virtually all possessed high molecular weight bands at 95-98, 127, or 149 kDa, regardless of whether the strains possessed type A or B LPS (Blaser et al., 1987). However, for strains possessing the 127 or 149 kDa bands, the crystalline structure of these S-layers was tetragonal. In addition to finding mutant strains that had lost the ability to produce S-layer proteins, we also found strains that were able to shift the size of the major S-layer protein which was produced. This size shift resulted in both a change in crystalline structure and in antigenicity as revealed by polyclonal and monoclonal antibodies (Fujimoto et al., 1991; Wang et al., 1990). Recent observations confirm that variation of the S-layer proteins occurs in vivo in the bovine vagina (M. Garcia, M.J. Blaser, unpublished) providing the basis for the antigenic variation of *C. fetus* cells that had been previously observed (Corbeil et al., 1975).

BIOLOGICAL PROPERTIES ASSOCIATED WITH THE S-LAYER PROTEINS

Comparison of the wild type S^+ strains and their spontaneous S^- mutants enabled insights into the properties conferred by the S-layer (Table 1).

Table 1. Biological differences associated with the presence of an S-layer on *C. fetus* cells

	S^+ (82-40LP,23D)	S^- (82-40HP,23B)
Log_{10} killing by serum[a]	<0.05	>1.0
Hydropathy of cell surface[b]	hydrophobic	hydrophilic
Lectin binding[c]	0	4+
IP LD_{50} for mice[d]	1.3×10^7	7.6×10^8
Log_{10} bacteremia after oral challenge[e]	4.2	<1.0
C3 binding[f]	0.2	4.0
C5 consumption (%)[g]	5	35
Alternative pathway activation (%)[h]	0-11	92-93
Log_{10} killing by PMN in NHS[i]	<0.05	2.3
Log_{10} killing by PMN in IHS[j]	1.8	ND^k

a   Killing by pooled normal human serum in standard assay for 60 minutes at 37°C (Blaser et al., 1985; 1987).
b   As determined from phase-partition assay (Blaser et al., 1987).
c   Binding of wheat germ, *Helix pomatia*, and *Bandeiraea simplicifolia* agglutinins (Fogg et al., 1990).
d LD_{50} for adult HA-ICR mice after intraperitoneal challenge (Pei and Blaser, 1990).
e   Bacteremia in adult HA-ICR mice 30 minutes after oral challenge (Pei and Blaser, 1990).
f   Binding of ^{125}I-C3 to pelleted bacterial cells after 30 minute incubation (Blaser et al., 1988).
g   Activation of C5 after 30 minute incubation of normal human serum with bacterial cells (Blaser et al., 1988).
h   Activation of the alternative complement pathway after incubation of normal human serum with bacterial cells (Washburn et al., 1991).
i   Killing by human neutrophils in the presence of normal human serum at 120 minutes (Blaser et al., 1988).
j   Killing by human neutrophils in the presence of immune serum at 120 minutes (Blaser et al., 1988).
k   ND, not determined.

The S^+ strains have hydrophobic surfaces (Blaser et al., 1987) that block lectin binding to LPS-polysaccharide side chains (Fogg et al., 1990). Serum resistance is associated with inability of complement component C3b to bind to the surface of S^+ cells even in the presence of immune serum; thus activation of the membrane attack complex (C5-9) does not occur (Blaser et al., 1988). The basis for the inability of C3b to bind to S$^+$ cells is not known, but may be related to the inefficiency with which these cells activate the alternative complement pathway (Washburn et al., 1991). Failure of C3b to bind to the cell surface also explains phagocytosis resistance in that S$^+$cells are not properly opsonized in the absence of immune serum (Blaser et al., 1985; 1988). In vivo, S$^+$ cells are more virulent than S$^-$ cells when injected intraperitoneally in mice (Pei and Blaser, 1990). Chemical removal of the S-layer by treatment of S$^+$ cells with pronase has little effect on cell viability, but the treated cells become susceptible to serum-mediated killing as well as phagocytosis, and virulence of the treated cells is lower after intraperitoneal challenge (Blaser and Pei, 1993). These findings indicate that presence of the S-layer is critical for virulence properties observed in both in vitro and in vivo systems (Pei and Blaser, 1990; Blaser and Pei, 1993). Oral challenge of mice with S$^+$ but not S$^-$ cells results in bacteremia, which in essence recapitulates the phenomenon of ingestion of the wild-type organisms by either sheep or humans (Pei and Blaser, 1990).

CHARACTERISTICS OF THE *C. FETUS* S-LAYER PROTEINS

Purification to homogeneity of the 97, 127, and 149 kDa S-layer proteins from *C. fetus* cells with type A LPS (Fig. 1) indicates that all are highly acidic, contain similar amino acid compositions, are not glycosylated, possess the same N-terminus, and are antigenically cross-reactive using polyclonal antisera (Pei et al., 1988). S-layer proteins from type B LPS strains share the same N-terminal amino acids with one another but differ in some of their amino acids from the type A strains and represent a parallel but separate protein family (Dubreuil et al., 1988; Yang et al., 1992). As

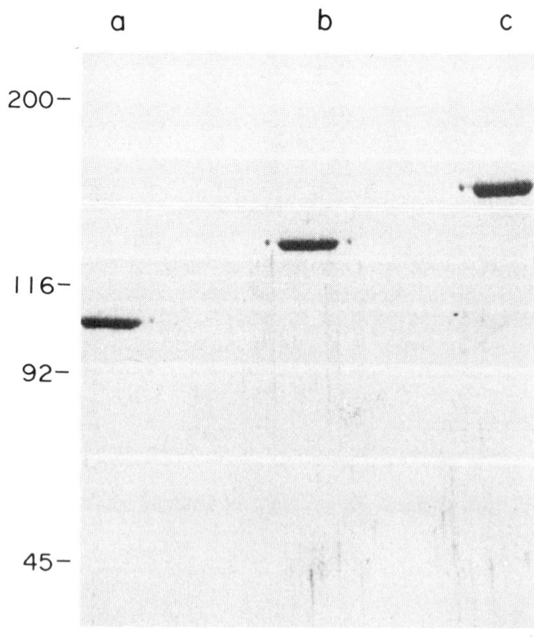

Figure 1. SDS-PAGE of purified S-layer proteins from *C. fetus*. Proteins (kDa) are: 97, lane a; 127, lane b; 149, lane c.

has been shown for other bacteria possessing S-layer proteins, extraction of *C. fetus* cells in hypotonic solutions permits removal of the S-layer, and addition of divalent, but not monovalent or trivalent, cations retards this process (Fogg et al., 1990; Yang et al., 1992). Conversely, type A S-layer proteins may be reattached to S⁻ strains possessing type A LPS but not to S⁻ strains possessing type B LPS (Yang et al., 1992). Type B S-layer proteins show a parallel specificity in reattachment. These interactions require divalent cations, and the ligand for each is the specific LPS molecule (Yang et al., 1992). Among a group of monoclonal antibodies raised to the 97 kDa S-layer protein purified from a strain possessing type A LPS, one class recognized all S-layer proteins present on type A strains regardless of their size, one recognized only 96-98 kDa proteins regardless of whether they were present on type A or B strains, and one group recognized only proteins of 96-98 kDa present on type A strains (Wang et al., 1990). Thus, S-layer proteins possess a diversity of epitopes, some of which are specific to molecular size and others to protein family. Elucidating the properties of and differences between the *C. fetus* S-layer proteins (Table 2) suggests that knowledge of the genes encoding these proteins would be of interest.

Table 2. Putative functional domains on *C. fetus* S-layer proteins

Complement interactive domain	Export signalling domain
Crystal-(multimer-) forming domain	Antigenic variability sites
Lipopolysaccharide binding site	Procoagulant domain (?)
Divalent cation binding site(s)	

STRUCTURE OF THE *sapA* GENE

We constructed a library in *Escherichia coli* of chromosomal fragments from wild-type strain 23D using λgt11 (Blaser and Gotschlich, 1990; Tummuru and Blaser, 1992) and the *sapA* gene, containing a 2799 bp open reading frame (ORF) encoding a protein of 96,758 Da, was cloned. The deduced sequence in the ORF begins with the same amino acids as those found experimentally in the mature protein, indicating that *C. fetus* secretes this S-layer protein (and two others for which the sequence has been determined; M.K. Tummuru, J. Dworkin, M.J. Blaser, unpublished) without the benefit of a cleaved signal sequence. An 18-residue stretch near the C-terminus (amino acids 672-689) has the characteristics of a membrane spanning structure, suggesting that it may play a role in export. An inverted repeat, strongly suggestive of a transcription terminator, is present immediately downstream of the ORF. Approximately 130 bp upstream of the ORF there is a putative sigma 70 promoter, with the mRNA +1 site several bases later, and a ribosomal binding site 6 bases before the ORF (Blaser and Gotschlich, 1990; Tummuru and Blaser, 1992). This arrangement suggests that the *sapA* operon contains only a single gene. Northern analysis indicates that the *sapA* transcript is approximately 2.8 kb (Tummuru and Blaser, 1992), which is consistent with this hypothesis.

CONSERVATION OF *sapA* HOMOLOGS

Considering the multiplicity of S-layer proteins that *C. fetus* can produce, an important question is whether there are other genes homologous to *sapA*. In Southern analysis of *Hin*dIII-restricted chromosomal DNA from two wild-type *C. fetus* strains (23D and 82-40LP), using a probe corresponding to the N-terminal region of the 97 kDa S-layer protein, we found 8 hybridizing bands for each, indicating that there are multiple *sapA* homologs (Tummuru and Blaser, 1992). Interestingly, for each of the spontaneous mutant strains (23B and 82-40HP), there were only 5 bands that hybridized with the probe. Thus, each of these mutants contains *sapA* homologs but there has clearly been deletion or rearrangement of the genome. In another Southern analysis, using a probe representing the putative promoter region for *sapA*, the two wild-type strains each had a single hybridizing band whereas none was present for the two mutants (Tummuru and Blaser, 1992). This result indicates that the *sapA* promoter region was deleted from the two strains, and Northern analysis did not reveal a transcript for either mutant. We now know that *C. fetus* strain 23B possesses a full-length *sapA* homolog (*sapA*1) which can express an immunoreactive and functional S-layer protein when cloned in an *E. coli* background (M.K. Tummuru, and M.J. Blaser, unpublished). These data indicate that the mutant has full-length *sapA* homologs but failure to express an S-layer protein is due to absence of a transcript, possibly because of deletion of a required promoter sequence.

Finally, Southern hybridization with probes proceeding toward the 3' end of the *sapA* ORF indicated that *sapA* contains sequences that are only present in single copy (Tummuru and Blaser, 1992). The hybridization data indicate that the *sapA* homologs share conserved regions at the 5' end but that sequences diverge by the middle of the ORF. Studies to exactly define the conserved and divergent regions are underway. It is presumed that the conserved regions encode important structures determined by the physiochemical constraints under which this molecule operates, whereas the divergent regions may be the basis for the antigenic variability observed.

CONCLUSIONS

The S-layer proteins of *C. fetus* are critical for virulence of these significant veterinary and human pathogens, and therefore their study is important. In addition, the *C. fetus* system has great utility for analysis of structure-function relationships pertinent to other S-layer proteins because of the: (i) diversity of S-layer protein sizes, (ii) presence of the two parallel families of proteins in the type A and type B strains, (iii) variation in paracrystalline (tetragonal and hexagonal) structure, (iv) flexibility with regard to divalent cation requirements for binding, (v) presence of conserved domains, and (vi) presence of a mechanism for generating appropriate diversity. The availability of cloned genes, mutant strains and a system for mutagenesis, monoclonal antibodies, and a suitable animal model to examine the fitness of engineered adaptations will facilitate analysis of a number of important biological and biophysical questions.

ACKNOWLEDGMENTS

The author thanks P.F. Smith, G.I. Pérez-Pérez, J. Hopkins, J.E. Repine, K.A. Joiner, Z. Pei, R.T. Ellison, III, B. Dunn, E. Gotschlich, G.C. Fogg, L. Yang, E.

Wang, S. Fujimoto, M. Tummuru, A. Labigne, and R. Washburn for their many contributions to the work presented. This research was supported in part by the Medical Research Service of the Department of Veterans Affairs, the U.S. Army Medical Research and Development Command, and the National Institutes of Health (AI-24145).

REFERENCES

Blaser, M.J., and Gotschlich, E.C., 1990, Surface array protein of *Campylobacter fetus*: cloning and gene structure, *J. Biol. Chem.*, 265:14529.

Blaser, M.,J., and Pei, Z, 1993, Pathogenesis of *Campylobacter fetus* infections, Critical role of high molecular weight S-layer proteins in virulence, *J. Infect. Dis.*, in press.

Blaser, M.J., and Reller, L.B., 1981, *Campylobacter* enteritis, *New England J. Med.* 305:1444.

Blaser, M.J., Smith, P.F., and Kohler, P.A., 1985, Susceptibility of *Campylobacter* isolates to the bactericidal activity in human serum, *J. Infect. Dis.* 151:227.

Blaser, M.J., Smith, P.F., Hopkins, J.A., Bryner, J., Heinzer, I., and Wang, W.L.L., 1987, Pathogenesis of *Campylobacter fetus* infections. Serum-resistance associated with high molecular weight surface proteins, *J. Infect. Dis.* 155:696.

Blaser, M.J., Smith, P.A., Repine, J.E., and Joiner, K.A., 1988, Pathogenesis of *Campylobacter fetus* infections, Failure to bind C3b explains serum and phagocytosis resistance, *J. Clin. Invest.* 81:1434.

Corbeil, L.B., Schurig, G.D., Pier, P.J., and Winter, A.J., 1975, Bovine venereal vibriosis: antigenic variation of the bacterium during infection, *Infect. Immun.* 11:240.

Dubreuil, J.,D., Logan, S.M., Cubbage, S., NiEidhin, D., McCubbin, W.D., Kay, C.M., Beveridge, T.J, Ferris, F.G., and Trust, T.J., 1988, Structural and biochemical analyses of a surface array protein of *Campylobacter fetus*, *J. Bacteriol.* 170:4165.

Fogg, G.C., Yang, L., Wang, E., and Blaser, M.J., 1990, Surface-array proteins of *Campylobacter fetus* block lectin-mediated binding to type A lipopolysaccharide, *Infect. Immun.* 89:464.

Fujimoto, S., Takade, A., Amako, K., and Blaser, M.J., 1991, Correlation between molecular size of surface-array protein and both morphology and antigenicity in the *Campylobacter fetus* S-layer, *Infect. Immun.* 59:2017.

McCoy, E.C., Doyle, D., Burda, .K, Corbeil, L.B., Winter, A.J.,1975, Superficial antigens of *Campylobacter (Vibrio) fetus*: characterization of antiphagocytic component, *Infect. Immun.* 11:517.

Pei, Z., and Blaser, M.J., 1990, Pathogenesis of *Campylobacter fetus* infections. Role of surface array proteins in virulence in a mouse model, *J. Clin. Invest.* 85:1036.

Pei, Z., Ellison, R.T., Lewis, R.V., and Blaser, M.J., 1988, Purification and characterization of a family of high molecular weight surface-array proteins from *Campylobacter fetus*, *J. Biol. Chem.* 263:6416.

Pérez-Pérez, G.I., and Blaser, M.J., 1985, Lipopolysaccharide characteristics of pathogenic campylobacters, *Infect. Immun.* 47:353.

Pérez-Pérez, G.I., Blaser, M.J., and Bryner, J., 1986, Lipopolysaccharide structures of *Campylobacter fetus* related to heat-stable serogroups, *Infect. Immun.* 51:209.

Smibert, R.M., 1984, Genus *Campylobacter*, *in*: "Bergey's Manual of Systematic Bacteriology. Vol. I",N.R. Krieg and H.G. Holt, eds., Williams & Wilkins, Baltimore.

Tummuru, M.K.R., and Blaser, M.J., 1992, Characterization of the *Campylobacter fetus sapA* promoter: evidence that the *sapA* promoter is deleted in spontaneous mutant strains, *J. Bacteriol.* 174:5916.

Wang, W.L.L., and Blaser, M.J., 1986, Detection of pathogenic *Campylobacter* species in blood culture systems. J. Clin. Microbiol. 23:709.

Wang, E., Pei, Z., and Blaser, M.J., 1990, Antigenic shift in surface array proteins of *Campylobacter fetus*. Proc. Amer. Soc. Microbiol., B-233, Anaheim, CA.

Washburn, R.G., Julian, N.C., Wang, E., and Blaser, M.J.,1991, Inhibition of complement by surface array proteins from *Campylobacter fetus* subsp. *fetus*, Proc. 31st Intersci. Conf. Antimicrob. Agents Chemotherapy, Chicago, IL.

Winter, A.J., McCoy, E.C., Fullmer, C.S., Burda, K., and Bier, P.J., 1978, Microcapsule of *Campylobacter fetus*: chemical and physical characterization, *Infect. Immun.* 22:963.

Yang, L., Pei, Z., Fujimoto, S., and Blaser, M.J., 1992, Reattachment of surface array proteins to *Campylobacter fetus* cells, *J. Bacteriol.* 174:1258.

Chapter 18

DEFINITION OF FORM AND FUNCTION FOR THE S-LAYER OF *CAULOBACTER CRESCENTUS*

Wade H. Bingle, Stephen G. Walker, and John Smit

Department of Microbiology
University of British Columbia
Vancouver, British Columbia, Canada

INTRODUCTION

Caulobacters are often cited as examples of the diverse group of microorganisms categorized as the prosthecate bacteria; bacteria that produce appendages that are an extension of the cell membrane and wall. As is true for other prosthecate bacteria, caulobacters are found widely in the open environment, inhabiting lakes, streams, soils and the oceans (Poindexter, 1981). Because of their abundance or at least frequent appearance in locales with limited amounts of available nutrients, they are often used as the example of an oligotrophic bacterium.

The notoriety accorded caulobacters, however, stems as much from the use of *Caulobacter crescentus* in laboratory research as a model for cell development and differentiation as it does from interest in their role or niche in the environment. Caulobacters exhibit a biphasic life style, alternating between a stalked cell and a non-stalked dispersal phase cell motile by means of a single flagellum. Thus stalked cells (normally attached to a surface in the environment) produce swarmer cells which later differentiate to stalked cells. The observation that the time of appearance and the precision of localization of the structures involved with the developmental progression (the stalk, flagellum, polar pili and other surface markers) is remarkably reproducible and precise, has led to a whole field of research endeavoring to determine the biochemical and genetic basis for this progression (Shapiro, 1985; Newton, 1989).

One consequence of using this bacterium as a model for cell development studies is that a considerable amount of information on the genetics of *C. crescentus* and the development of molecular genetics techniques, as applied to this species, has accumulated. Thus, as we have considered investigation of surface structures on caulobacters and the role of such structures in their native environment, we have an ability to characterize and manipulate the genetics of these structures to an extent equal to or greater than that possible for any other bacterium found in aquatic or soil environments. This impacts not only on our capabilities for analysis, but also on the possibilities for development of biotechnological applications of caulobacters.

Caulobacters adhere to surfaces using an adhesive organelle found at the tip of the cellular stalk, the holdfast. This adhesion capability sugests that caulobacters are an important component of microbial biofilm or biofouling communities in the environment. We have committed some of our efforts to define the nature and genetic control of this adhesive organelle, which, parenthetically, is another of the developmentally regulated structures of caulobacters (Merker and Smit, 1988; Mitchell and Smit, 1990; Ong et al., 1990; Kurtz and Smit, 1992). This adhesiveness enables caulobacters to spontaneously form fixed-cell biofilms and leads to one major avenue of development of biotechnological applications for caulobacters.

The cell surface of *C. crescentus* is completely covered with an S-layer, which we noted as early as 1976 (J. Smit, unpublished observations). A comparable S-layer is seen on most isolates of caulobacters collected by our laboratory and as we detail below, most S-layers show numerous similarities to that of *C. crescentus*. Because of the ubiquity of S-layer-containing caulobacters in the environment as well as the possibility for biotechnology applications, we have considered the S-layer in the intervening years; analyzing its form, functional regions, aspects of S-layer biogenesis, genetic control and its role for the bacterium in its natural setting.

THE *CAULOBACTER* S-LAYER STRUCTURE AND BIOGENESIS

The *C. crescentus* S-layer is a hexagonal lattice that forms on the surface of the outer membrane of this gram-negative eubacterium. The most prominent feature in electron micrographs of negatively-stained patches of S-layer material is a circular structure, arranged at 22 nm intervals (Fig. 2), which is itself composed of six subunits. In side-view the structures form an upright structure sitting on the outer membrane surface. The connections between the circular structures has always been difficult to discern in negatively stained images, but the existence of rigid connections

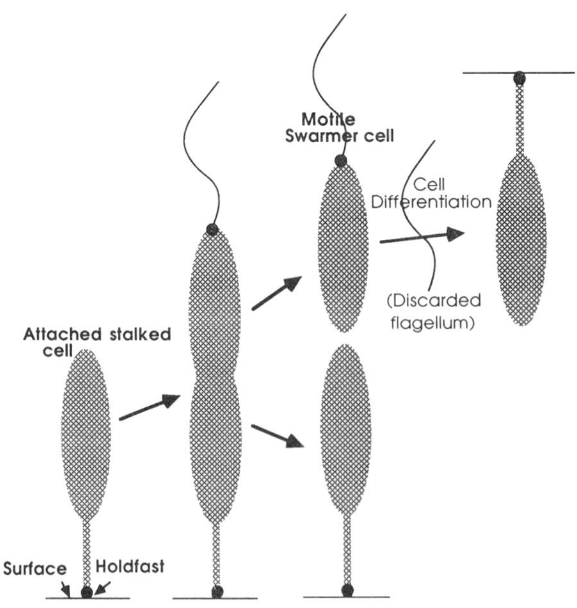

Figure 1. Schematic diagram of the *Caulobacter* life cycle.

has been assumed since optical diffractograms always show a high degree of structural regularity and resolution.

Recently, a three dimensional reconstruction of the *C. crescentus* S-layer has been produced by W. Baumeister and collaborators (Smit et al., 1992). Using preparations of S-layer material that are occasionally shed by some strains in high density cultures (Smit et al., 1981), a structure with resolution of 2 nm has been generated. In these, connections are now readily discerned and the linker "arms" enter the circular structure about midway at its vertical height.

The protein that comprises the S-layer (termed RsaA) can be isolated with considerable purity in a single step process by treatment of cells with either low pH (pH 2.0) or ethyleneglycol-bis(ß-aminoethylether)tetraacetic acid (EGTA). The single protein that results migrates at about 100 kDa on sodium dodecyl sulfate-polyacrylamide gels (SDS-PAGE). Although not rigorously examined, there is no indication that the protein is glycosylated; Schiff staining of SDS-PAGE gels gives a negative reaction and the expression product of the cloned gene (*rsaA*) in *Escherichia coli* gives rise to a band that co-migrates precisely with *Caulobacter*-derived protein.

Studies with the cloned gene (see below) have shown that the protein is expressed from a single copy of the gene which is transcribed at relatively high efficiency from a longlived mRNA (Smit and Agabian, 1984; Fisher et al., 1988). There is little indication of developmental regulation during the cell cycle. Since individual cells from a culture are found to be completely covered by S-layer (including the stalk) the question has arisen as to how the S-layer is enlarged on a growing cell and whether the stalk (which is produced during a short interval in the cell cycle) is immediately covered by S-layer. Using a combination of antibody/colloidal gold labeling and cell synchronization techniques, we learned that stalk growth occurs at the precise juncture of the cell body and the stalk and that the S-layer is indeed immediately applied during growth (Smit and Agabian, 1982). In perhaps an analogous fashion, the site of the incipient division is also a region of new S-layer production. In contrast, the main cell body appeared to be covered becaue of the diffuse intercalation of new subunits. The experimental procedure allowed only a low resolution view of this process, since it involved labeling cells with S-layer specific antibody and monitoring the development of unlabeled regions over time. Another and perhaps better means to monitor the transport of new subunits to the cell surface is best done by attaching a regulated promoter to *rsaA* and monitoring protein appearance after promoter induction. So far we have not yet found a regulated promoter that is suitably repressed in the uninduced condition in *C. crescentus* to allow this experiment; it remains an important future goal. We are particularly interested in learning whether the S-layer monomer arrives at the surface via fusion regions (i.e., Bayer junctions) between the inner and outer membrane, or via some other route.

S-LAYER ASSEMBLY AND INTERACTION WITH THE OUTER MEMBRANE

In our search for mechanisms of S-layer surface attachment and clues to the nature of the crystallization process we have gradually learned the importance of calcium ions for both processes.

The S-layer can be efficiently extracted from the cell surface using 10 mM EGTA or by adjusting the pH to 2.0 (treatments that interrupt salt-bridging). The extracted S-layer protein will recrystallize in the presence of calcium but not magnesium or strontium (Walker et al., 1992).

Calcium is required by *C. crescentus* for viability, a relatively rare occurrence among bacteria. We isolated "calcium-independent" mutants which grow at normal rates in the absence of calcium, no longer attach the S-layer and fail to produce a cell surface molecule termed the S-layer associated oligosaccharide (SAO). We believe SAO mediates the calcium specific attachment to the outer membrane of the S-layer. (For a more detailed account see S.G. Walker and J. Smit, this book).   When grown on calcium-containing solid medium the shed RsaA protein of the mutants is still capable of subunit-subunit crystallization, producing large sheets of highly organized S-layer (Smit et al., 1992) (Fig. 2). Image analysis of such crystalline patches indicates that they are composed of mirror-image double layers.

Single-layer sheets, with attached outer membrane material, can be isolated from high density cultures of wild-type cells (Smit et al., 1981) and double-layer sheets form spontaneously from this material when detergent is used to solublize the membrane material (Smit et al., 1992). We hypothesize in both cases of double layer formation that bridging of two S-layers is accomplished by divalent Ca^{2+}, in the absence of a suitable outer membrane surface.

When colonies of calcium-independent mutants are carefully embedded and examined by thin-section electron microscopy, we see that the double-layer sheets accumulate into stacks, often at the upper surface of the colony. Colonies of calcium-independent mutants grown in calcium-free medium do not exhibit S-layer sheet production. Instead, an amorphous layer is detected on the cell surfaces (Smit et al., 1992).

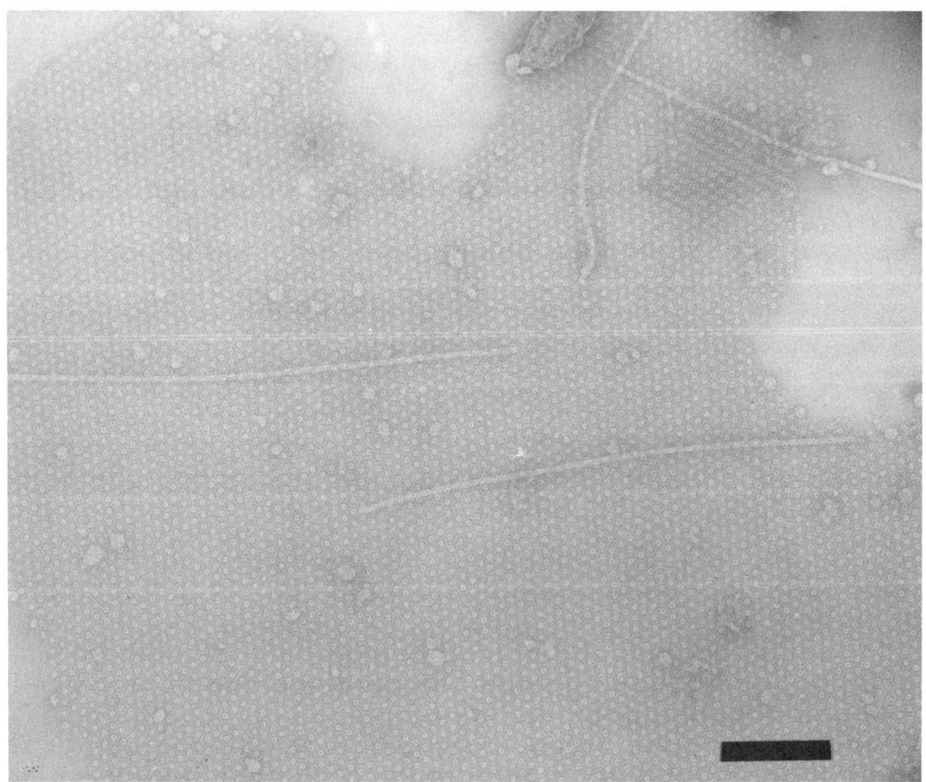

Figure 2. A portion of a patch of shed S-layer produced by a calcium-independent mutant of *C. crescentus*. This image is typical of the large patches seen in gently suspended colonies of these "shedding" mutants. Image analysis indicates that this is a mirror-image double layer. The bar indicates 0.2 μm.

The importance of calcium in subunit-subunit interactions was reinforced by experiments showing that treatment of the shed double layers, produced by calcium-independent mutants, with 800 μM EGTA resulted in complete disassembly. The same concentration of ethylene diaminetetraacetic acid (EDTA) has no effect (J. Smit, unpublished observations) presumably because EGTA is more selective in the chelation of calcium ions than EDTA.

Finally, there are indications that the RsaA protein contains calcium binding regions (see next section). Taken together, these observations indicate a specific role for calcium in the attachment of the S-layer to the cell surface as well as in subunit-subunit interactions.

As a postscript to the studies with calcium-independent mutants, it appears that the shed double layers are displayed at higher resolution than the standard preparations from wild-type cells, in part due to the absence of attendant attached membrane material, and they crystallize into much larger array patches. With specialized negative stain techniques designed to minimize the distortion effects of drying, a three dimensional reconstruction at a resolution of about 1 nm seems feasible.

S-LAYER GENE CLONING AND THE SEARCH FOR FUNCTIONAL REGIONS

The gene for the S-layer protein (*rsaA*) was cloned initially from a library of DNA fragments in lambda phage by antibody detection of protein produced in *E. coli* during a lytic cycle of the phage (Smit and Agabian, 1984). Subcloning and restriction mapping localized the *rsaA* gene to a 4.4 kb *Hind*III/*Bam*HI DNA fragment and indicated that the gene was present as a single copy in the chromosome. More recently, pulsed field gel electrophoretic analysis has placed the *rsaA* gene on a physical map of the *C. crescentus* chromosome (Ely and Gerardot, 1988).

We have also been able to easily manipulate the *rsaA* gene in both *C. crescentus* and *E. coli*. Complementation experiments have shown that when the 4.4 kb *Hind*III/*Bam*HI fragment carrying *rsaA* (and its native promoter) is cloned into a broad-host range vector and introduced into an S-layer negative mutant of *C. crescentus*, the cells synthesize an S-layer protein which assembles on the cell surface (Edwards and Smit, 1991).

The entire 4.4 kb *Hind*III/*Bam*HI fragment carrying the *rsaA* gene has been sequenced (Fisher et al., 1988; Gilchrist et al., 1992) and a centrally-located open reading frame (ORF) encoding a polypeptide of 1026 amino acids (98 kDa) was identified. This ORF extends for approximately 3.1 kb and is followed by a palindromic sequence which resembles a transcription terminator; the presence of a transcription terminator was anticipated based on earlier Northern and S1 nuclease analysis which indicated that the RsaA mRNA was not polycistronic. It is clear that the 3.1 kb ORF encodes RsaA because N-terminal amino acid sequencing of RsaA indicates perfect alignment to the predicted beginning of the protein, the predicted amino acid composition agrees well with the actual amino acid composition of RsaA, the third position codon bias is consistent with other sequenced genes of *C. crescentus* and several peptide fragments prepared from RsaA align with the predicted amino acid sequence.

Analysis of the amino acid composition of RsaA revealed that 60% of the protein is composed of alanine, glycine, threonine and serine and like most other S-layer proteins, no cysteines are present. Although acidic residues are not abundant, the low amount of balancing basic residues yields a protein with predicted acidic pI of 3.46.

Comparison of the amino acid sequence with other S-layer proteins has revealed some homology, but not especially significant in our view. Interestingly, the closest match is with a *Campylobacter fetus* S-layer protein, the only other gram-negative organism displaying a hexagonal S-layer for which a sequence is available.

An additional discovery has been of a region of the RsaA protein containing aspartate and glycine residues which shows homology with a region found in other calcium-stabilized proteins, including secreted proteins of gram-negative bacteria such as hemolysins; these regions are believed to serve as calcium binding motifs. In the case of the RsaA protein, the motif is a nine amino acid repeat, centered with an aspartic acid residue, and is repeated at least 4 times, compared to 3-16 times in other calcium-requiring proteins. Because of the apparent involvement of calcium in attachment and assembly, these aspartate/glycine repeats may serve a similar calcium binding function. One focus of current attention is to determine directly whether the region serves as the RsaA component of the surface attachment mechanism or subunit-subunit interactions or both.

The site of transcription initiation for *rsaA* has been determined by S1 nuclease mapping of *C. crescentus* RNA (Fisher et al., 1988); it lies approximately 60 bp upstream of the initiation codon. Within this region lies a probable Shine Dalgarno (SD) sequence centered about 12 bp upstream of the initiation codon (Fig. 3). Upstream of the transcription start site at -10 and -35 bp lies a sequence showing significant homology to the consensus promoter sequence of *E. coli* (Fig. 3), however S1 nuclease analysis and expression studies have indicated the promoter is not active in *E. coli*. Because the *rsaA* promoter is not similar to developmentally regulated promoters of *C. crescentus*, it was not surprising to find that the promoter is not developmentally regulated (Fisher et al., 1988).

In an effort to confirm that the putative promoter/SD regions of the *rsaA* gene were indeed active in transcriptional and translational initiation, we isolated the 135 bp *Msp*I/*Hae*III fragment expected to contain this information and linked it to a cellulase-encoding reporter gene (Bingle et al., 1992). This construction yields high levels of intracellular cellulase activity in *C. crescentus* when carried on a low copy number plasmid (Fig. 3). Indeed, the level of cellulase activity found is comparable to *lacZ*-directed expression of a similar construction carried on a pUC plasmid in *E.*

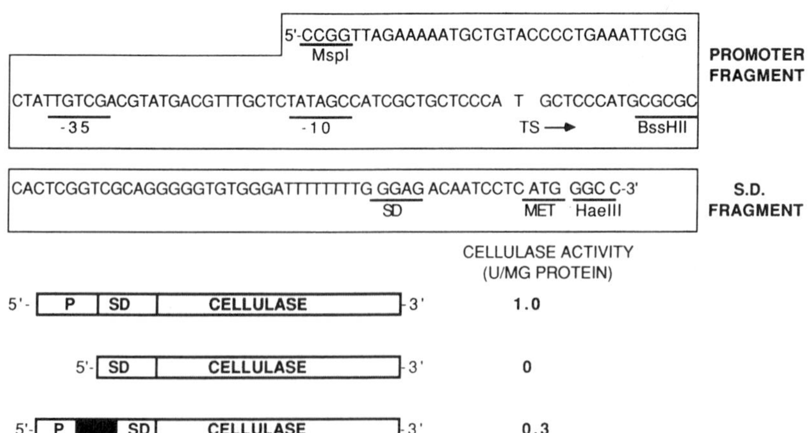

Figure 3. The transcription/translation initiation region of the *rsaA* gene and analysis using a cellulase-encoding reporter gene.

coli (Bingle and Smit, 1990). Considering the plasmid copy number differences between the two systems, it is clear that the 135 bp *MspI/Hae*III fragment directs high levels of cellulase synthesis in *C. crescentus*. Using conveniently placed restriction sites (Fig. 3), we examined this region of the *rsaA* gene further. Not surprisingly, deletion of the putative promoter region (leaving the SD region intact) abolishes cellulase activity. However, insertion of an additional 100 bp of DNA between the putative promoter and SD sequences reduces cellulase levels by 70%, suggesting that the high level of RsaA synthesis is partly linked to the 5'-untranslated region of the *rsaA* mRNA. This region may be important for mRNA stability; the *rsaA* mRNA is known to possess a relatively long half-life (Fisher et al., 1988).

Using the information from the studies outlined above, we have constructed a number of high level expression vectors for *C. crescentus*, incorporating either the *rsaA* transcription initiation region alone or the same region in conjunction with the translation initiation region (Bingle and Smit, 1990). These vectors are useful for general heterologous gene expression needs in a research field where there are few promoters available from non-regulated "housekeeping" genes since emphasis is placed on the characterization of developmentally-regulated promoters. The vectors have been particularly useful to us for expressing promoterless versions of the *rsaA* gene carrying linker insertions (see below).

RsaA is essentially a secreted protein for *C. crescentus*, quite possibly the only protein of the species that is secreted (other than the flagellum and pili). It was interesting to learn that the secretion process occurs without any of the secretion-related sequences or processing events associated with secreted proteins of other species. There is no cleaved signal leader peptide on RsaA; only the terminal methionine is cleaved at some point following translation. There is also no post-translational C-terminal processing; amino acid sequencing of a number of peptides resulting from a proteolytic digest of the purified, mature RsaA fortuitously identified a fragment that extends from a predicted protease cleavage site to the site predicted as the C-terminus by DNA sequencing (Gilchrist et al., 1992). Although there are other secreted proteins that have a similar lack of processing (notably, the hemolysins of some *E. coli* strains), there is no strong homology to RsaA. In effect, we are apparently presented with a new class of secreted proteins and so we are only able to search for secretion-related signals in a general fashion.

Although the N-terminal region of the RsaA protein bears no striking resemblance to signal leader regions of other proteins, it is somewhat hydrophobic. For this reason we suggested that it may have a role in export/secretion (Fisher et al., 1988). In order to investigate this possibility we created, using gene fusion technology, hybrid proteins consisting of various stretches of the RsaA N-terminus followed by C-terminal domains derived from two export-competent reporter molecules, a cellulase and an alkaline phosphatase (W.H. Bingle and J. Smit, unpublished). The cellulase reporter was used because it is active both intra- and extracellularly, allowing an evaluation of the efficiency of export or secretion which is not easily done using alkaline phosphatase. Preliminary data has indicated that fusion of the first 35 to 52 N-terminal amino acids of RsaA to either reporter results in their export to the periplasm; the efficiency of export appears to be on the order of 10%. This data suggests that the N-terminus could function in secretion of RsaA to the cell surface. Indeed when this region of the N-terminus is deleted from the native RsaA protein, secretion of the protein to the cell surface is abolished. The protein does not appear to accumulate intracellularly however, suggesting it is either degraded or its production is regulated in some manner. Other support for the importance of the N-terminus in RsaA secretion has been derived from linker insertion mutagenesis (described below); to date, we have been unable to introduce any additional amino acids insertions within the first 50 N-terminal amino acids and still achieve secretion of RsaA to the cell surface. Other regions of the RsaA molecule have tolerated such insertions with no effect on secretion of the molecule.

FUNCTIONS FOR THE S-LAYER

As is the case for many other S-layers, prime functions are difficult to ascertain (Smit, 1986; Messner and Sleytr, 1992). However, at least one function for the *C. crescentus* S-layer appears to be protection of the cell from environmental factors. Koval and Hynes (1991) have clearly shown that a *Bdellovibrio*-like organism, parasitic for *C. crescentus*, is prevented from infecting by the presence of an S-layer. This observation also suggests that attack from some bacteriophage may be prevented by the S-layer. We found, however, that phage ØCR30 (a transducing phage for *C. crescentus*) uses the S-layer as a receptor (Edwards and Smit, 1991); S-layer minus mutants are resistant to the phage.

The three dimensional reconstruction analysis has indicated that no hole or crevice in the S-layer structure appears to be larger than 2.5-3.5 nm, suggesting a porosity that would permit the entrance of molecules no larger than about 17 kDa. This suggests that most lytic enzymes would be excluded; it may be interesting to examine the sensitivity of *C. crescentus* with and without an S-layer to the digestive efforts of an aggressive lytic enzyme-producing bacterium such as *Myxococcus*. In any case, a direct determination of the porosity of the assembled S-layer could be attempted using techniques developed by Sleytr and Sára (1986).

We have also determined that the S-layer served as an impediment to the introduction of plasmids by electroporation, reducing efficiency by about 10-fold (Gilchrist and Smit, 1991). Whether this is suggestive of a degree of screening of foreign DNA prior to transformation or conjugal events in natural settings can only be speculated upon.

EXAMINATION OF S-LAYERS IN OTHER CAULOBACTERS

The molecular and biochemical studies of the *Caulobacter* spp. S-layer have centered on the protein produced by *C. crescentus* NA1000 (previously named CB15A). We were interested in the degree of conservation of S-layer proteins among a sizable group of independent caulobacter isolates, in part to learn something of the cohesiveness of the caulobacter group and also to determine if the hexagonal S-layers, which all appeared quite similar by electron microscopy, were in fact ancestrally related. Colony lift hybridization experiments, using the NA1000-derived *rsaA* gene as a probe and under moderate stringency conditions, could differentiate between caulobacter and non-caulobacter isolates (MacRae and Smit, 1991). This suggested that the S-layer gene may be a group-specific character. However, Southern blot analysis of the environmental caulobacter isolates indicated that the S-layer genes and their various genomic contexts were not conserved adequately to permit the clustering of isolates by restriction fragment length polymorphism analysis.

Walker et al. (1992) examined the same *Caulobacter* spp. environmental isolates using biochemical and immunological methods. Strains that gave a positive response in the colony lift hybridization studies were subjected to the low pH or EGTA extraction procedure used to extract RsaA from *C. crescentus* NA1000. SDS-PAGE analysis showed that in most cases a single major protein species was produced and the M_r of the extracted proteins ranged from 100 to 193. Western blot analysis, using antisera raised against the NA1000 S-layer protein, gave a positive signal for all these proteins; we presume them to be the S-layer protein for these various strains. All but one of the environmental isolates that produced an S-layer also produced an SAO-like molecule in that the antiserum that recognizes the NA1000 SAO showed various degrees of recognition to a similar molecule by Western blot analysis.

Appended to these approaches is a recent study of the comparison of 16S rRNA sequences among caulobacters, both marine and freshwater (Stahl et al., 1992). In it we determined that the entire caulobacter group appears to be a diverse assemblage with an early evolutionary divergence of marine and freshwater caulobacters. Of relevance to the present topic, the most common type of freshwater isolate we encountered (see MacRae and Smit, 1991) has an S-layer and belongs to the same discrete subgroup as *C. crescentus* strains. Those freshwater strains without an S-layer (a relatively rare occurrence) were significantly divergent from these typical S-layer-producing strains.

Overall, the several studies indicate that the most common caulobacters in freshwater environments form a cohesive subgroup of bacteria that produce an S-layer. In turn, the S-layer proteins show a detectable degree of conservation as judged by the two dimensional crystal structure, immunological crossreactivity, gene cross-hybridization and, apparently, the general mechanism by which outer membrane attachment is mediated. We hypothesize that the limits to the degree of structural conservation are due to common mechanisms of assembly, surface attachment and export. Portions of the protein that are not involved in these functions may have evolved independently, resulting in a considerable degree of M_r heterogeneity between isolates.

DEVELOPMENT OF BIOTECHNOLOGICAL APPLICATIONS FOR THE S-LAYER

One of the major projects in this laboratory over the last couple of years has been to develop the molecular genetic information to do an incisive analysis of the functional regions of the *C. crescentus* S-layer protein. That is, what regions of the protein are important for secretion, attachment to the cell surface and self-assembly process. From this approach comes a corollary question which has a number of practical sequelae. Are there regions of the molecule that are not involved with any of the primary functions and can these regions be exploited to expose useful activities (e.g., enzyme activities, vaccine epitopes on other binding activities) of practical value on the surface of this group of bacteria?

S-layer proteins are ideal carrier proteins to expose such activities on the cell surface. S-layer proteins typically constitute 5-10% of cell protein (Smit, 1986) and little, if any, of the protein is embedded in the outer membrane. Thus S-layer proteins present a large surface area to the environment and this should provide numerous surface exposed sites for insertion of various activities. The S-layer protein of *C. crescentus* is one of the larger S-layer proteins (98 kDa) which should serve to increase the likelihood of identifying such sites. In the case of vaccine epitope insertions, the geometrical arrangement S-layer proteins may be a distinct advantage over other carrier proteins (see the chapter by Malcolm et al.). Since S-layer proteins form a two dimensional array, it is difficult to imagine that an epitope could be packed at a higher density. And, because S-layers are repetitive structures, any foreign epitope will be presented to the immune system as a repeating unit. There is some evidence that this is advantageous for immunogenicity (Feldman and Basten, 1971).

We have identified potential sites for foreign epitope insertion into the *C. crescentus* S-layer protein using a modified random linker mutagenesis method developed in this laboratory that allows selection for linker insertion events (Bingle and Smit, 1991; Bingle et al., this book). Classical linker mutagenesis involves inserting a double-stranded oligonucleotide encoding for a unique restriction site in-frame into a plasmid-borne target gene using sites created at random by partial digestion with an endonuclease. Because the inserted oligonucleotides are usually in the order of 12 bp in length, major perturbations to the resulting protein are avoided. In fact, many mutant genes express protein products indistinguishable from wild-type. Such insertions pin-point so-called "permissive" sites, sites in the protein which can tolerate insertions of extra amino acids without loss of most (or any) biological properties and therefore without major changes in cellular location and structure. These sites then become the candidates for the insertion of larger peptides.

Using the methodology outlined above we have examined the effect of such linker insertions on the secretion of the *C. crescentus* S-layer protein. Of the 240 linker insertions examined so far, 55 have permitted secretion of the S-layer protein to the cell surface. Excepting the N-terminus and a region encompassing amino acids 650-750, these permissive insertions map throughout the RsaA protein (Fig. 4), suggesting there may indeed be many regions of the S-layer protein capable of accepting additional amino acids without deleterious effects on S-layer secretion. We are currently examining the effect of these insertions on the assembly of the S-layer but preliminary results suggest that most will not produce any gross distortion of S-

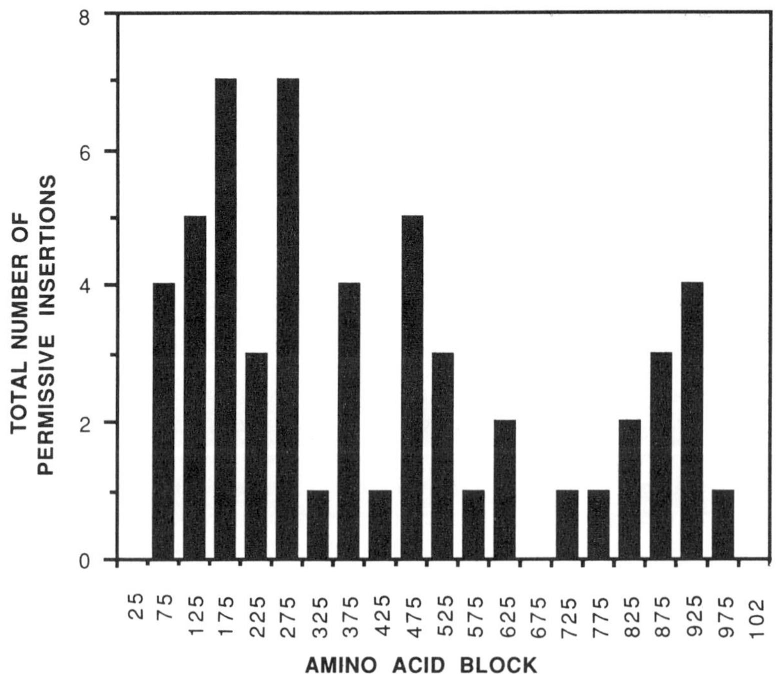

Figure 4. Regions of the *C. crescentus* S-layer protein containing permissive sites. The protein has been arbitrarily divided into blocks of 50 amino acids and the number of linker insertions found in each block is indicated.

190

layer assembly. At the moment we have successfully introduced a metal-binding peptide into the S-layer in at least one of the permissive sites are currently investigating the resulting metal binding capabilities.

We have viewed the *C. crescentus* S-layer gene as appropriate for the expression of additional activities or epitopes from at least two perspectives. First, S-layers in general are likely to be at least as good a vehicle and probably better in many instances to comparable insertion attempts of others involving outer membrane or flagellar proteins of other bacteria. Moreover, the extent of genetics development for *C. crescentus* make its S-layer particularly attractive. Second and more important, the underlying characteristics of *C. crescentus* make its S-layer uniquely suitable for several applications involving insertions of additional activities into an S-layer. The ability of caulobacters to form a monolayer of cells on surfaces (as a monoculture), their stability and widespread natural presence in aquatic environments, their tolerance to low nutrient stress, and their completely non-pathogenic nature are all useful characteristics, from the perspective of a fermentation engineer. Thus the potential to produce fixed-cell bioreactors, for the enzymatic conversion of substrates or the binding of specific reagents (heavy metals for example), the possibilities for dispersal of caulobacters with specific surface-presented capabilities in environments and the potential of using engineered caulobacters with their altered S-layer as vaccination agents for fish and other aquatic life are all obvious examples of combining modified S-layers with the inherent capabilities of the underlying bacterium.

REFERENCES

Bingle, W.H., and Smit J., 1990, High level plasmid expression vectors for *Caulobacter crescentus* incorporating the transcription and transcription-translation initiation regions of the paracrystalline surface layer protein gene, *Plasmid* 24:143.

Bingle, W.H., and Smit, J., 1991, Linker mutagenesis using a selectable marker: A method for tagging specific-purpose linkers with an antibiotic-resistance gene, *BioTechniques* 10:150.

Bingle, W.H., Kurtz, H.D., and Smit, J., 1992, An "all-purpose" cellulase reporter for gene fusion studies and application to the paracrystalline surface (S)-layer protein of *Caulobacter crescentus, Can. J. Microbiol.*, in press.

Edwards, P., and Smit, J., 1991, A transducing bacteriophage for *Caulobacter crescentus* uses the paracrystalline surface layer protein as receptor, *J. Bacteriol.* 173:5568.

Ely, B., and Gerardot, C.J., 1988, Use of pulsed-field gel electrophoresis to construct a physical map of the *Caulobacter crescentus* genome, *Gene* 68:323.

Ely, B., and Johnson, R.C, 1977, Generalized transduction in *Caulobacter crescentus*, *Genetics* 87:391.

Feldman, M., and Basten, A., 1971, The relationship between antigenic structure and the requirement for thymus-derived cells in the immune response, *J. Experimental Med.* 143:103.

Fisher, J., Smit, J., and Agabian, N., 1988, Transcriptional analysis of the major surface array gene of *Caulobacter crescentus, J. Bacteriol.* 170:4706.

Gilchrist, A., and Smit, J., 1991, Transformation of freshwater and marine caulobacters by electroporation, *J. Bacteriol.* 173:921.

Gilchrist, A., Fisher, J.A., and Smit, J., 1992, Nucleotide sequence analysis of the gene encoding the *Caulobacter crescentus* paracrystalline surface layer protein, *Can J. Microbiol.* 38:193.

Koval, S.F., and Hynes, S.H.,1991, Effect of paracrystalline protein surface layers on predation by *Bdellovibrio bacteriovorus*, *J. Bacteriol.* 173:2244.

Kurtz, H.D., and Smit, J., 1992, Analysis of a *Caulobacter crescentus* gene cluster involved in attachment of the holdfast to the cell, *J. Bacteriol.* 174:687.

MacRae, J.D., and Smit, J., 1991, Characterization of Caulobacters isolated from wastewater treatment systems, *Appl. Environ. Microbiol.* 57:751.

Merker, R.I., and Smit, J., 1988, Analysis of the adhesive holdfast of marine and freshwater Caulobacters, *Appl. Environ. Microbiol.* 54:2078.

Messner, P., and Sleytr, U.B., 1992, Crystalline bacterial cell-surface layers. *Adv. Microbial Physiol.*   33:213.

Mitchell, D., and Smit, J., 1990, Identification of the genes affecting production of the adhesion organelle of *Caulobacter crescentus* CB2, *J. Bacteriol.* 172:5425.

Newton, A., 1989, Differentiation in *Caulobacter*: flagellum development, motility, and chemotaxis, *in*: "Genetics of Bacterial Diversity", D.A. Hopwood, and K.F. Chater, eds., p.199-220, Academic Press, New York.

Ong, C., Wong, M.L.Y., and Smit, J., 1990, Attachment of the adhesive holdfast organelle to the cellular stalk of *Caulobacter crescentus*, *J. Bacteriol.* 172:1448.

Poindexter, J.S., 1981, The caulobacters: ubiquitous unusual bacteria, *Microbiol. Rev.* 45:123.

Shapiro, L., 1985, Generation of polarity during *Caulobacter* differentiation, *Ann. Rev. Cell Biol.* 1:173.

Sleytr, U.B., and Sára, M., 1986, Ultrafiltration membranes with uniform pores from crystalline bacterial cell envelope layers, *Appl. Microbiol. Biotechnol.* 25:83.

Smit, J., Grano, D.A., Glaeser, R.M., and Agabian, N., 1981, A periodic surface array in *Caulobacter crescentus*: Chemical and fine structure analysis, *J. Bacteriol.* 146:1135.

Smit, J., and Agabian, N., 1982, Cell surface patterning and morphogenesis: biogenesis of a periodic surface array during *Caulobacter* development, *J. Cell Biol.* 95:41.

Smit, J., and Agabian, N., 1984, Cloning the major protein of the *Caulobacter crescentus* periodic surface layer: detection and characterization of the cloned peptide by protein expression assays, *J. Bacteriol.* 160:1137.

Smit, J., 1986, Protein surface layers of bacteria, *in*: "Outer Membranes as Model Systems", M. Inouye, ed., pp. 343-376, J. Wiley and Sons, New York.

Smit, J., Engelhardt, H., Volker, S., Smith, S.H., and Baumeister, W., 1992, The S-layer of *Caulobacter crescentus*: Three-dimensional image reconstruction and structure analysis by electron microscopy, *J. Bacteriol.* 174:6527.

Stahl, D.A., Key, R., Flesher, B., and Smit, J., 1992, The phylogeny of marine and freshwater caulobacters reflects their habitat, *J. Bacteriol.* 174:2193.

Walker, S.G., Smith, S.H., and Smit, J., 1992, Isolation and comparison of the paracrystalline surface layer proteins of freshwater caulobacters, *J. Bacteriol.* 174:1783.

V. APPLICATIONS FOR S-LAYERS

Chapter 19

S-LAYERS AS IMMOBILIZATION AND AFFINITY MATRICES

Margit Sára, Seta Küpcü, Christian Weiner, Stefan Weigert,
and Uwe B. Sleytr

Center for Ultrastructure Research and
Ludwig Boltzmann Institute for Molecular Nanotechnology
University of Agriculture
Vienna, Austria

INTRODUCTION

The procedures for immobilizing macromolecules can be classified into cross-linking, entrapping and carrier-binding and these can be subdivided into physical adsorption, ionic, and covalent binding (Mosbach 1987, 1988). Among the various methods, covalent binding is most frequently applied since the forces between the macromolecules and the carrier are strong and leakage of the immobilized molecules from the support matrix should be negligible under disrupting conditions, such as high salt concentration and low or high pH. The carriers usually used for immobilization are particles made of different polymers, such as agarose or poly-acrylamide. Despite their different chemical composition, all carriers are heteroporous meshworks with a random distribution of the polymer chains and a random orientation of the functional groups. Due to the presence of differently sized pores, immobilization of macromolecules does not only occur on the surface of the gel particles but also in the interior of the polymer network. Although higher binding capacities can be achieved if immobilization is possible in the interior of the gels, it can be of disadvantage in that the diffusion of subsequently applied reaction partners may be a limiting factor for the speed of reaction.

In comparison to the heteroporous beads (Mosbach, 1987, 1988) generally used for immobilization purposes, S-layers can be considered as ideal matrices. As isoporous two-dimensional monomolecular protein crystals composed of identical subunits, their functional groups have a well defined position and orientation (Sleytr and Messner, 1988; Messner and Sleytr, 1992; Sára et al., 1992a).

S-LAYERS AS MATRICES FOR THE IMMOBILIZATION OF MACROMOLECULES

Activation Procedures

On the outer surface of S-layers from most mesophilic and thermophilic Bacillaceae, amino and carboxyl groups are found in equimolar amounts. They are arranged in close proximity with one another, thus neutralizing themselves by direct electrostatic interactions (Pum et al., 1989; Sára et al., 1992b). In glycosylated S-layer proteins, the carbohydrate chains were also shown to be exposed on the S-layer surface (Sára et al., 1989). Consequently, both amino and carboxyl groups from the S-layer protein and hydroxyl groups from the carbohydrate chains are available for activation and immobilization of molecules.

To increase the stability of the S-layer lattice against acidic and alkaline pH conditions, as well as against hydrogen bond breaking agents, the S-layer subunits are usually crosslinked with glutaraldehyde before activating the functional groups for immobilization. If the carboxyl or hydroxyl groups are activated, crosslinking is done under conditions where at least 80% of the amino groups are modified due to their reaction with glutaraldehyde (Küpcü, 1991). On the other hand, if the foreign molecules had to be linked to the amino groups of the S-layer protein, crosslinking with glutaraldehyde is performed in such a way that up to 75% of the amino groups remained unmodified (Sára and Sleytr, 1992).

After crosslinking the S-layer lattice with glutaraldehyde, the carboxyl groups from the acidic amino acids can be either activated with 1-ethyl-3,3' (dimethylaminopropyl) carbodiimide (EDC) (Carraway and Koshland Jr., 1972) or with 1-ethoxy-carbonyl-2-ethoxy-1,2-dihydroquinoline (EEDQ). The advantage of using EDC lies in the fact that the reaction can be carried out in water, whereas EEDQ is only soluble in organic solvents which frequently induces changes to the regular structure even with glutaraldehyde-treated S-layer lattices. Vicinal hydroxyl groups from the carbohydrate chains of S-layer glycoproteins are generally activated with cyanogen bromide (Axen and Ernback, 1971). An alternative method is to succinylate the hydroxyl groups and activate the introduced carboxyl groups with EDC. In such modified S-layer glycoproteins, both the carboxyl groups from the acidic amino acids and those introduced by succinylation were available for EDC-activation. More detailed studies revealed that the pH optimum for EDC-activation strongly depends on the type of carboxyl group. For example, the pH-optimum for the β- and γ-carboxyl groups from either aspartic or glutamic acid was between pH 4.2 - 4.7 thus approximating their pK values, whereas the pH-optimum for activating the carboxyl groups from the succinic acid residues was between pH 3.2 - 3.5 (Küpcü, 1991). If the carboxyl groups from the acidic amino acids are amidated before succinylation of the hydroxyl groups, immobilization of the macromolecules is restricted to the carbohydrate chains.

In addition to the direct immobilization of macromolecules to S-layer lattices, the foreign proteins are also bound to S-layers after introducing spacer molecules of different length. 4-aminobutyric acid, 6-aminocaproic acid, glycine, glutamic acid, poly-D-glutamic acid, poly-L-serine and poly-L-lysine which was succinylated before further activation can be employed as spacer molecules (Sára et al., 1992a).

For obtaining an immobilization matrix which could be used for reversible covalent immobilization, sulfhydryl groups can also be introduced into the S-layer lattice (Sára and Sleytr, 1992). For this purpose, glutaraldehyde-treated S-layers in

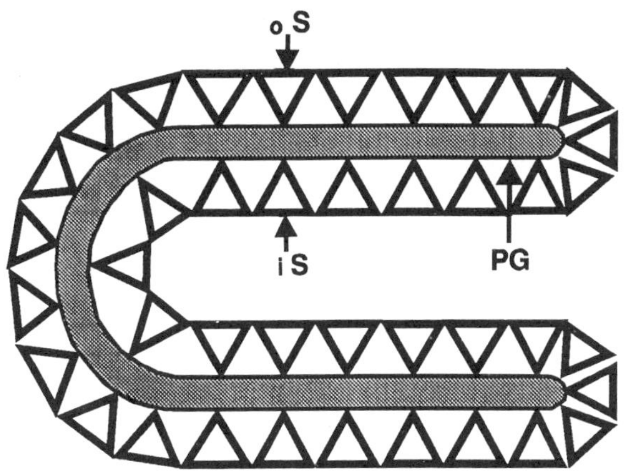

Figure 1. Schematic drawing of "cup-shaped" cell wall fragments. The peptidoglycan-containing layer (PG) is completely covered with mirror-symmetrically arranged S-layer lattices (OS is the outer S-layer whereas iS is the inner S-layer).

which 75% of the amino groups are still available, were modified with iminothiolane (Traut et al., 1973). After activation of the sulfhydryl groups with 2,2'-dipyridyl-disulfide, enzymes containing thiol-groups such as β-D-galactosidase could be immobilized. The advantage of immobilizing functional molecules via disulfide bonds is seen in the fact that the unaltered native S-layer matrix can be regenerated by cleaving the disulfide bonds with dithiothreitol.

S-Layer Material used as an Immobilization Matrix

Depending on the field of application, S-layers can be used in different forms for immobilizing macromolecules; as self-assembly products, as "cup-shaped" cell wall fragments (Fig.1) or as S-layer ultrafiltration membranes (SUMs) (Sára and Sleytr, 1987). Regardless of the type of material, the S-layer protein is first crosslinked with glutaraldehyde before modification or activation steps are performed. In the case of S-layer self-assembly products, only double layers are used as an immobilization matrix. Due to the mirror symmetrical orientation of the constituent layers, either the outer or the inner S-layer faces are exposed (Pum et al., 1989). Consequently carboxyl groups from both surfaces can be activated at identical pH optima.

In comparison to double layer self-assembly products, the advantage of using cell wall fragments is their higher stability towards mechanical, centrifugal and shear forces which is due to the presence of the rigid peptidoglycan-containing layer. Since the peptidoglycan-containing layer from the *Bacillus* strains used in our immobilization studies is completely covered with an outer and an inner S-layer lattice, covalent attachment of macromolecules can only occur on the outer face of both S-layer lattices (Fig.1).

In addition to self-assembly products and cell wall fragments, SUMs can be employed for immobilizing enzymes. These were used to design and develop a new type of biosensor (Pum et al., 1992; Pum and Sleytr, this book). For producing SUMs, cell wall fragments were deposited on microfiltration membranes and the S-layer protein was subsequently crosslinked with glutaraldehyde (Sára and Sleytr, 1987). After immobilizing the different oxidases, the enzyme monolayer is coated with a thin metal layer for transferring the electrons released during the enzymatic reaction to the electrodes (Pum et al., 1992). The advantage of using SUMs as enzyme carriers in biosensor development is that the microfiltration membrane and their S-layers are capable of rejecting foreign particles and macromolecules. At the same time both membranes allow a rapid diffusion of substrates to the site of enzymatic reaction due to their high porosities.

Immobilization of Ferritin

For visualizing the density and distribution of a protein molecule bound to an S-layer lattice by electron microscopy, ferritin with a molecular weight of 440000 and a molecular size of 12 nm can be used. For comparison, the amount of ferritin bound per mg S-layer protein can also be determined by UV measurement. Quantitative determination showed that in the case of the hexagonally ordered S-layer lattice from *Clostridium thermohydrosulfuricum* L111-69, 700 µg ferritin were bound per mg S-layer protein. Since the apparent molecular weight of the S-layer subunits from *C. thermohydrosulfuricum* L111-69 is 120000, approximately one ferritin molecule was attached per 14 nm S-layer subunit leading to the formation of a monolayer of ferritin on the surface. These results were in good accordance with freeze-etching images of whole cells, which showed ferritin molecules densely arranged on the outer face of the S-layer lattice (Fig.2). Only small areas of immobilized molecules reflected the hexagonal lattice from the crystalline matrix. Thin-sections of cell wall fragments also showed a monolayer of ferritin molecules attached to the S-layer surface (Fig.3).

In order to bind more than one monolayer of ferritin on an S-layer surface, immobilization can be done on the carbohydrate chains of the S-layer glycoprotein from *C. thermohydrosulfuricum* L111-69. The carbohydrate chains consist of up to 30 repeating disaccharides of rhamnose and mannose (Christian et al., 1988), and their hydroxyl groups can be succinylated and subsequently activated with EDC. On such modified S-layer lattices, 1500 µg ferritin were bound per mg S-layer protein. This

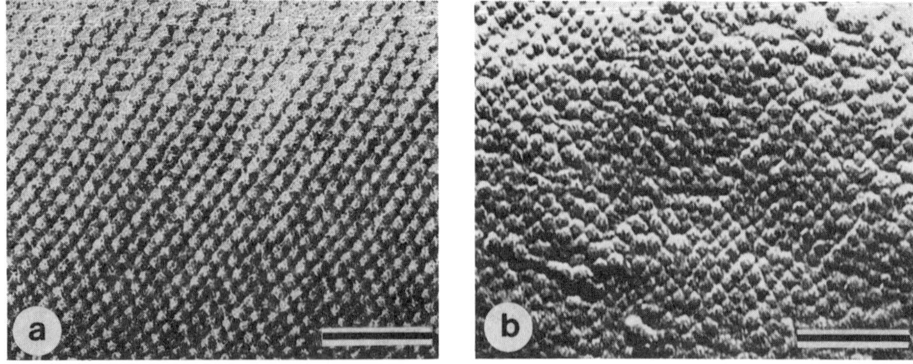

Figure 2. Freeze-etched preparations of glutaraldehyde-treated whole cells from *C. thermohydrosulfuricum* L111-69 incubated with ferritin (a) before and (b) after activation of the carboxyl groups with carbodiimide. In (b) ferritin molecules were bound to the S-layer surface in the maximum dense packing order. Bars = 100 nm.

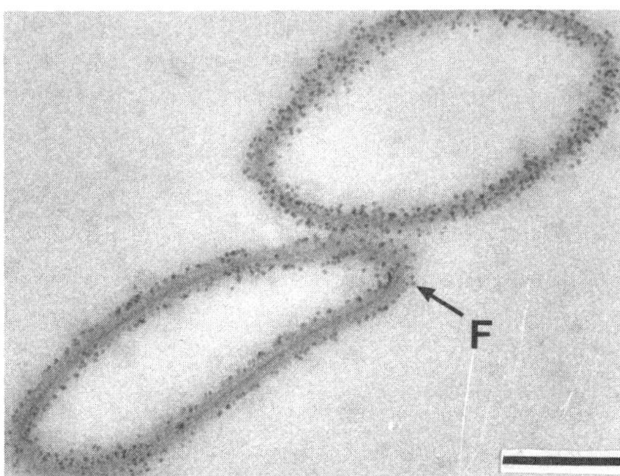

Figure 3. Thin section of glutaraldehyde-treated cell wall fragments from *C. thermohydrosulfuricum* L111-69 after immobilization of ferritin to the carbodiimide-activated carboxyl groups of the outer and inner S-layer lattice. Bar = 200 nm.

corresponded to approximately 2 ferritin molecules per morphological unit. Considering the size of the morphological units (14 nm) and the size of the ferritin molecules (12 nm), it can be calculated that by exploiting the exposed carbohydrate moities as immobilization sites two separate monolayers of ferritin had been formed on the S-layer surface. These calculations could also be confirmed by thin-sections of cell wall fragments which exhibited two superimposed ferritin monolayers (Sára et al., 1989).

Immobilization of Ovalbumin and Glucose Oxidase to the Protein Moiety of the S-Layer Lattice from *C. thermohydrosulfuricum* L111-69 Before and After Introduction of Spacer Molecules

If ovalbumin is directly linked to the EDC activated carboxyl groups of the S-layer lattice of *C. thermohydrosulfuricum* L111-69, 370 µg could be immobilized per mg S-layer protein. This corresponds to six ovalbumin molecules per morphological unit of the hexagonal array. Such a binding capacity clearly showed that at least one activated carboxyl group per S-layer subunit was accessible to the spherical, 5 nm-sized ovalbumin molecules. In this context one has to consider that the surface of the S-layer lattice is corrugated. Consequently, it could not be excluded that additional carboxyl groups are actually present on the S-layer surface but, due to steric hindrance, these are not accessible to foreign molecules such as ovalbumin. Therefore, low molecular weight spacer molecules of different length were linked to the EDC-activated S-layer lattice from *C. thermohydrosulfuricum* L111-69. After introduction of either 4-aminobutyric acid, 6-aminocaproic acid or glycine as spacer molecules, the S-layer lattice was capable of binding up to 760 µg ovalbumin per mg S-layer protein which was approximately twice the amount immobilized without a spacer. This increase clearly demonstrated that by introducing the spacer molecules at least one additional carboxyl group, by spacer extension, became available on the S-layer lattice for the immobilization of ovalbumin molecules.

Glucose oxidase (M_r 150000) was chosen to investigate the advantage of spacers attached to S-layer lattices for the binding capacity for molecules with a size comparable to that of individual S-layer subunits. Independent of the type and length of the introduced spacer molecules, the S-layer lattice was capable of binding 550 µg enzyme protein per mg S-layer protein. The same binding capacity was observed without spacers, corresponding to 3-4 immobilized glucose oxidase molecules per morphological unit. It was concluded that one free accessible carboxyl group per S-layer subunit is sufficient for achieving the maximum possible packing density. However, the introduction of the spacers had a significant influence on the retained activity of the immobilized enzymes. Whereas glucose oxidase directly linked to the S-layer lattice retained 35% of its activity, up to 60% of the original activity could be preserved if the immobilization was done via spacer molecules.

Immobilization of Differently Sized Proteins to the S-Layer Lattice from *B. stearothermophilus* PV72

In order to investigate the influence of the molecular weight and the size of protein molecules on their binding capacity per unit cell of an S-layer lattice, proteins differing in their molecular weights by up to a factor of 10 were immobilized on the hexagonal array from *B. stearothermophilus* PV72. These were ferritin (M_r 440000), invertase (M_r 270000), glucose oxidase (M_r 160000), ß-D-galactosidase (M_r 116000), bovine serum albumin (M_r 67000) and ovalbumin (M_r 43000). The morphological units of the S-layer have a dimension of 22.5 nm and consist of six identical S-layer subunits with a molecular weight of 130000. As can be seen from Table 1 the number of protein molecules bound per morphological unit strongly depended on their molecular weights. In the case of the 12 nm ferritin, only one to two molecules were bound per morphological unit, whereas up to 6 molecules were immobilized per morphological unit if the molecular weight of the protein approached M_r 70000 (Table 1 and Fig.4).

Table 1. Covalent attachment of different proteins to the S-layer lattice from *B. stearothermophilus* PV72. Carboxyl groups from the S-layer protein were activated with EDC so that they could then react with the free amino groups on the protein molecules.

Protein	Molecular weight	µg protein bound per mg S-layer protein	molecules bound per unit cell
Ferritin	440000	1,000	1-2
Invertase	270000	960	2-3
Glucose Oxidase	150000	660	3
ß-D-Galactosidase	116500	460	3
Bovine Serum Albumin	67000	500	6
Ovalbumin	43000	340	6

Immobilization of Protein A for Generating Affinity Microstructures

For studying the suitability of S-layer lattices as affinity matrices, Protein A was linked to cell wall fragments from *C. thermohydrosulfuricum* L111-69 which

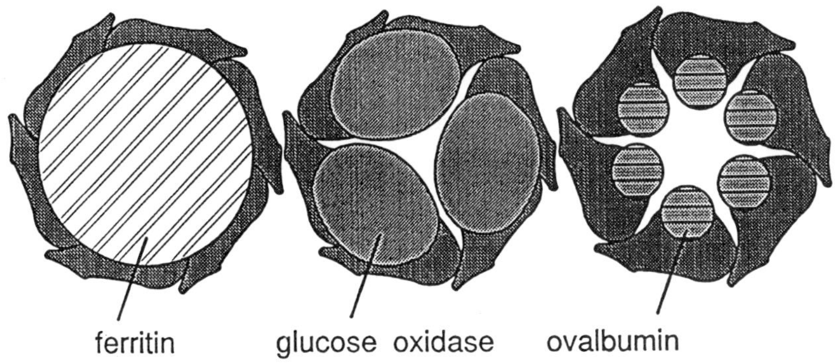

Figure 4. Schematic drawing showing the immobilization of differently sized protein molecules on the hexagonally ordered S-layer lattice from *B. stearothermophilus* PV72.

predominantly had the shape of half cells (Fig.1). These structures, composed of a peptidoglycan layer completely covered on both surfaces with an S-layer lattice and loaded with Protein A, will subsequently be referred to as affinity microstructures. Protein A is known for its binding to the F_c regions of most mammalian immunoglobulins. The F_c regions of the IgG molecules are bivalent for Protein A, whereas Protein A itself has four to five potential binding sites for IgG. Since this "pseudo-immune binding" is reversible under acidic or high salt conditions, Protein A has found many applications, particularly in downstream-processing for the isolation and purification of antibodies (Langone, 1982).

At present, two types of Protein A are commercially available, one is isolated from the cell wall of *Staphylococcus aureus* and has a molecular weight of 42000, and the other is recombinant Protein A from *Escherichia coli* with a molecular weight of 32000.

The hexagonal S-layer lattice from *C. thermohydrosulfuricum* L111-69 is capable of binding 700 µg Protein A from *S. aureus* and 500 µg recombinant Protein A per mg S-layer protein; in both cases two Protein A molecules were bound per S-layer subunit. This clearly showed that at least two activated carboxyl groups per S-layer subunit must have been accessible for the Protein A molecules. Such a binding capacity is in remarkable contrast to the results obtained with ovalbumin (see above), where only one carboxyl group was available for a molecule with a comparable molecular weight. The different binding capacity of the S-layer lattice for Protein A and ovalbumin can be explained by differences in shape and surface topography of the foreign protein molecules. Protein A is extremely elongated and thin (Björk et al., 1972; Deisenhofer et al., 1978) and thus could reach carboxyl groups on the corrugated S-layer surface not accessible to spherical ovalbumin molecules.

To obtain affinity matrices with long shelf-lifes, the S-layer protein of the affinity microstructures can first be crosslinked with glutaraldehyde. Since Schiff's bases are susceptible to hydrolysis during storage, the affinity microstructures can also be reduced with sodium borohydride prior to the immobilization of Protein A. On the glutaraldehyde-treated and reduced S-layer material only 500 µg instead of 700 µg of Protein A (M_r 42000) could be immobilized per mg S-layer protein. This corresponded to 1 - 2 Protein A molecules per S-layer subunit (Fig.5). The reason for the lower binding capacity of S-layer lattices reduced with sodium borohydride may be explained by inter- and intramolecular crosslinking reactions between secondary amino groups and the EDC-activated carboxyl groups.

The Protein A affinity microstructures crosslinked only with glutaraldehyde or further treated with borohydride each bound 660 μg of human polyclonal antibodies per mg S-layer protein. Such a binding capacity corresponds to 3 IgG molecules per morphological S-layer unit. Since the amount of IgG bound to the affinity microstructures was not affected by the amount of immobilized Protein A, it seems that the space available on the Protein A-treated S-layer surface is the limiting factor for IgG binding (Fig.5). Further experiments with reduced amounts of Protein A showed that on average 2 Protein A molecules per morphological S-layer unit were sufficient for generating a tightly packed IgG monolayer on the S-layer surface. In comparison to soluble Protein A, where all potential IgG-binding sites are accessible, the retained IgG-binding capacity of the immobilized Protein A was in the range of 10-25% and strongly depended on the number of Protein A molecules present per morphological unit.

CONCLUSION

The data presented in this chapter clearly demonstrate that S-layers represent a remarkably good immobilization matrix for enzymes and ligands. Functional molecules can be either bound to the protein moiety or to the carbohydrate chains of S-layer glycoproteins. In comparison to amorphous polymers commonly used as immobilization matrices, S-layers have higher binding capacities. This was attributed to the fact that S-layers are crystalline lattices composed of identical subunits.

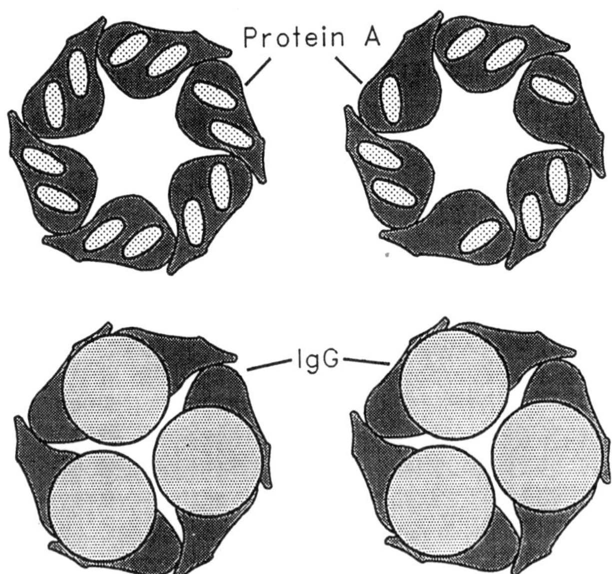

Figure 5. Schematic drawing showing the immobilization of Protein A (M_r 42,000) on the hexagonally ordered S-layer lattice from *C. thermohydrosulfuricum* L111-69. On S-layers crosslinked with glutaraldehyde, 12 Protein A molecules could be immobilized per morphological unit, whereas only up to 9 Protein A molecules were bound if the crosslinked S-layer lattice had been subsequently reduced with borohydride. In both cases 3 IgG molecules were bound per morphological unit of the S-layer lattice.

Consequently, the same number of functional groups with identical orientation is present on each S-layer subunit leading to a matrix with well-defined repetitive physiochemical surface properties down to the nanometer range. Depending on the specific application, different types of S-layers can be used. So far, the broad application of S-layers as immobilization matrices has been demonstrated with such diverse products as enzyme membranes, biosensors and affinity microstructures applied in cross flow systems.

ACKNOWLEDGMENTS

This work was supported by the Fonds zur Förderung der Wissenschaftlichen Forschung in Österreich, project S50/02 and by the Bundesministerium für Wissenschaft und Forschung.

REFERENCES

Axen, R., and Ernback, S., 1971, Chemical fixation of enzymes to cyanogen halide activated polysaccharide carriers, *Eur.J.Biochem.* 18:351.

Björck, I., Peterson, B.A., and Sjöquist, J., 1972, Some physiochemical properties of Protein A from *Staphylococcus aureus, J. Biochem.* 29: 579.

Carraway, K.L., and Koshland, Jr. D.E., 1972, Carbodiimide modification of proteins, "Methods in Enzymology", vol. 25, S.P. Colowick, and N.O. Kaplan, eds., Academic Press, Inc., London.

Christian, R., Messner, P., Weiner, C., Sleytr, U.B., and Schulz, G., 1988, Structure of a glycan from the surface-layer glycoprotein of *Clostridium thermohydrosulfuricum* L 111-69, *Carbohydr. Res.* 176:160.

Deisenhofer, J., Jones, T.A., and Huber, R., 1978, Crystallization, crystal structure analysis and atomic model of the complex formed by a human Fc fragment and fragment B of Protein A from *Staphylococcus aureus, Hoppe-Seyler's Z. Physiol. Chem.* 359: 975.

Küpcü, S., 1991, Chemische Modifikation von S-Schicht Ultrafiltrationsmembranen mit aliphatischen Nucleophilen und Versuche zur Immobilisierung von Makromolekülen an 2 D Glykoproteinkristallen, Ph. D. thesis, Universität für Bodenkultur, Wien.

Langone, J.J., 1982, Protein A of *Staphylococcus aureus* and related immunoglobulin receptors by Streptococci and Pneumonococci, *Adv. Immunol.* 32:157.

Messner, P., and Sleytr, U.B, 1992, Crystalline bacterial cell surface layers, *Adv. Microbial Physiol.* 33: 213.

Mosbach, K., 1987, Immobilized enzymes and cells (Part C), *in*: "Methods in Enzymology", Vol. 136, S.P. Colowick, and N.O. Kaplan, eds., Academic Press, Inc., London.

Mosbach, K., 1988, Immobilized enzymes and cells (Part D), *in*: "Methods in Enzymology", Vol. 137, S.P. Colowick, and N.O. Kaplan, eds., Academic Press, Inc, London.

Pum, D., Sára, M., and Sleytr, U.B., 1989, Structure, surface charge and self-assembly of the S-layer lattice from *Bacillus coagulans* E38-66, *J. Bacteriol.* 171:5296.

Pum, D., Sára, M., and Sleytr, U.B. 1992, Two-dimensional (glyco)protein crystals as patterning elements and immobilisation matrices for the development of biosensors, *in*: "Immobilized Macromolecules: application potential", U.B. Sleytr, P. Messner, M. Sára, and D. Pum, eds., Springer, London (in press).

Sára, M., and Sleytr, U.B., 1987, Production and characteristics of ultrafiltration membranes with uniform pores from two-dimensional arrays of proteins, *J. Membrane Sci.* 33:27.

Sára, M., and Sleytr, U.B., 1992, Introduction of sulfhydryl groups into the crystalline bacterial cell surface layer protein from *Bacillus stearothermophilus* PV72 and its application as immobilization matrix, *Appl. Microbiol. Biotechnol.* (in press).

Sára, M., Küpcü, S., and Sleytr, U.B., 1989, Localization of the carbohydrate residue of the S-layer glycoprotein from *Clostridium thermohydrosulfuricum* L 111-69, *Arch. Microbiol.* 151: 416.

Sára, M., Küpcü, S., Weiner, C., Weigert, S., and Sleytr, U.B., 1992a, Crystalline protein layers as isoporous molecular sieves and immobilization and affinity matrices, *in*: "Immobilized Macromolecules: Application Potential", U.B. Sleytr, P. Messner, M. Sára, and D. Pum, eds., Springer, London (in press).

Sára, M., Pum, D., and Sleytr, U.B., 1992b, Permeability and charge dependent adsorption properties of the S-layer from *Bacillus coagulans* E38-66, *J. Bacteriol.* 174:3487.

Sleytr, U.B. and Messner, P., 1988, Crystalline surface layers in procaryotes, *J. Bacteriol.* 170:2891.

Traut, R.R., Bollen, A., Sun, T.T., Hershey, J.W.B., Sundberg, J., and Pierce, L.R., 1973, Methyl-4-mercapto butyrimidate as a cleavable crosslinking reagent and its application to the *Escherichia coli* 30S ribosome, *Biochem.* 12:3266.

Chapter 20

MOLECULAR NANOTECHNOLOGY WITH S-LAYERS

Dietmar Pum and Uwe B. Sleytr

Center for Ultrastructure Research and
Ludwig Boltzman Institute for Molecular Nanotechnology
University of Agriculture
Vienna, Austria

INTRODUCTION

Nanotechnology is not only a new word it is a new way of thinking since it combines knowledge from such diverse disciplines as physics, biology, and chemistry. In an interdisciplinary way it comprises the techniques and properties of molecular manufacturing. Molecular nanotechnology originated from the efforts made by the electronic industry in miniaturizing integrated circuits. Increased processing speed, decreased energy consumption as well as decreased size and weight of the products should all lead to greatly improved computer performance. Nevertheless, scientists already predict that this conventional technology will lead to a dead end since severe physical limits imposed by quantum and thermal fluctuation phenomena will be rapidly approached. When the size of conventional transistors approaches the wavelength of electrons the ability of silicon chips to function will deteriorate because electrons will tunnel through the insulating layers of the switching elements. As a consequence the operation of these miniature devices will become unreliable.

With these ultimate limits in mind, scientists have started to study biological self-assembly systems which are appealing models for investigating the structural principles of miniaturized functional units and devices. These systems operate with functional elements of molecular size. Of course, utilizing biological or organic materials for a new age of information processing demands the exploitation of their chemistry as well as the development of completely new computer architectures and languages. This new way of thinking led to a new science called *Molecular Electronics* and, in a more general sense, all those phenomena and techniques at the molecular level are referred to as *Molecular Nanotechnology*. Today the entire field of molecular nanotechnology has become an interdisciplinary science with a working framework using theory, practice, and computer simulation directed to developing nanometer scale devices.

Advances in Bacterial Paracrystalline Surface Layers
Edited by T.J. Beveridge and S.F. Koval, Plenum Press, New York, 1993

Until recently, biomaterials were not seriously considered as patterning elements for the fabrication of molecular devices since they were thought to be too fragile and inaccessible for conventional processing techniques. Now it is becoming evident that proteins are worth exploiting as new nanotechnology materials since they have already acquired significant degrees of molecular complexity and function in the course of evolution (Nagayama, 1992). New techniques using two-dimensional protein crystals appear to be an especially effective alternative approach for constructing usable structures in the nanometer range. Unlike conventional approaches which use focussed particle beams to make complicated etchings in minute devices, self-assembly of S-layer proteins is an exacting and efficient method of fabricating precise arrays at molecular resolution.

As will be shown in this chapter all concepts of nanofabrication depend strongly on the availability of a patterning structure which is precisely defined in terms of its geometry and physicochemistry and on the controlled immobilization of functional molecules on such a patterning structure. This demand has already been met by nature in the form of crystalline bacterial cell surface layers (for reviews see Sleytr, 1978; Smit, 1987; Beveridge, 1981; Sleytr and Messner, 1983, 1988; Hovmöller et al., 1988; Koval, 1988; Sleytr et al., 1987, 1988a; Messner and Sleytr, 1992). Since S-layer lattices are composed of identical protein or glycoprotein subunits, functional groups are located in an identical position and orientation on the constituent protein or carbohydrate moities (Sára and Sleytr, 1989). It is not only their well defined physicochemical properties which make S-layers so ideally suited as patterning structures in the nanofabrication of molecular machines and bioelectronic devices, it is also their ability to self-assemble in suspension, to recrystallize on solid surfaces and to form large scale monolayers at various interfaces.

SURFACE PROPERTIES AND INTERMOLECULAR FORCES

A spectrum of S-layers has shown considerable asymmetry in the topography and charge distribution of the outer and inner S-layer faces. As revealed by three-dimensional image reconstructions, the outer face is generally less textured than the inner face (Baumeister and Engelhardt, 1987; Hovmöller et al., 1988). Chemical modification and adsorption studies have shown that in S-layers of different Bacillaceae the outer face and the pore areas are charge neutral due to an equal number of amino and carboxyl groups whereas the inner face is generally negatively charged because of an excess of free carboxyl groups (Sára and Sleytr, 1987). In addition, the outer face seems to be more hydrophobic (Gruber and Sleytr, 1991). In the case of glycosylated S-layer proteins, the carbohydrate chains are located on the outer face (Sára et al., 1989; Messner and Sleytr, 1991).

In most S-layer lattices the constituent subunits are held together and to the supporting surface (e.g., outer membrane or peptidoglycan) by non-covalent forces such as hydrogen or ionic bonds, and hydrophobic or electrostatic interactions. Dissolution experiments clearly demonstrated that the intermolecular forces between subunits are stronger than those binding the crystalline array to the supporting envelope layer (Sleytr, 1978; Beveridge, 1981; Koval, 1988; Sleytr and Messner 1989; Beveridge and Graham, 1991).

SELF-ASSEMBLY IN SUSPENSION

S-layers isolated from a broad spectrum of species have shown the inherent ability to reassemble into two-dimensional arrays after removal of the disrupting agent used in the dissolution procedure. Studies on the self-assembly process have shown that the initial phase is determined by a rapid nucleation of the subunits into oligomeric precursors consisting of several unit cells which aggregate together; after this, growth into larger crystalline arrays is much slower (Jaenicke et al., 1985). These self-assembly products may have the form of flat sheets, open-ended cylinders or closed vesicles (for reviews see Sleytr, 1983; Sleytr and Messner, 1989). Size and structural form may depend on several environmental parameters such as temperature, pH, ion concentration, or ionic strength in the course of dialysis. Self-assembly products from *Bacillus* or *Clostridium* spp. possessing a size of several square micrometers are normally found. The morphology and bonding properties of the subunits also determine the assembly route, the final reassembly form, and the capacity to form mono- or double layers. In the case of double layered structures, the two constituent monolayers are facing each other either with their inner or their outer sides together. Spiral growth is initiated in a reassembly when a screw dislocation is formed at a very early stage in the assembly process.

RECRYSTALLIZATION ON SOLID SURFACES

Numerous S-layer proteins have shown the ability to recrystallize on solid surfaces such as glass, silicon, mica, carbon or synthetic polymers (Pum et al., 1989a; Sleytr et al., 1993). The environmental conditions are frequently the same as those used to induce self-assembly without supporting layers. In general, as a precondition for many S-layer assembly systems, the surface of the supporting layer has to be chemically modified in order to favour the recrystallization at the interface. Depending on the surface properties of the supporting layer and on the chemical properties of S-layer proteins, the subunits can bind either with their inner or their outer face (Pum et al., 1989a). This preferential orientation is a consequence of the asymmetry in the physicochemical surface properties of each face. Electrostatic and hydrophobic interactions are the main forces responsible for the orientation of the subunits with respect to the solid support surface. Positively charged surfaces can be obtained either by treatment with alcian blue or poly-L-lysine. Silanization with different compounds can be used to produce either positively or negatively charged surfaces on glass or silicon. An example of a reconstituted oblique S-layer lattice from *Bacillus coagulans* E38-66 on a poly-L-lysine coated film is shown in Fig.1 (Pum et al., 1989a). Since the S-layer proteins from this bacterium do not bind with their negatively charged inner face but, instead, with their charge neutral outer face to the positively charged support, it was concluded that hydrophobic interactions dominated the adherence of the S-layer to the surface. Numerous crystal boundaries, which are seen as line imperfections, indicate that crystal growth on the supporting layer is initiated at several distant nucleation points. These points can be the result of randomly oriented monomers, oligomers or small crystallites. The individual crystalline patches grow isotropically in all directions until the front end of a neighbouring crystalline area is reached.

S-layers which have recrystallized on solid surfaces in monolayers are the basis for all our nanomanufacturing investigations (Pum et al., 1991; Sleytr et al., 1992). The choice of support is critical and must be selected to fulfill certain requirements such as stiffness, flatness or the availability of functional groups. The mechanical properties are particularly important when a scanning force microscope (SFM) or atomic force microscope (AFM) is used to manipulate elements on the molecular level. Silicon has proven to be one of the best materials since it is atomically flat, is chemically well characterized, and can be produced with almost perfect crystallinity. Precious metals such as platinum or gold which are chemically inert are also good choices since they allow the investigation of redox reactions in molecular assemblies by electrochemical measurements.

FORMATION OF MONOLAYERS AT INTERFACES

Recrystallization of isolated subunits and oligomeric precursors into coherent S-layer lattices at an air/water interface and on a phospholipid film has proven to be an easy and reproducible way for generating large scale S-layer sheets (Pum et al., 1991; Sleytr et al., 1992, 1993). The idea for building composite structures from S-layers and biomembranes or Langmuir-Blodgett (or LB) films was derived from the observation that many archaeobacteria which grow under extreme environmental conditions (low pH, high temperatures, concentrated salt solutions) possess cell envelopes composed of an S-layer attached to the cytoplasmic membrane. Thus, the structural principle of a porous crystalline protein network combined with a lipid membrane containing additional proteins must have been optimized during billions of years of evolution.

Recrystallization experiments with the S-layer protein from *B. coagulans* E38-66 have shown that the S-layer subunits were oriented with their negatively charged inner face against the hydrophilic head groups of an LB film generated from dipalmitoylphosphatidylcholine (DPPC). This led to the assumption that electrostatic interactions were primarily responsible for the defined orientation of the subunits.

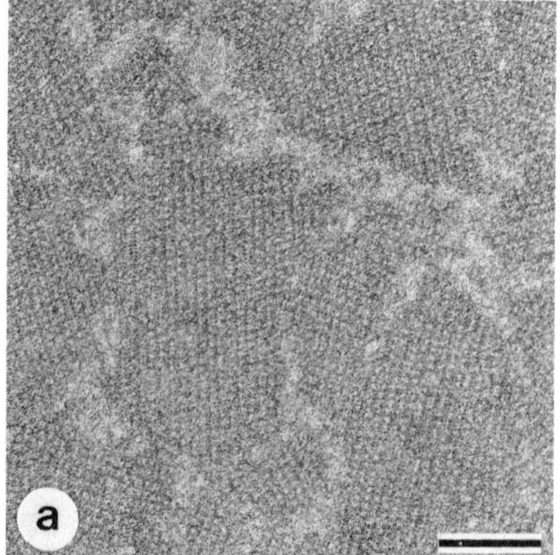
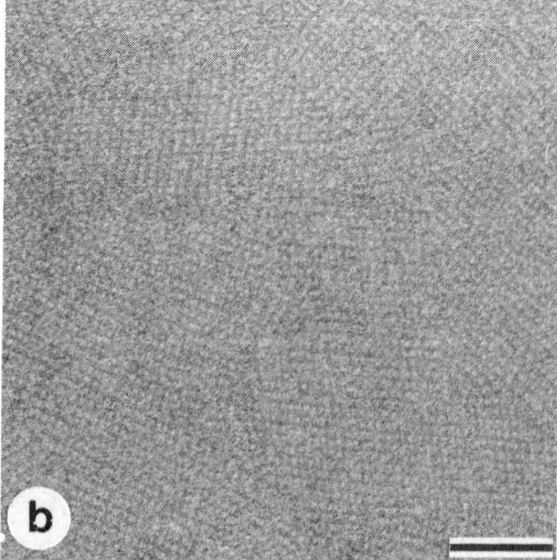

Figure 1. Electron micrographs illustrating the oblique S-layer lattice from *B. coagulans* E38-66 reconstituted on a poly-L-lysine coated film. (a) Lattice formation is initiated at several distant randomly oriented nucleation points. Bar = 100 nm. (b) The individual (mono)crystalline patches grow isotropically in all directions until the front end of a neighbouring crystalline area is reached. The crystal boundaries are seen as line imperfections between them. Bar = 100 nm.

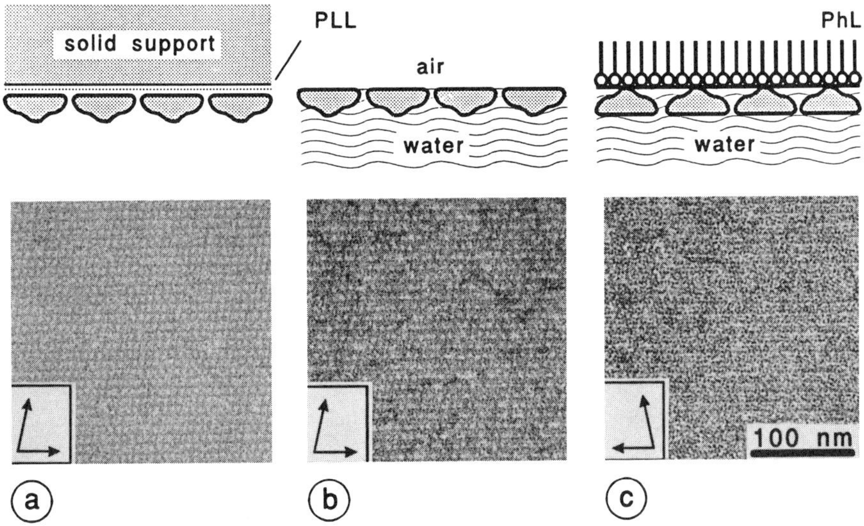

Figure 2. Summary of the specific orientations of the oblique S-layer lattice of *B. coagulans* E38-66. (a) on solid surfaces coated with poly-L-lysine, (b) at the air/water interface, and (c) on DPPC lipid films (modified after Sleytr et al., 1993).

For these experiments the lipid film had been compressed within the barriers in a LB trough. S-layer subunits in solution were injected beneath the trough barrier from the non-lipid side into the water phase. The S-layer crystals which formed at the interface had a consistently elongated morphology because of non-coherent crystalline areas. One explanation for this may be because of imperfections in the lipid film at which the S-layer proteins cannot fill the gaps by lateral growth.

The major driving force for the recrystallization of S-layer proteins at an air/water interface are hydrophobic interactions (Israelachvili, 1985). Preliminary experiments with the S-layer protein of *B. coagulans* E38-66 have shown that at this interface large-scale closed monolayers may be obtained. Electron microscopical examinations revealed that the S-layer is oriented with its outer, or charge neutral, face towards the air/water interface. This is in perfect agreement with the experiments on solid supports. S-layer protein of this organism was able to recrystallize on surfaces coated with poly-L-lysine. Poly-L-lysine is a positively charged, rather hydrophobic molecule and thus should be capable of undergoing electrostatic and hydrophobic interactions with the S-layer surface. Although proteins generally exhibit a strong tendency to unfold upon adhering to hydrophobic interfaces and to expose their hydrophobic parts, the S-layer subunits were not denatured under these experimental conditions and maintained their ability to form two-dimensional arrays. The orientation of S-layers generated at different interfaces was inspected by electron microscopy after transferring them onto holey grids. A summary of the specific orientations of the oblique S-layer of *B. coagulans* E38-66 is shown in Fig.2.

Preliminary results demonstrate that the formation of large scale crystals is directly related to the physicochemical properties of the proteins and the interface. The two dominant forces determining the intermolecular interactions are hydrophobic and electrostatic forces. The strength of the hydrophobic interaction may be manipulated by adjusting the solvent composition. Methanol and glycerol may be added to the aqueous buffer in order to decrease or increase, respectively, the contribution of the hydrophobic effect on the protein interactions. Methanol preferentially solvates non-polar species, thereby hindering the tendency of water molecules to orient at the interfacial region. This is due to the single hydroxyl group which prevents methanol from entering into a hydrogen-bonded three-dimensional network. In contrast, glycerol, which possesses three hydroxyl groups, can form co-operative structures with water and thus enhances hydrophobic interactions.

S-layers which have either formed monolayers at the air/water interface or at phospholipid films can easily be transferred onto solid surfaces by dip-coating using standard LB techniques. The surface properties of the support will determine the orientation of the crystalline arrays (Fig.3). This technology will play a key role in the fabrication of mono- or multilayer protein arrays attached to solid supports. In addition, the combination of S-layers with mono- or multilayer LB films, in which functional molecules such as molecular carriers, pore-forming proteins, proton pumps, light harvesting complexes, receptors and other biologically active molecules can be incorporated, opens a broad spectrum of applications in molecular nanotechnology and in the development of biosensors.

S-LAYERS AS MATRICES FOR THE CONTROLLED IMMOBILIZATION OF EXOGENOUS MACROMOLECULES

S-layers in suspension as well as S-layers attached to solid surfaces have proven to be ideal matrices for the controlled immobilization of macromolecules. Unlike other carriers where the location, local density, orientation of functional groups, the porosity, and pore size are known only approximately, the properties of a single constituent unit for crystalline layers are replicated with the periodicity of the lattice. This defines the characteristics of the whole two-dimensional array. Carboxyl, amino and hydroxyl groups are located on the constituent protein or carbohydrate moities in an identical position and orientation and this allows for a controlled immobilization of exogenous functional molecules (Sára and Sleytr, 1989; Sára et al., 1992b). These can be bound to S-layer lattices either through electrostatic, hydrophobic or covalent bonds. The distribution of net negatively

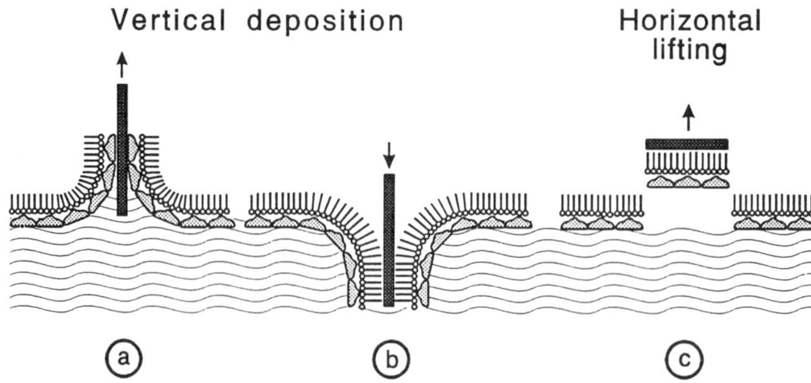

Figure 3. S-layers and S-layer/lipid film composite structures can be transferred onto solid substrates by vertical deposition (a,b) or by horizontal lifting (c). Depending on the surface properties of the support, the deposition may either start (a) with the S-layer or (b) with the lipid film. In the horizontal lifting method (c) the substrate is removed vertically from but parallel to the surface.

charged areas on S-layers can be visualized by electron microscopical methods after labelling with positively charged marker molecules such as polycationized ferritin (PCF). For example, the regular arrangement of free carboxyl groups on the hexagonally-ordered S-layer lattice from *Thermoproteus tenax* could be clearly demonstrated in this way (Messner et al., 1986). With glycosylated S-layers not only the protein but also the carbohydrate chains which are exposed on the outer face are available for the binding of molecules. For example, hydroxyl groups of glycan chains can be converted to carboxyl groups by succinylation thereby producing a strongly electronegative surface (Sára et al., 1989).

For a covalent attachment of molecules carboxyl groups can be activated with carbodiimide whereas hydroxyl groups can be treated with cyanogen bromide (Sára and Sleytr, 1989; Sára et al., 1988, 1992a). A broad spectrum of functional macromolecules, including ferritin, can be used to demonstrate the high binding capacity of S-layers. The spatial distribution of immobilized molecules which were smaller than the constituent S-layer subunits, themselves, has revealed that more than one carboxyl group per subunit must be available for binding (Sára et al., 1992a). Covalently linked molecules frequently exhibit a less ordered arrangement on the lattice than electrostatically bound molecules. This may be explained by the fact that most activation reactions lead to an increase in the number of potential binding sites over each subunit. Consequently, the exogenous molecule may not bind to the exact same spot on each subunit.

The possibility of immobilizing different types of molecules on the protein or on the carbohydrate opens a broad spectrum of applications and a new dimension in the specific binding capacity of S-layers. The first application of S-layers as an immobilization matrix for an ordered binding of biologically active molecules was the development of bioanalytical sensors (Pum et al., 1992).

S-LAYER BIOSENSORS

Biosensors transform a hardly detectable signal into an easily measurable entity by means of a biological process (Turner et al., 1988). Bioanalytical sensors, developed at our research center, were the first molecular machines involving S-layers as the principle structural and immobilization components. In these devices enzymes are covalently linked to S-layer fragments or S-layer self-assemblies. Since the periodicity of the S-layer lattice and, consequently, the location of the functional groups is in the same scale range as the size of the immobilized enzymes, molecules can be covalently bound as monolayers at a maximum packing order with preferential orientations. For the fabrication of a single enzyme-based sensor, such as a glucose sensor, glucose oxidase (GOD) molecules are covalently bound to the exposed S-layer surface of S-layer ultrafiltration membranes (SUM) which are made by depositing and crosslinking S-layer fragments onto hydrophilic microfiltration membranes (Fig.4a) (Sára and Sleytr, 1988). GOD was bound to the carbodiimide activated carboxyl groups of the S-layer protein. In order to facilitate the electrochemical oxidation of hydrogen peroxide in close vicinity to the glucose oxidase molecules, the enzyme layer was metal-coated by sputtering with platinum or gold. The high degree of preservation of enzyme activity (50%-70%) of the immobilized enzyme, activity, the high packing density and the short diffusion distances yielded high signal levels ($150nA/mm^2$/mmol glucose), a broad linear range of reactivity (up to 20 mmol glucose) and a fast response time (10-30 seconds).

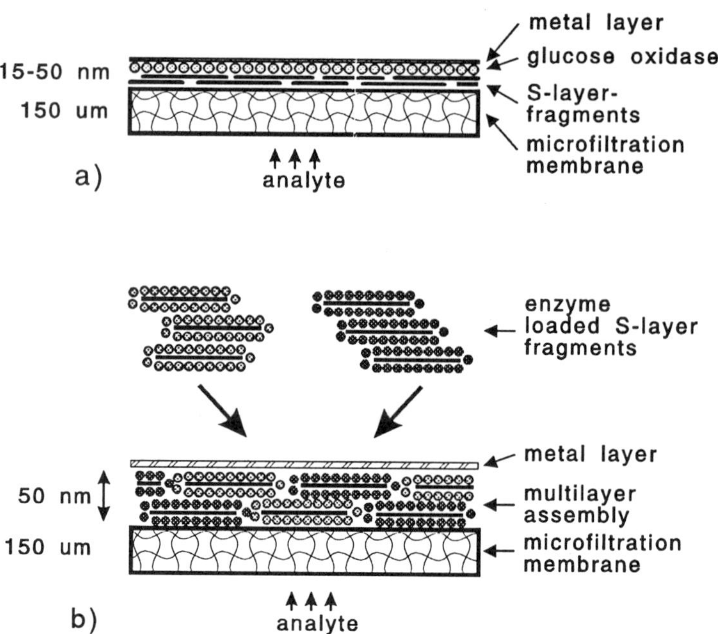

Figure 4. Schematic illustration of an S-layer biosensor. (a) Single enzyme-based biosensor. The sensing layer consists of a thin layer of enzyme molecules which have been covalently bound to a SUM. The electrical contact to the enzyme is established by depositing a platinum or gold layer on the sensing layer. The analyte reaches the sensing layer through the open structure of the microfilter. (b) Schematic illustration of the fabrication process of a S-layer biosensor: The multilayered assembly can be produced either by sequentially depositing the individual enzyme-loaded S-layers or by mixing them (b) prior to deposition on the microporous support. In both cases the multilayer assembly can be crosslinked with glutaraldehyde before coating with platinum or gold. The analyte reaches the multienzyme assembly through the microfilter (modified after Pum et al., 1991).

For the construction of multienzyme sensors a different construction principle was developed since simultaneous immobilization of different enzymes generally leads to an uncontrollable competition for the available activated sites. Individual enzymes were first immobilized to different S-layer fragments in suspension, and then deposited and fixed on a porous support (i.e. microfiltration membrane) (Fig.4b). In this way the immobilization parameters could be optimized for each enzyme and the eventual ratio of each enzyme accurately controlled. This method produces a well structured sandwich of thin enzyme monolayers to which protective S-layers, charged or uncharged isoporous molecular sieves, can easily be integrated. Although such a multienzyme sensor may consist of many superimposed S-layers the total thickness of the sensing unit remains very small in comparison to those produced by conventional manufacturing techniques. Using this technique a sucrose sensor with three enzymes (fructosidase-mutarotase-glucose oxidase) was developed.

The method of depositing enzyme-loaded S-layer fragments onto microfiltration membranes can also be used for the production of single enzyme sensors. In comparison to the SUM-technique (see above) this method has the advantage of binding a higher amount of enzyme per (projected) unit area. Based on this technique prototypes of the mono- and multienzyme biosensors have been developed (Table 1).

S-layers can also be used to interface optical fibers or waveguides with molecules of the sensing layer (e.g., enzymes, fluorescent markers). For this type of biosensor S-layer protein monolayers are recrystallized on transparent supports since S-layer fragments cannot be used. Recently, a glucose sensor using glucose oxidase and a pH-sensitive fluorescent dye has been developed. Detection is based on changes in pH at the vicinity of the enzyme due to the production of gluconic acid during the course of the enzyme reaction.

Table 1. Mono- and multi-enzyme S-layer biosensors

Sensor type	Enzymes	Signal($\mu A \cdot l/cm^2 \cdot mmol$)	Linear Range	Application
glucose	glucose oxidase	15	0 - 20 mmol	blood analysis
ethanol	alcohol oxidase	2.5	0 - 7 mmol	blood analysis, microbiology, food technology
xanthine	xanthine oxidase	30	0 - 0.6 mmol	food technology
maltose	maltase + glucose oxidase	1.5	0 - 1.5 mmol	microbiology
sucrose	ß-fructosidase + mutarotase + glucose oxidase	1	0 - 12 mmol	microbiology
cholesterol	cholesterol esterase + cholesterol oxidase	6	0 - 0.5 mmol (cholesteryl palmitate)	blood analysis

INFORMATION PROCESSING WITH THE SFM

The SFM emerges as the most promising tool for molecular manufacturing because it can be used for "writing" with atoms (Eigler and Schweitzer, 1990; Eigler et al., 1991). Although SFMs were originally designed to image surfaces with sub-Ångstrom resolution the capability to be used as a molecular assembler has extended their application range dramatically. From a general point of view, writing with molecules is conceptually simple but requires a tool which has enough precision to select, drag and finally drop the molecules at selected position. In the conventional approach of down-scaling lithographic procedures, the microfabrication of very large-scale integrated circuits and of miniaturized mechanical devices (e.g., pumps) was only possible by engraving the required structures into a silicon surface (Shedd and Russel, 1990). Although these methods have reached an enormously high level of sophistication, structuring in the third dimension is still a problem. In the same way the assembly of different fucntional parts at the nanoscale level is almost impossible.

Molecular nanotechnology and manufacturing practices have devoted themselves to a different approach because molecular structures should only be constructed from single molecules or molecular assemblies. This seems to be the only way to fabricate multifunctional molecular machines by assembling simple functional units into devices of higher complexity. It is a fundamental principle which has been optimized by nature over billions of years. A controlled assembly of functional devices is only possible when a physicochemically well defined base structure or "work bench" is available. Surface properties of such immobilization matrices must be precisely defined over the entire area. Based on our knowledge of the ultrastructure, chemistry, assembly, function, and binding capacity of S-layers, these two-dimensional crystals are perfectly tailored natural patterning elements for use in nanotechnology. Recently it has been demonstrated that under specific experimental conditions which eliminated uncontrolled movement on a surface (i.e., ultrahigh vacuum and low temperature (4°K)), it is possible to place gas atoms at specific positions on metal surfaces (Eigler and Schweitzer, 1990). Unlike this approach, S-layers allow us to bind macromolecules at well defined locations in buffer solution and at room temperature (Sleytr et al., 1992). The type of interaction (electrostatic, hydrophobic or ligand type interactions) define the strength of the binding forces involved. As work progresses, advanced tip chemistry allow molecules to be picked up more specifically from the S-layer matrix. Binding ligands or antibodies onto the end of the tip, or modifying the surface properties of the probe to make it more hydrophobic or hydrophilic will increase the capability of SFM. If necessary, functional molecules can also be immobilized on activated S-layer surfaces thereby, introducing covalent linkage to the support. Charged molecules (e.g., PCF or cytochrome c) are bound by electrostatic interactions to the S-layer lattice in a perfect regular fashion as shown by electron microscopical studies (Fig.5) (Messner et al., 1986; Beveridge et al., 1988; Pum et al., 1989a; Sleytr et al., 1992). This will allow us to write information with start/end marks, and to read stored information as well (Fig.6a). The writing process is performed by either (i) picking up or (ii) depositing individual molecules on the underlying S-layer (Fig.6b). The first technique is referred to as positive writing and the second as negative writing. In the reading process these morphological units are scanned in order to detect the presence or absence of molecules (i.e., "bits"). The information storage capacity of this type of memory is determined by the size of the molecules and the lattice constant of the underlying S-layer. If the unit cell is 10 nm square, then 10^{12} bits or 1 terabit can be stored in a square centimeter. If the S-layer allows storage of more than 1 bit molecule per morphological unit, a much higher storage capacity could be achieved (e.g., $2^6 = 64$ levels with p6 symmetry). High resolution replication

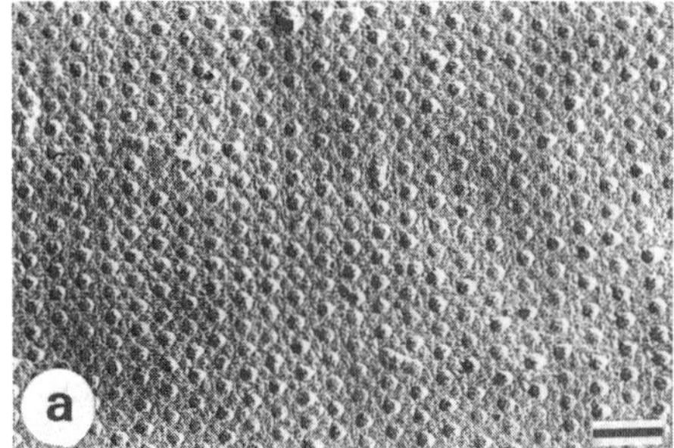
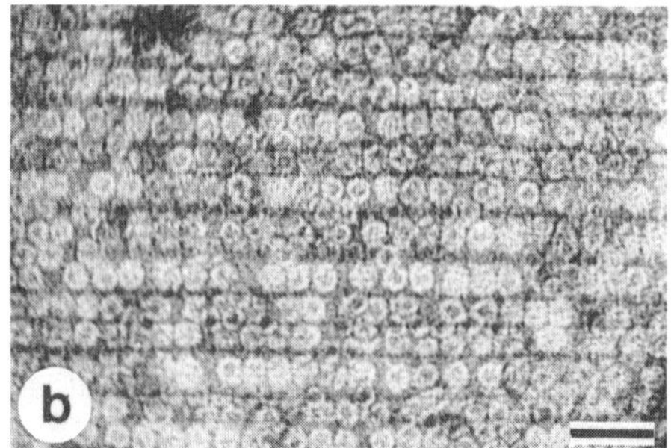

Figure 5. PCF molecules are immobilized in a regular fashion by electrostatic interactions on different types of S-layer lattices. (a) Hexagonal array with a 30 nm lattice spacing. Bar = 100 nm. (b) Linear arrangement with a 12 nm spacing. Bar = 100 nm.

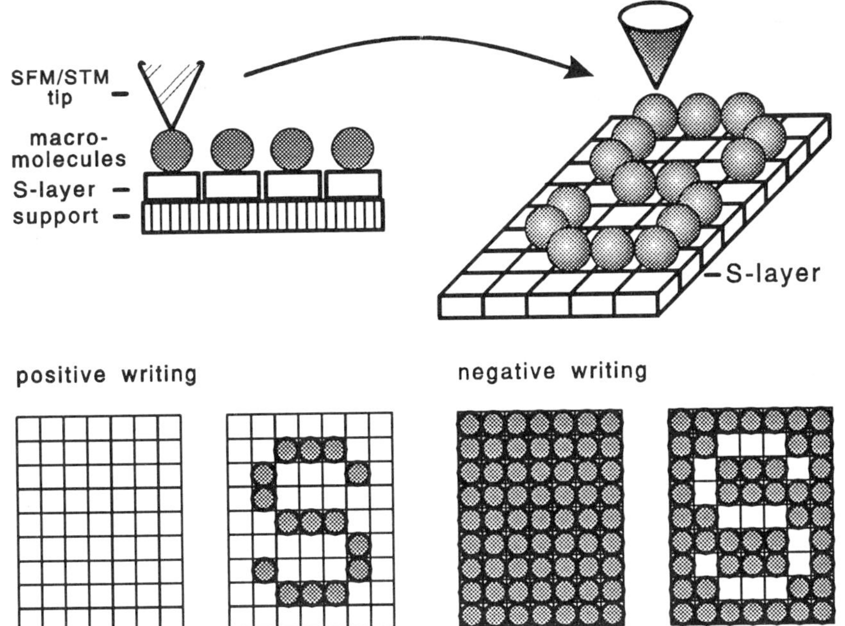

Figure 6. Schematic representation of writing with molecules on a square S-layer lattice. (a) Molecules which are bound by electrostatic forces onto the S-layer may be transferred with the SFM tip in a controlled way to a blank S-layer lattice. (b) It depends on the prospective application whether a positive or a negative writing process will be employed (modified after Sleytr et al., 1992).

techniques involving carbon sublimation or tantalum/tungsten evaporation, as used in specimen preparation for electron microscopy, could be suitable procedures to make replicas of the stored information. After replication the biological material (S-layer protein with attached molecules) could be dissolved following the replica cleaning procedures used in freeze-etching. The high resolution imprints obtained could subsequently act as replication matrices.

CONCLUSIONS

Currently we are investigating the application potential of S-layers as structural and patterning elements for different aspects of molecular nanotechnology, nanomanufacturing and bioelectronics. Based on our knowledge of the ultrastructure, chemistry, morphogenesis, and function of S-layers it has been shown that these two-dimensional protein crystals are appealing model systems for studying several aspects of molecular self-organization at the nanometer level. Their simple construction involving a single constituent subunit, allows a deeper understanding of the "molecular logic" involved in the generation of these two-dimensional lattices. The distribution and orientation of functional groups is also precisely defined. This makes S-layers ideal immobilization matrices with high binding capacity for most biologically active molecules. Since the arrangement of immobilized molecules frequently mimics the crystalline structure of the S-layer lattice we expect that this unique property will lead to new developments in nanotechnology. S-layers which have been recrystallized on solid supports or which have been transferred to them from interfaces may be used as patterning elements for structuring microelectronic or microoptical devices

(Douglas and Clark, 1986; Fisher et al., 1990; Pum et al., 1989b; Sleytr et al., 1988b, 1992; Utsugi, 1990).

Since the surface properties of S-layers can be rendered more hydrophilic or hydrophobic by a broad spectrum of chemical modification reactions, tenside and lipid films (e.g., LB films) can either bind with their hydrophilic or hydrophobic side. The combination of S-layers with bimolecular lipid membranes or membranes composed of tetraether lipids where functional molecules (e.g., carriers, ion channels, ionophores, proton pumps, light harvesting complexes and receptor molecules) are incorporated, opens wide range of new applications in diagnostics, biosensor development and molecular nanotechnology.

ACKNOWLEDGEMENTS

This work was supported by grants from the "Österreichischer Fonds zur Förderung der wissenschaftlichen Forschung" Projekt S5705 and the Österreichisches Bundesministerium für Wissenschaft und Forschung.

The assistance of K. Gruber, C. Hödl, U. Kainz, A. Neubauer, A. Scheberl and M. Weinhandel is gratefully acknowledged.

REFERENCES

Baumeister, W., and Engelhardt, H., 1987, Three-dimensional structure of bacterial surface layers, *in*: "Electron microscopy of proteins", Vol. 6, J.R. Harris, and R.W. Horne, eds., Academic Press Inc., London.

Beveridge, T.J., 1981, Ultrastructure, chemistry, and function of the bacterial wall, *Int. Rev. Cytol.* 72:229.

Beveridge, T.J., and Graham, L.L., 1991, Surface layers of bacteria, *Microbiol. Rev.* 55:684.

Beveridge, T.J., Sprott, G.D., and Stewart, M., 1988, The structure, chemistry and physicochemistry of the *Methanospirillum hungatei* GP1 sheath, *in*: "Crystalline Bacterial Cell Surface Layers", U.B. Sleytr, P. Messner, D. Pum, and M. Sára, eds., Springer, Berlin.

Douglas, K., and Clark, N.A., 1986, Nanometer molecular lithography, *Appl. Phys. Lett.* 48:676.

Eigler, D.M., and Schweizer, E.K., 1990, Positioning single atoms with a scanning tunneling microscope, *Nature* 344:524.

Eigler, D.M., Lutz, C.P., and Rudge, W.E., 1991, An atomic switch realized with the scanning tunneling microscope, *Nature* 352:600.

Fisher, K.A., Yanagimoto, K.C., Whitfield, S.L., Thomson, R.E., Gustafsson, M.G.L., and Clarke, J., 1990, Scanning tunneling microscopy of planar biomembranes, *Ultramicroscopy* 33:117.

Gruber, K., and Sleytr, U.B., 1991, Influence of an S-layer on surface properties of *Bacillus stearothermophilus*, *Arch. Microbiol.* 156:181.

Hovmöller, S., Sjögren, A., and Wang, D.N., 1988, The structure of crystalline bacterial surface layers, *Prog. Biophys. Molec. Biol.* 51:131.

Israelachvili, J., 1985, "Intermolecular and Surface Forces", Academic Press Inc., London.

Jaenicke, R., Welsch, R., Sára, M., and Sleytr, U.B., 1985, Stability and self-assembly of the S-layer protein of the cell wall of *Bacillus stearothermophilus*, *Physiol. Chem. Hoppe-Seyler* 366:663.

Koval, S.F., 1988, Paracrystalline protein surface arrays on bacteria, *Can. J. Microbiol.* 34:407.

Messner, P., and Sleytr, U.B., 1991, Bacterial surface layer (S-layer) glycoproteins, *Glycobiol.* 1:545.

Messner, P., and Sleytr, U.B., 1992, Crystalline bacterial cell-surface layers, *Adv. Microbial Physiol.* 33:213.

Messner, P., Pum, D., Sára, M., Stetter, K.O., and Sleytr, U.B, 1986, Ultrastructure of the cell envelope of the archaebacteria *Thermoproteus tenax* and *Thermoproteus neutrophilus, J. Bacteriol.* 166:1046.

Nagayama, K., 1992, Protein array: an emergent technology from biosystems, *Nanobiology* 1:25.

Pum, D., Sára, M., and Sleytr, U.B. ,1989a, Structure, surface charge and self-assembly of the S-layer lattice from *Bacillus coagulans* E38-66, *J. Bacteriol.* 171:5296.

Pum, D., Sára, M., Sleytr, U.B., 1989b, Use of two-dimensional protein crystals from bacteria for nonbiological applications, *J. Vac. Sci. Technol. B* 6:1391.

Pum, D., Sára, M., Messner, P., and Sleytr, U.B., 1991, Two-dimensional (glyco)protein crystals as patterning elements for the controlled immobilization of functional molecules, *Nanotechnology* 2:196.

Pum, D., Sára, M., and Sleytr, U.B., 1992, Two-dimensional (glyco)protein crystals as patterning elements and immobilisation matrices for the development of biosensors, *in*: "Immobilised Macromolecules: Application Potential", U.B. Sleytr, P. Messner, M. Sára, and D. Pum, eds., Springer, London (in press).

Sára, M., and Sleytr, U.B., 1987a, Production and characteristics of ultrafiltration membranes with uniform pores from two-dimensional arrays of proteins, *J. Membr. Sci.* 33:27.

Sára, M., and Sleytr, U.B., 1987b, Charge distribution on the S-layer of *Bacillus stearothermophilus* NRS 1536/3c and the importance of charged groups for morphogenesis and function, *J. Bacteriol.* 169:2804.

Sára, M., and Sleytr, U.B., 1988, Membrane biotechnology: Two-dimensional protein crystals for ultrafiltration purposes, *in*: "Biotechnology", Vol. 6b, H.J. Rehm, ed., VCH Verlagsgesellschaft, Weinheim.

Sára, M., and Sleytr, U.B., 1989, Use of regularly structured bacterial cell envelope layers as matrix for the immobilization of macromolecules, *Appl. Microbiol. Biotechnol.* 30:184.

Sára, M., Wolf, G., Küpcü, S., Pum, D., Sleytr, U.B., 1988, Use of crystalline bacterial cell envelope layers as ultrafiltration membranes and supports for the immobilization of macromolecules, *in*: "Dechema Biotechnology Conferences", Vol. 2, VCH Verlagsgesellschaft, Weinheim.

Sára, M., Küpcü, S., and Sleytr, U.B., 1989, Localization of the carbohydrate residue of the S-layer glycoprotein from *Clostridium thermohydrosulfuricum* L111-69, *Arch. Microbiol.* 151:416.

Sára, M., Pum, D., and Sleytr, U.B., 1992a, Permeability and charge-dependent adsorption properties of the S-layer lattice from *Bacillus coagulans* E38-66, *J. Bacteriol.* 174:3487.

Sára, M., Küpcü, S., Weiner, C., Weigert, S., and Sleytr, U.B., 1992b, Crystalline protein layers as isoporous molecular sieves and immobilization and affinity matrices. *in*: "Immobilised Macromolecules: Application Potential", U.B. Sleytr, P. Messner, M. Sára, and D. Pum, eds., Springer, London (in press).

Shedd, G.M., and Russell, P.E., 1990, The scanning tunneling microscope as a tool for nanofabrication, *Nanotechnol.* 1:67.

Sleytr, U.B., 1978, Regular arrays of macromolecules on bacterial cell walls: structure, chemistry, assembly, and function, *Int. Rev. Cytol.* 53:1.

Sleytr, U.B., and Messner, P., 1983, Crystalline surface layers on bacteria, *Annu. Rev. Microbiol.* 37:311.

Sleytr, U.B., and Messner, P., 1988, Crystalline surface layers in procaryotes, *J. Bacteriol.* 170:2891.

Sleytr, U.B., and Messner, P., 1989, Self assembly of crystalline bacterial cell surface layers (S-layers), *in*: "Electron Microscopy of Subcellular Dynamics", H. Plattner, ed., CRC Press Inc., Boca Raton.

Sleytr, U.B., Messner, P., and Pum, D., 1987, Analysis of crystalline bacterial surface layers by freeze-etching, metal shadowing, negative staining, and ultra-thin sectioning, *in*: "Methods in Microbiology", Vol. 20, F. Mayer, ed., Academic Press Inc., London.

Sleytr, U.B., Messner, P., Pum, D., and Sára, M., 1988a, "Crystalline Bacterial Cell Surface Layers", Springer, Berlin.

Sleytr, U.B., Sára, M., and Pum, D., 1988b, Application potentials of two dimensional protein crystals. *in*: "Microcircuit Engineering '88", F. Paschke, W. Fallmann, and H. Löschner, eds., Elsevier Science Publishers B.V., Amsterdam.

Sleytr, U.B., Pum, D., Sára, M., and Messner, P., 1992, Two-dimensional protein crystals as patterning elements in molecular nanotechnology, *in*: "AIP conference proceedings 262, Molecular Electronics - Science and Technology", A. Aviram, ed., American Institute of Physics, New York.

Sleytr, U.B., Messner, P., Pum, D., and Sára, M., 1993, Crystalline bacterial cell surface layers: General principles and application potential, *in*: "The Cell Envelope - Structure and Function", Special Issue of *J. Appl. Bacteriol.* (in press).

Smit, J., 1987, Protein surface layers of bacteria, *in*: "Bacterial Outer Membranes as Model Systems", M. Inouye, ed., John Wiley and Sons, New York.

Turner, A.P.F., Karube, I., and Wilson, G.S., 1988, "Biosensors: Fundamentals and Applications", Oxford University Press, Oxford.

Utsugi, Y., 1990, Nanometre-scale chemical modification using a scanning tunneling microscope, *Nature* 347:747.

Chapter 21

SURFACE LAYERS FROM *BACILLUS ALVEI* AS A CARRIER FOR A *STREPTOCOCCUS PNEUMONIAE* CONJUGATE VACCINE

Andrew J. Malcolm*, Michael W. Best*, Roderick J. Szarka*, Zina Mosleh* and Frank M. Unger

Chembiomed Ltd.
Edmonton, Alberta, Canada

Paul Messner and Uwe B. Sleytr

Center for Ultrastructure Research and Ludwig Boltzman
Institute for Molecular Nanotechnology
University of Agriculture
Vienna, Austria

INTRODUCTION

Approaches to the use of crystalline bacterial cell surface layers (S-layers) as carrier/adjuvants for conjugate vaccines have been reported recently (Sleytr et al., 1987; Sleytr et al., 1988; Malcolm, et al., 1993; Messner and Sleytr, 1992; Messner et al., 1992; Smith et al., in press). In this chapter, we report on some of the results we have obtained using an S-layer derived from *Bacillus alvei* CCM 2051 (Altman et al., 1991) as a carrier for a prototype conjugate vaccine against *Streptococcus pneumoniae*, serotype 8.

We have developed methods to produce oligosaccharides from bacterial polysaccharides and chemical coupling procedures which have resulted in *S. pneumoniae* oligosaccharide hapten-S-layer conjugates effective in eliciting immunoprotective cellular and antibody responses to the pathogenic organism. This technology is currently being utilized to develop other conjugate vaccines to several bacterial strains.

---

*Present Address: Alberta Research Council, P. O. Box 8330,
Edmonton, Alberta, Canada, T6H 5X2

Advances in Bacterial Paracrystalline Surface Layers
Edited by T.J. Beveridge and S.F. Koval, Plenum Press, New York, 1993

CURRENT PNEUMOCOCCAL VACCINE

Pneumococci are currently divided into 84 serotypes based on their capsular polysaccharides. Serotypes 1,3,4,7,8 and 12 are more clinically prevalent in the adult population; serotypes 6,14,19 and 23 often cause pneumonia in children (Mandell, 1990). The most widely used anti-pneumococcal vaccine is composed of purified capsular polysaccharides from 23 strains of pneumococi (Pneumovax®23, Merck Sharp & Dohme). The 23 pneumococcal capsular types included in Pneumovax®23 are 1, 2, 3, 4, 5, 6B, 7F, 8, 9N, 9V, 10A, 11A, 12F, 14, 15B, 17F, 18C, 19F, 19A, 20, 22F, 23F, 33F (Danish nomenclature). These serotypes are said to be responsible for 90 percent of serious pneumococcal disease in the world. Some controversy exists in the literature over the efficacy of the Pneumovax®23 vaccine (Borgano et al.,1978; Broome et al., 1980; Sloyer et al., 1981; Shapiro and Clemens, 1984; Bolan et al., 1986; Simberkoff et al., 1986; Forester et al., 1987; Shapiro, 1987; Sims et al., 1988; Simberkoff, 1989; Shapiro, 1991). The pneumococcal vaccine is effective for induction of an antibody response in healthy young adults (Hilleman et al., 1981; Mufson et al., 1985). The antibodies have been shown to have *in vitro* opsonic activity (Chudwin et al., 1985). However, there is marked variability in the response and in the persistence of antibody titers to the different serotypes (Hilleman et al., 1981; Mufson et al., 1987). Children under 2 years of age are the group at highest risk of systemic disease and otitis media caused by pneumococci, but they do not respond to this vaccine (Sell et al., 1981; Douglas et al., 1983). Furthermore, elderly patients with a variety of conditions and immunosuppressed patients have impaired or varied responses to Pneumovax®23 (Siber et al., 1978; Schildt et al., 1983; Forester et al., 1987; Simberkoff, 1989; Shapiro, 1991). These population groups do not respond well to the thymus independent polysaccharide antigens of this vaccine. Typical of thymus independent antigens, antibody class switching from an IgM to IgG isotype or an anamnestic response to a booster immunization is not usually observed (Borgono et al., 1978). Clearly, new vaccines against pneumoccocci are needed, especially for high risk groups and children.

CONJUGATE VACCINES

Several protein carrier conjugates have been developed which elicit thymus dependent responses to a variety of bacterial polysaccharides and oligosaccharides. To date the development of conjugate vaccines to *Hemophilus influenzae* type b (Hib) has received the most attention. Connaught Laboratories has adopted the technology of Schneerson et al. (1980) by covalently coupling Hib polysaccharides (polyribitol-phosphate) to diphtheria toxoid. This group has also developed a Hib vaccine by derivatizing the polysaccharide with an adipic acid dihydrazide spacer and coupling this material to tetanus toxoid with carbodiimide (Schneerson et al., 1986). A similar procedure was used to produce conjugates containing diphtheria toxoid as the carrier (Gordon, 1986 and 1987). Merck Sharp & Dohme utilized a bifunctional spacer to couple the outer membrane protein of group B *Neisseria meningitidis* to Hib polysaccharides (Marburg et al., 1986 and 1989). Finally, Anderson (1983 and 1987) has produced a conjugate vaccine using Hib oligosaccharides coupled by reductive amination to a nontoxic, cross-reactive mutant diphtheria toxin CRM_{197}. This technology is being used by Praxis Biologics, Inc. Reports in the literature differ on the efficacy of these vaccines and many studies are still in progress. However, the oligosaccharide conjugates are immunogenic in infants and elicit a thymus dependent response (Anderson et al., 1985a, 1985b).

Other conjugate vaccines have been developed by Jennings et al. (1985 and 1989) who utilized periodate activation to couple polysaccharides of *Neisseria meningitidis* to tetanus or diphtheria toxoid carriers. Porro (1987) defined methods to couple esterified *N. meningitidis* oligosaccharides to carrier proteins.

Miles Laboratories developed conjugate vaccines containing polysaccharides of *Pseudomonas aeruginosa* coupled by the periodate procedure to detoxified protein from the same organism (Tsay and Collins, 1987). Cryz and Furer (1988) used adipic acid dihydrazide as a spacer arm to produce conjugate vaccines against *P. aeruginosa*.

Polysaccharides of specific serotypes of *S. pneumoniae* have also been coupled to classical carrier proteins such as tetanus or diphtheria toxoids (Schneerson et al., 1984; Fattom et al., 1988), *N. meningitidis* membrane protein (Marburg et al., 1987) and, more recently to a pneumolysin mutant carrier (Paton et al., 1991; Lock et al., 1992). American Cyanamid Company has patented technology for coupling *S. pneumoniae* oligosaccharides to CRM_{197} protein (Porro, 1992). Again, these conjugate vaccines have variable or as yet undetermined immunopotentiation properties.

In this paper, we present previously unreported data which demonstrates that S-layers derived from *B. alvei* CCM 2051 may be useful as a carrier for pneumonia prototype vaccines. We will discuss the immunological and chemical properties of S-layers which we feel may make S-layers better vaccine carriers than other widely used carriers such as tetanus or diphtheria toxoids. We will briefly describe our oligosaccharide preparation, conjugation technologies and the effectiveness of our conjugate in animal models to elicit antibody and cellular mediated immunoprotective responses.

IMMUNE RESPONSES TO S-LAYER CONJUGATES

At the outset of our investigation, the following requirements of a carrier for a conjugate vaccine were established.
(a)   Multiple reactive groups for coupling a variety of haptens.
(b)   Ease of handling and good recovery after coupling.
(c)   Ability to form stable hapten-carrier conjugates.
(d)   Injectable form does not cause noticeable trauma.
(e)   Ability to elicit a cellular response.
(f)   Ability to elicit immunoprotective antibodies to otherwise weakly immunogenic carbohydrate antigens.
(g)   Ability to induce B and T cell memory.

In an initial study, both glutaraldehyde fixed and non-fixed forms of S-layer proteins were injected into mice at various concentrations and by different immunization routes. All non-fixed S-layers tested to date (*Clostridium thermohydrosulfuricum* L111-69, Christian et al., 1988; *C. thermosaccharolyticum* D120-70, Altman et al., 1990; *B. alvei* CCM 2051, Altman et al., 1991; *B. stearothermophilus* NRS 2004/3a, Christian et al., 1986 and Messner et al., 1987; and *B. stearothermophilus* PV72, Sleytr and Messner, 1983) elicited anamnestic and long lasting antibody responses without the necessity of an extraneous adjuvant. Typical of a thymus dependent response, the secondary IgG antibodies were much higher (approximately 10 times) than those seen after a primary response (Malcolm et al., 1993).

Although glutaraldehyde-fixed S-layers have been reported to be immunogenic carriers in experimental delayed-type hypersensitivity assays (Smith et al., in press), this form of S-layer generally elicited poor antibody responses. Adjuvants (Freund's complete and incomplete, dimethyldioctadecyl-ammonium bromide, 2, 6, 10 ,14-tetramethylpentadecane, or Adjuvax™) did not markedly enhance any of the responses to fixed S-layer. We believe that glutaraldehyde treatment results in conformational changes which alter important immunogenic epitopes. Fig. 1 clearly illustrates the superiority of unfixed versus fixed S-layers as carrier/adjuvant material.

EXPERIMENTAL ANTI-PNEUMOCCOCAL VACCINE

We have focused our research efforts on the development of S-layer carrier conjugates which are capable of eliciting immunoprotective antibody responses to *S. pneumoniae*. For the design of an universally effective anti-pneumococcal vaccine, we selected *S. pneumoniae* type 8 as the first target organism. This strain is associated with bacteremic disease and high case fatality rates. Oligosaccharide haptens were prepared by acid-catalyzed hydrolysis of serotype 8 polysaccharides and were separated by size-exclusion chromatography into fractions comprising from one to eight repeat units. Oligosaccharides derived from hyaluronic acid were used to standardize the chromatographic system. The number of repeat units in each oligosaccharide fraction was confirmed by thin-layer chromatorgraphy. Oligosaccharides of selected size were then coupled to unfixed S-layer 2051 (i.e., the S-layer from *B. alvei* CCM 2051) by modified carbodiimide and reductive amination procedures (Davis and Preston, 1981; Roy et al., 1984; Malcolm et al., manuscript in preparation). Fig. 2 illustrates (a) the type 8 oligosaccharide repeating unit, (b) the diol group activated by periodate for reductive amination coupling and (c) the -COOH reactive group for carbodiimide coupling.

Methods for conjugate formation have been developed for the utilization of different reactive groups on the hapten, or on the glycan chains or protein of the S-layer 2051. This provides flexibility in preparing immunogenic compounds and the option to design multi-hapten vaccines. Furthermore, the degree of haptenation of the S-layer protein can be closely controlled.

Oligosaccharides of various repeat unit length have been coupled to the S-layer 2051 and injected into mice. Conjugates comprising relatively small oligosaccharide haptens have elicited good antibody responses, as measured by ELISA (Fig. 3). No adjuvant material other than the S-layer carrier is necessary to elicit specific antibodies. Injection of the polysaccharide material alone did not induce significant antibody titers. However, injection of the type 8 oligosaccharide-S-layer 2051 conjugate into mice

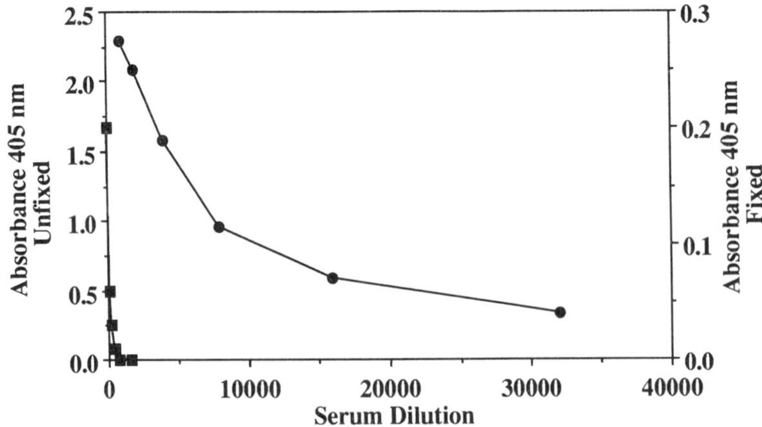

Figure 1. Tertiary antibody (IgA, IgG and IgM) response to unfixed (●) or glutaraldehyde fixed (■) S-layer 2051.

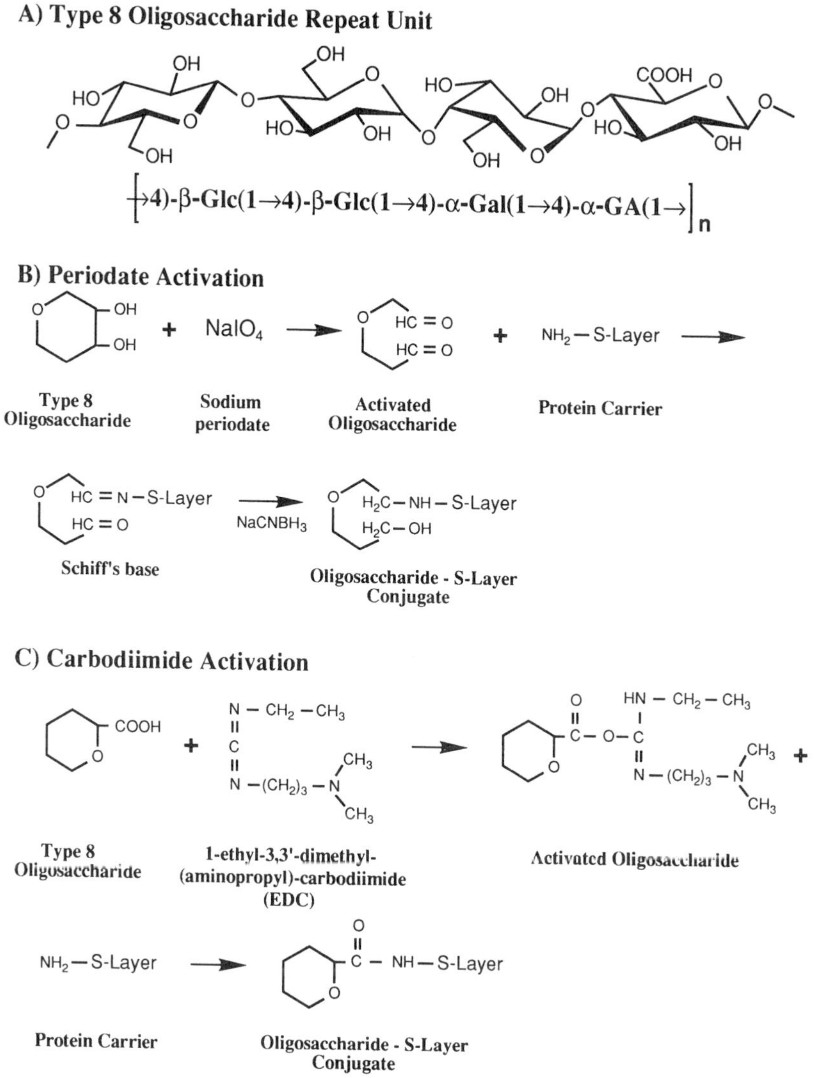

Figure 2. Coupling procedures for *S. pneumoniae* conjugate vaccines.
A) The capsular polysaccharide of *S. pneumoniae* serotype 8 is composed of repeating oligosaccharide units of: $\rightarrow 4$)-β-glucose $(1\rightarrow4)$-β-glucose $(1\rightarrow4)$-α-galactose $(1\rightarrow4)$-α-glucuronic acid $(1\rightarrow$.
B) Oxidative amination by sodium periodate activates diol groups in the oligosaccharide and with the addition of an amino group found on the protein carrier, an intermediate (Schiff's base) is formed. Reductive amination by sodium cyanoborohydride protonates the Schiff's base, resulting in the formation of a stable oligosaccharide-S-layer conjugate.
C) Carbodiimide activation of carboxyl groups in the oligosaccharide reacts with amino groups found on the protein carrier to form carbohydrate-protein conjugates.

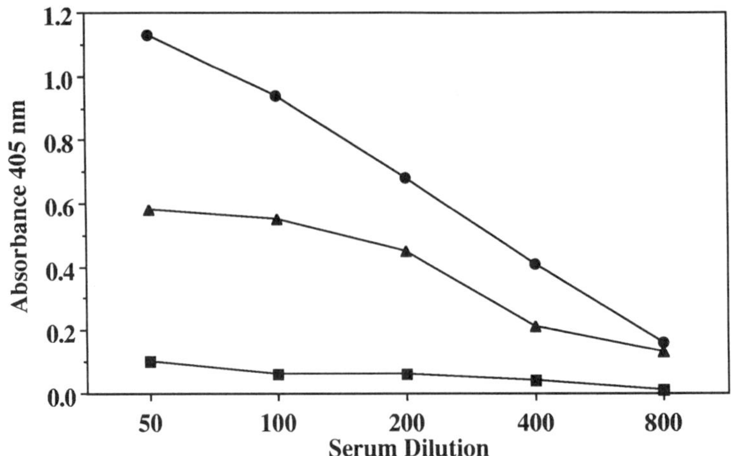

Figure 3. Tertiary antibody (IgA, IgG and IgM) response using an ELISA assay of BALB/c mice injected with either *S. pneumoniae* serotype 8 oligosaccharide-S-layer 2051 conjugate (●), serotype 8 polysaccharide (■) or with two injections of polysaccharide followed by a single injection of the type 8-2051 conjugate (▲).

which had previously received the type 8 polysaccharide alone induced an increase in the antibody titer.

Serum from mice immunized with type 8 oligosaccharide-S-layer 2051 conjugates were found to be immunoprotective. Results from bactericidal assays show that these antibodies greatly reduced (by 99%) serotype 8 pathogenic colony forming units of *S. pneumoniae* grown on blood agar plates (Table 1, #1a). Immunization with the

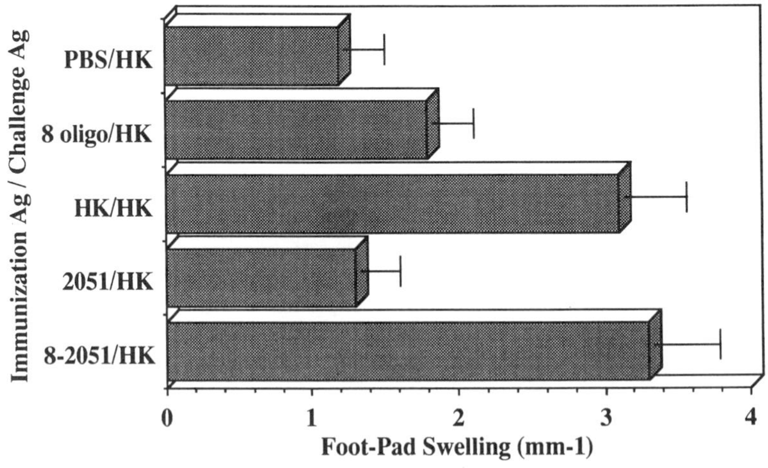

Figure 4. The immunogenicity of *S. pneumoniae* type 8 oligosaccharide S-layer 2051 conjugate as measured by delayed-type hypersensitivity in BALB/c mice. One week after intramuscular immunization of *S. pneumoniae* type 8 oligosaccharide alone or coupled to S-layer 2051, or with heat-killed (HK) *S. pneumoniae*, S-layer 2051 alone or PBS as controls, all mice were footpad challenged with heat-killed *S. pneumoniae* type 8. The delayed inflammatory reaction at the challenge site was measured by micrometer 24 hours later.

with the polysaccharide alone elicited no protection (Table 1, #1b). Injection of 8 oligosaccharide-S-layer 2051 conjugates into mice previously administered the 8 polysaccharide (Table 1, #1c) induced some immuno-protective antibodies (65% reduction of colonies).

Table 1. Bactericidal Assay*

	Immunogen		Percent Reduction
1.	Oligosaccharide Conjugate/ Polysaccharide Immunizations		
	a. 8-2051/8-2051/8-2051		99
	b. 8 poly/8 poly/8 poly		0
	c. 8 poly/8 poly/8-2051		65
	d. PBS		0
2.	Rabbit Immunizations		
	a. 8-2051		88
	b. 8 oligo		49
	c. 8 poly		55
	d. NRS		0
3.	Carrier Pre-Immunizations	Immunogen	
	a. Tetanus Toxoid	8-2051	99
	b. S-layer 2051	8-2051	0
	c. S-layer PV72	8-2051	99
	d. PBS	8-2051	99
	e. PBS	PBS	0

*The ability of mouse and rabbit antisera to inhibit the growth of *S. pneumoniae* serotype 8 as measured by the blood bactericidal assay. Three sets of antisera from experiments shown in Figures 3, 5, and 6 were mixed with heparinized, naive mouse blood containing live *S. pneumoniae*. After an incubation of one hour at 37°C, samples were plated onto blood agar. Results were expressed as percent reduction of colony forming units when compared with PBS control serum or normal rabbit serum (NRS). 8=type 8 oligosaccharide, 2051=S-layer 2051 and PBS=phosphate buffered saline.

These results demonstrate that our experimental S-layer conjugate vaccine elicits bactericidal antibodies against the serotype 8 *S. pneumoniae* pathogen. Polysaccharide material, which is used in the commercially available pneumonia vaccine, did not induce a bactericidal antibody response in our experimentation. These findings indicate that vaccination with the type 8 oligosaccharide-S-layer 2051 conjugate will result in immunoprotection. As well, the conjugate material may be useful for enhancing the immune response to specific serotypes in patients previously administered Pneumovax®23. Specific oligosaccharide serotypes coupled to S-layers may be beneficial as a booster to augment the immunoprotection of high risk groups.

These conjugates could also elicit a cellular response. Mice immunized with 8 oligosaccharide-S-layer 2051 conjugate exhibited a delayed-type hypersensitivity response on challenge with heat killed serotype 8 *S. pneumoniae* bacteria (Fig. 4). This response was equivalent to the reaction of mice primed with heat-killed *S. pneumoniae*. Immunization with the 8 oligosaccharide alone or with the S-layer 2051 alone generated only a marginal delayed-type hypersensitivity response above the PBS control. Previously, Smith et al. (in press) reported that fixed S-layers could prime for a hapten specific delayed-type hypersensitivity response, when both the immunogen and challenge material contained a fixed S-layer carrier. In this article we demonstrate a further use of unfixed S-layer conjugates in that they have the ability to generate a specific cellular response to the native pathogen.

Good antibody titers were also elicited in rabbits immunized with the 8 oligosaccharide-S-layer 2051 conjugate (Fig. 5). Compared with the conjugate material, injection of type 8 oligosaccharides or of type 8 polysaccharides elicited reduced or low titers, respectively. Serum from rabbits immunized with 8 oligosaccharide-S-layer 2051 conjugates had good bactericidal activity (88% colony reduction; Table 1, #2a.). Bactericidal antibody levels were also observed in the sera of rabbits immunized to the 8 oligosaccharide and polysaccharide (49% and 55% colony reduction respectively; Table 1, #2b and c). However, the oligosaccharide-S-layer 2051 conjugate elicited long lasting and the best bactericidal responses, reinforcing the findings of our mouse studies.

UNIQUENESS OF S-LAYER CARRIERS

Although our methodologies can be adapted to conjugate haptens to a variety of carrier molecules, experimental results to date lead us to believe that S-layers may act as superior vaccine carriers, for the following reasons:

1. We obtain more reproducible and better coupling efficiency with S-layers than with other protein carriers such as diphtheria or tetanus toxoid. We believe this may be due to the repeating crystalline glycoprotein nature of S-layers, with the amino, carboxyl or hydroxyl groups available for hapten binding occurring on each protomer in identical position and orientation. Different hapten to S-layer ratios may be achieved by varying the reaction time and the amounts of hapten or S-hapten or S-layer in the reaction vial. In our laboratory, hapten incorporation is consistently greater with S-layer glycoprotein than with toxoids.

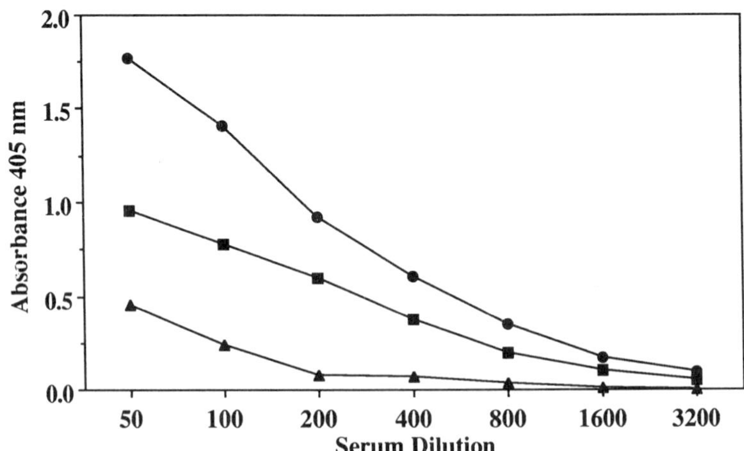

Figure 5. Tertiary IgG antibody response in rabbits to *S. pneumoniae* type 8 oligosaccharide either coupled to S-layer 2051 (●) or alone (■), and to type 8 capsular polysaccharide (▲).

2.   Multi-hapten vaccines can also be produced with S-layers. The amino, carboxyl or hydroxyl groups of S-layers can be activated by different methods to bind haptens. Alternately, reactive groups of haptens can be activated and then bound to S-layers.

3.   S-layer carriers elicit delayed-type hypersensitivity and immunoprotective antibody responses without the use of extraneous adjuvants. The aggregate nature of the S-layer carrier may endow it with an intrinsic adjuvant property. No injection site trauma or granuloma formation has been observed in our animal studies. S-layer conjugates are not toxic or pyrogenic. Other vaccine formulations such as amphiphilic adjuvant molecules (Quil A™), immunogenic complexes containing lipid micelles (Iscoms) or other lipids are reported to be pyrogenic.

4.   S-layer carrier conjugates can be administered by several different immunization routes. Intramuscular or nasal/oral administration of our hapten-S-layer conjugates can elicit a delayed-type hypersensitivity response (Smith et al., in press.) Subcutaneous, intramuscular or intraperitoneal injection of S-layer conjugates elicit a good immunoprotective antibody response. Nasal/oral administration results in a reduced level of bactericidal antibody. We are currently modifying our immunization regime in an attempt to elicit higher antibody titers by the nasal/oral inoculation route.

5.   The most important property of the S-layer carriers is their immunological uniqueness. Antibody and delayed-type hypersensitivity responses to each S-layer are specific and not cross-reactive. Immunization of mice with the S-layer 2051 elicits antibodies which are inhibited from binding to an S-layer 2051-coated ELISA plate by free S-layer 2051 molecules, whereas S-layer PV72 or L111 molecules do not inhibit this binding (results not shown - Malcolm et al., manuscript in preparation). Similarly, antisera of other S-layers are inhibited by only homologous S-layer molecules (Malcolm et al., 1993; Malcolm et al., manuscript in preparation). As well, delayed-type hypersensitivity responses are elicited only when mice are challenged with the same immunizing S-layer (Malcolm et al., manuscript in preparation).

CARRIER-INDUCED EPITOPE SUPPRESSION

Repeated administration of vaccine carriers, such as tetanus or diphtheria toxoid, may cause a phenomenon called carrier-induced epitope suppression (Herzenberg and Tokuhisa, 1982). Antibody responses to a hapten coupled to a carrier protein can be reduced or absent when the recipient has been previously immunized with the carrier protein. Tetanus and diphtheria toxoids are widely used as vaccine carriers both for the immunization of children and adults, and as a prophylactic measure following trauma. Epitope suppression has been described in the literature with synthetic peptides coupled to tetanus toxoid (Schutze et al., 1985; Etlinger et al., 1988; Gaur et al., 1990). A carrier suppressive effect has also been observed with saccharide-toxoid conjugate vaccines (Granoff et al., 1984; Cryz et al., 1986; Di John et al., 1989; Peeters et al., 1991). Investigators have attempted to overcome this problem by using material produced by a mutant *Corynebacterium diphtheriae* (CRM_{197}) (Porro et al., 1980). The administration of toxin or endotoxin from *Bordetella pertussis*, or the adjuvant-active nonpyrogenic derivative of muramyl dipeptide reduced epitopic suppression of anti-peptide responses (Vogel et al., 1987).

Vaccines based on the immunologically distinct S-layer carrier molecules could be an efficacious way to avoid the carrier suppression phenomenon.

Experimentally, we have demonstrated that carrier suppression may occur. This suppression can be circumvented by using different S-layer carriers (Fig. 6). Groups of mice were injected three times with tetanus toxoid, with S-layer 2051, with S-layer PV72 or with PBS. All groups were subsequently injected three times with type 8 oligosaccharide-S-layer 2051 conjugates. Mice pre-sensitized to S-layer 2051 did not exhibit a significant antibody titer to the 8 oligosaccharide coupled to the S-layer 2051 carrier, indicative of a carrier suppression phenomenon. However, mice pre-immunized to immunologically unrelated carriers (tetanus toxoid and the S-layer PV72) did produce antibodies to the 8 oligosaccharide. Bactericidal activity of antisera towards the serotype 8 pathogen was measured. Sera from mice pre-immunized with the S-layer 2051 carrier had no ability to reduce *S. pneumoniae* colony growth (Table 1, #3b). However, sera from mice pre-sensitized to tetanus toxoid (Table 1, #3a) or PV72 (Table 1, #3c) had excellent bactericidal antibody capacity (99% reduction of colonies).

SUMMARY

The information presented in this chapter indicates that S-layers have good potential as a unique carrier/adjuvant for conjugate vaccines. We have developed reproducible, efficient and effective coupling technologies. We have produced a hapten which contains the epitopes necessary to elicit a specific and immunoprotective response to the native pathogen. We demonstrate that a thymus independent response to a bacterial *S. pneumoniae* polysaccharide can be changed to an immunoprotective thymus dependent response with S-layer 2051 conjugate material.

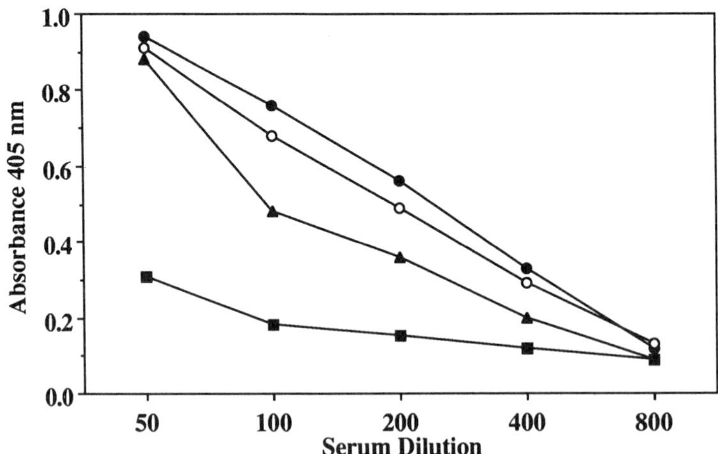

Figure 6. The effect of carrier pre-immunization on the antibody (all isotypes) response to *S. pneumoniae* serotype 8 oligosaccharides coupled to S-layer 2051. The protein carriers tetanus toxoid (●), S-layer 2051 (■) and S-layer PV72 (▲) were injected three times. PBS (O)was administered to one group of BALB/c mice as controls. All groups then received a series of three injections of the 8-oligosaccharide-S-layer 2051 conjugate.

The S-layer carrier induced specific cellular and immunoprotective antibody responses to otherwise weakly immunogenic haptens without the use of extraneous adjuvants. The aggregate nature of the S-layer may endow our vaccine prototype with an intrinsic adjuvant property. Additionally, the glycoprotein nature of S-layers allows us to activate specific reactive groups on the protein portion or on the glycan chains. Haptens can also be activated by various methods and bound to the S-layer molecules. This affords the possibility of producing multi-hapten vaccines. A multi-hapten pneumonia vaccine containing oligosaccharides of 3 to 4 serotypes may be very useful to augment the response to Pneumovax®23 in high risk patients. It may be possible to develop multi-hapten conjugate vaccines for different population groups.

We have identified several S-layer glycoproteins which elicit non-cross reactive antibody and cellular responses. Vaccines to a variety of diseases can be developed using S-layers isolated from various bacterial strains, thereby avoiding the carrier suppression phenomenon observed with other widely used vaccine carriers.

Unresponsiveness of children, the immunosuppressed or the elderly to polysaccharides remains a great medical problem. We have demonstrated that mice which are unresponsive to polysaccharide material will exhibit an immunoprotective response to oligosaccharide haptens when coupled to a S-layer carrier. To examine whether S-layer conjugates can induce immunoprotection and memory responses in populations with immature or suppressed immune systems, we are currently testing the immune responses of athymic and SCID mice to S-layer conjugates. Should the conjugates elicit immunoprotective responses in these mice, S-layer carrier conjugates may well become powerful new agents for human vaccination.

ACKNOWLEDGEMENTS

S-layers were provided under a licence agreement by U. B. Sleytr, W. Mundt and P. Messner of Universität für Bodenkultur, Wien, Austria with Chembiomed Ltd., Edmonton, Alberta. Repeating disaccharide units from hyaluronic acid was a gift from Dr. Josef Glössl, Universität für Bodenkultur, Wien. The vaccine research and development program at Chembiomed Ltd. was partially funded by the Canadian Bacterial Diseases Network, one of the Networks of Centres of Excellence of Canada. This work was also supported in part by the "Fonds zur Förderung der Wissenschaftlichen Forschung in Österreich" project P7757-BIO and the "Österreichisches Bundesministerium für Wissenschaft und Forschung". P. Messner was a recipient of the Erwin Schrödinger Stipendium while on sabbatical at Chembiomed Ltd.

REFERENCES

Altman, E., Brisson, J-R., Messner, P., and Sleytr, U. B., 1990, Chemical characterization of the regularly arranged surface layer glycoprotein of *Clostridium thermosaccharolyticum* D120-70, *Eur. J. Biochem.* 188:73.

Altman, E., Brisson, J-R., Messner, P., and Sleytr, U. B., 1991, Chemical characterization of the regularly arranged surface layer glycoprotein of *Bacillus alvei* CCM 2051, *Biochem. Cell. Biol.* 69:72.

Anderson, P., 1983, Antibody responses to *Haemophilus influenzae* type b and diphtheria toxin induced by conjugates of oligosaccharides of the type b capsule with the non-toxic protein CRM_{197}, *Infect. Immun.* 39:233.

Anderson, P., Pichichero, M. E., and Insel, R. A., 1985a, Immunogens consisting of oligosaccharides from the capsule of *Haemophilus influenzae* type b coupled to the diphtheria toxoid or the toxin protein CRM_{197}, *J. Clin. Invest.* 76:52.

Anderson, P., Pichichero, M. E., and Insel, R. A., 1985b, Immunization of 2-month-old infants with protein-coupled oligosaccharides derived from the capsule of *Haemophilus influenzae* type b, *J. Pediatr.* 107:346.

Anderson, P. W., 1987, Immunogenic conjugates, U.S. patent number 4,673,574.

Bolan, G., Broome, C. V., Facklam, R. R., Pitkaytis, B. D., Fraser, D. W., and Schlech, W. F. III, 1986, Pneumococcal vaccine efficacy in selected populations in the United States, *Ann. Internal. Med.* 104:1.

Borgono, J. M., McLean, A. A., Vella, P. P., Canepa, I., Davidson, W. L., and Hilleman, M. R., 1978, Vaccination and revaccination with polyvalent pneumococcal polysaccharide vaccines in adults and infants, *Proc. Soc. Exp. Biol. Med.* 157:148.

Broome, C. V., Facklam, R. R., and Fraser, D. W., 1980, Pneumococcal disease after pneumococcal vaccination: An alternate method to estimate the efficacy of pneumococcal vaccine, *N. Engl. J. Med.* 303:549.

Christian, R., Schulz, G., Unger, F. M., Messner, P., Küpcü, Z., and Sleytr, U. B., 1986, Structure of a rhamnan from the surface layer glycoprotein of *Bacillus stearothermophilus* strain NRS 2004/3a, *Carbohydr. Res.* 150:265.

Christian, R., Messner, P., Weiner, C., Sleytr, U. B., and Schulz, G., 1988, Structure of a glycan from the surface-layer glycoprotein of *Clostridium thermohydrosulfuricum* strain L111-69, *Carbohydr. Res.*176:160.

Chudwin, D. S., Artrip, S. C., Korenbilt, A., Schiffman, G., and Rao, S., 1985, Correlation of serum opsonins with in vitro phagocytosis of *Streptococcus pneumoniae*, *Infect. Immun.* 50:213.

Cryz, S. J., Sadoff, J. C., Fürer, E., and Germanier, R., 1986, *Pseudomonas aeruginosa* polysaccharide-tetanus toxoid conjugate vaccine: Safety and immunogenicity in humans, *J. Infect. Dis.* 154:682.

Cryz, S. J., and Fürer, E., 1988, Conjugate vaccine against infections by gram-negative bacteria, method for its preparation and use, U.S. patent number 4,771,127.

Davis, M-T. B., and Preston, J. F., 1981. A simple modified carbodiimide method for conjugation of small-molecular-weight compounds to immunoglobulin G with minimal protein crosslinking, *Anal. Biochem.* 116:402.

Di John, D., Torres, J. R., Murillo, J., Herrington, E. A., Wasserman, S. S., Lasonsky, M. J., Stüker, D., and Levine, M. M., 1989, Effect of priming with carrier on response to conjugate vaccine, *Lancet* ii:1415.

Douglas, R. M., Patton, J. C., Duncan, S. J., and Hansman, D. J., 1983, Antibody response to pneumococcal vaccination in children younger than five years of age, *J. Infect. Dis.* 148:131.

Etlinger, H. M., Felix, A. M., Lillessen, D., Haemer, E. D., Just, M., Pink, J. R. L., Sinigaglia, F., Stürchler, D., Tacaks, B., Trzeciak, A., and Matile, H., 1988, Assessment in humans of a synthetic peptide-based vaccine against the sporozoite stage of human malaria parasite, *Plasmodium falciparum*, *J. Immunol.* 140:626.

Fattom, A., Vann, W. F., Szu, S. C., Schneerson, R., Robbins, J. B., Chu, C., Sutton, A, Vickers, J. C., London, W. T., Curfman, B., Hardagree, M. C., and Shiloach, J., 1988, Synthesis and physicochemical and immunological characterization of pneumococcus type 12F polysaccharide-diphtheria toxoid conjugates, *Infect. Immun.* 56:2292.

Forester, H. L., Jahnigen, D. W., and LaForce, F. M., 1987, Inefficacy of pneumococcal vaccine in a high-risk population, *Amer. J. Med.* 83:425.

Gaur, A., Arunan, K., Singh, O., and Talwar, G. P., 1990, Bypass by an alternate carrier of acquired unresponsiveness to hCG upon repeated immunization with tetanus-conjugated vaccine, *Int. Immunol.* 2:151.

Gordon, L. K., 1986, Polysaccharide exotoxoid conjugate vaccines, U. S. patent number 4,619,828.

Gordon, L. K., 1987, *Haemophilus influenzae* b polysaccharide-diphtheria toxoid conjugate vaccine, U.S. patent number 4,644,059.

Granoff, D. M., Boies, E. G., and Munson, R. S., 1984, Immunogenicity of *Haemophilus influenzae* type b polysaccharide diphtheria conjugate vaccines in adults, *J. Pediatr.* 105:22.

Herzenberg, L.A., and Tokuhisa, T., 1982, Epitope-specific regulation. I. Carrier-specific induction of suppression for IgG anti-hapten antibody responses, *J. Exp. Med.* 155:1730.

Hilleman, M. R., Carlson, Jr., A. J., McLean, A. A., Vella, P. P., Weibel, R. E., and Woodhour, A. F., 1981, *Streptococcus pneumoniae* polysaccharide vaccine: Age and dose responses, safety, persistence of antibody, revaccination, and simultaneous administration of pneumococcal and influenza vaccines, *Rev. Infect. Dis.* 3 (suppl):S31.

Jennings, H. J., and Lugowski, C., 1985, Immunogenic polysaccharide-protein conjugates, Canadian patent number 1,181,344.

Jennings, H. J., Roy, R., and Gamian, A. J., 1989, Modified meningococcal group b polysaccharide for conjugate vaccine, Canadian patent number 1,261,320.

Lock, R. A., Hansman, D., and Paton, J. C., 1992, Comparative efficacy of autolysin and pneumolysin as immunogens protecting mice against infection by *Streptococcus pneumoniae, Microbial. Pathogen.* 12:137.

Malcolm, A. J., Messner, P., Sleytr, U. B., Smith, R. H., and Unger, F. M., 1993, Crystalline bacterial cell surface layers (S-layers) as combined carrier/adjuvants for conjugate vaccines, *in*: "Immobilised Macromolecules: Application Potentials," U. B. Sleytr, P. Messner, D. Pum and M. Sára, eds., Springer-Verlag, London.

Mandell, G. L., 1990, "Principles and Practise of Infectious Diseases", Churchill Livingston, New York.

Marburg, S., Jorn, D., Tolman, R. L., Arison, B., McCauley, J., Kniskern, P. J., Hagopian, A., and Vella, P.O., 1986, Biomolecular chemistry of macromolecules: Synthesis of bacterial polysaccharide conjugates with *Neisseria meningitidis* membrane protein, *J. Amer. Chem. Soc.* 108:5282.

Marburg, S., Tolman, R. L., and Kniskern, P. J., 1987, Covalently-modified polyanionic bacterial polysaccharides, stable covalent conjugates of such polysaccharides and immunogenic proteins with bigeneric spacers, and methods of preparing such polysaccharides and conjugates and of confirming covalency, U.S. patent number 4,695,624.

Marburg, S., Kniskern, P. J., and Tolman, R. L., 1989, Covalently-modified bacterial polysaccharides, stable covalent conjugates of such polysaccharides and immunogenic proteins with bigeneric spacers and methods of preparing such polysaccharides and conjugates and of confirming covalency, U. S. patent number 4,882,317.

Messner, P., and Sleytr, U.B., 1992, Crystalline bacterial cell-surface layers, *Adv. Microbial Physiol.* 33:213.

Messner, P., Sleytr, U. B., Christian, R., Schulz, G., and Unger, F. M., 1987, Isolation and structure determination of a diacetamidodideoxyuronic acid-containing glycan from S-layer glycoprotein of *Bacillus stearothermophilus* NRS 2004/3a, *Carbohydr. Res.* 168:211.

Messner, P., Mazid, M. A., Unger, F. M., and Sleytr, U. B., 1992, Artificial antigens. Synthetic carbohydrate haptens immobilized on crystalline bacterial surface layer glycoproteins, *Carbohydr. Res.* 233:175.

Mufson, M. A., Hughey, D., and Lydick, E., 1985, Type-specific antibody responses of volunteers immunized with 23-valent pneumococcal polysaccharide vaccine, *J. Infect. Dis.* 151:749.

Mufson, M. A., Krause, H. E., Schiffman, G., and Hughey, D. E., 1987, Pneumococcal antibody levels one decade after immunization of healthy adults, *Amer. J. Med. Sci.* 293:279.

Paton, J. C., Lock, R. A., Lees, C-J., Li, J. P., Berry, A. M., Mictchell, T. J., Andrew, P. W., Hansman, D., and Boulnois, G. J., 1991, Purification and immunogenicity of genetically obtained pneumolysin toxoids and their conjugation to *Streptococcus pneumoniae* type 19F polysaccharide, *Infect. Immun.* 59:2297.

Peeters, C. C. A. M., Tenbergen-Meekes, A-M., Poolman, J. T., Beurret, M., Zegers, B. J. M., and Rijkers, G. T., 1991, Effect of carrier priming on immunogenicity of saccharide-protein conjugate vaccines, *Infect. Immun.* 59:3504.

Porro, M., Saletti, M., Neucioni, L., Tagliaferi, L., and Marsili, L., 1980, I. Immunogenic correlation between cross-reacting material (CRM_{197}) produced by a mutant *C. diphtheria* and diphtheria toxoid, *J. Infect. Dis.* 142:716.

Porro, M., and Costantino, P., 1987, Glycoproteinic conjugates having trivalent immunogenic activity, U. S. patent number 4,711,779.

Porro, M., 1990, Oligosaccharide conjugate vaccines, Canadian patent number 2,052,323.

Roy, R., Katzenellenbogen, E., and Jennings, H. L., 1984, Improved procedures for the conjugation of oligosaccharides to protein by reductive amination, *Can. J. Biochem.* 62:270.

Schidt, R. A., Boyd, J. F., McCracken, J. D., Schiffman, G., and Giolma, J. P., 1983, Antibody response to pneumococcal vaccine in patients with solid tumors and lymphomas, *Med. Pediatr. Oncol.* 11:305.

Schneerson, R., Barrera, O., Sutton, A., and Robins, J. B., 1980, Preparation, characterization and immunogenicity of *Haemophilus influenzae* type b polysaccharide-protein conjugates, *J. Exp. Med.* 152:361.

Schneerson, R., Robbins, J. B., Chu, C., Sutton, A., Vann, W., Vickers, J. C., London, W. T., Curfman, B., and Hardegree, M. C., 1984, Serum antibody responses to juvenile and infant rhesus monkeys injected with *Haemophilus influenzae* type b and pneumococcus type 6a capsular polysaccharide protein conjugates, *Infect. Immun.* 45:582.

Schneerson, R., Robbins, J. B., Parke, J. C., Bell, C., Schlesselman, J. J., Sutton, A., Wang, Z., Schiffman, G., Karpas, A., and Shiloach, J., 1986, Quantitative and qualitative analyses of serum antibodies elicited in adults by *Haemophilus influenzae* type b and pneumococcus type 6a capsular polysaccharide-tetanus toxoid conjugates, *Infect. Immun.* 52:519.

Schutze, M. P., LeClerc, C., Jolivet., M., Audibert, F., and Chedid, L., 1985, Carrier-induced epitopic suppression, a major issue for synthetic vaccines, *J. Immunol.* 135:2319.

Sell, S. H., Wright, P. F., Vaughn, W. K., Thompson, J., and Schiffman, G., 1981, Clinical studies of pneumococcal vaccines in infants: I. Reactogenicity and immunogenicity of two polyvalent polysaccharide vaccines, *Rev. Infect. Dis.* 3 (suppl):S97.

Shapiro, E.D., and Clemens, J. D., 1984, A controlled evaluation of the protective efficacy of pneumococcal vaccine for patients at high risk for serious pneumococcal infections, *Ann. Intern. Med.* 101:325.

Shapiro, E. D., 1987, Pneumococcal vaccine failure, *New Engl. J. Med.* 316:1272.

Shapiro, E. D., 1991, Pneumococcal vaccine, *in:* "Vaccines and Immunotherapy," S. J. Fryz Jr., ed., Pergamon Press, New York.

Siber, G. R., Weitzman, S. A., Aisenberg, A. C., Weinstein, H. J., and Schiffman, G., 1978, Impaired antibody response to pneumococcal vaccine after treatment for Hodgkin's disease, *New Engl. J. Med.* 299:442.

Simberkoff, M. S., Cross, A. P., Al-Ibrahim, M., Baltch, A. L., Geiseler, P. J., Nadler, J., Richmond, A. S., Smith, R. P., Schiffman, G., and Shephard, D. S., 1986, Efficacy of pneumococcal vaccine in high-risk patients: Results of a veterans administration cooperative study, *New Engl. J. Med.* 315:1318.

Simberkoff, M. S., 1989, Pneumococcal vaccine in adults, *in:* "Immunization," M. A. Sande, and R. K. Root, eds., Churchill Livingstone, New York.

Sims, R. V., Steinman, W. C., McConville, J. H., King, L. K., Zwick, W. C., and Schwartz, J. C., 1988, The clinical effectiveness of pneumococcal vaccine in the elderly, *Ann. Intern. Med.* 108:653.

Sleytr, U.B., and Messner, P., 1983, Crystalline surface layers on bacteria, *Annu. Rev. Microbiol.* 37:311.

Sleytr, U. B., Messner, P., Pum, D., and Sára, M., eds., 1988, "Crystalline bacterial cell surface layers", Springer-Verlag, Berlin.

Sleytr, U. B., Mundt, W., and Messner, P., 1987, Pharmazeutische Struktur. European Patent Application No. 0306473A1.

Sloyer, J. L., Jr., Ploussard, J. H., and Howie, V. M., 1981, Efficacy of pneumococcal polysaccharide vaccine in preventing acute otitis media in infants in Huntsville, Al., *Rev. Infect. Dis.* 3 (suppl.):S19.

Smith, R. H., Messner, P., Lamontagne, L. R., Sleytr, U. B., and Unger, F. M., Induction of T-cell immunity to oligosaccharide antigens immobilized on crystalline bacterial surface layers (S-layers), *Vaccine*, in press.

Tsay, G. C., and Collins, M. S., 1987, Vaccines for gram-negative bacteria, U. S. patent number 4,663,160.

Vogel, F. R., LeClerc, C., Schutze, M. P., Jolivet, M., Audibet, F., Klein, T. W., and Chedid, L., 1987, Modulation of carrier-induced epitopic suppression by *Bordetella pertussis* components and muramyl peptide, *Cell Immunol.* 107:40.

Chapter 22

SCALE-UP OF S-LAYER PROTEIN SECRETION BY *BACILLUS BREVIS* 47

A.J. Daugulis, G.K. Whitney, A. Wong, and C.P. Wight

Department of Chemical Engineering
Queen's University
Kingston, Ontario, Canada

B.N. White

Department of Biology
McMaster University
Hamilton, Ontario, Canada

INTRODUCTION

Bacillus brevis 47 is a soil bacterium which produces two S-layer wall proteins which are 115 and 104 kDa in size. This strain can, under appropriate cell cultivation conditions, be induced to secrete up to about 14 g/L of these S-layer proteins into the culture medium. This situation has prompted researchers to evaluate this organism as a host for the production and secretion of large quantities of heterologous proteins of commercial value. Proteins which are secreted into the medium are potentially easier to recover and purify than are intracellular proteins. Indeed, the production of recombinant proteins by this organism has been reported (Takao et al., 1989; Yamagata et al., 1989; Konishi et al., 1990), although at substantially lower levels than the native S-layer proteins.

Strategies to encourage *B. brevis* to produce large quantities of protein have largely focused on culture medium formulation, particularly on supplementation with amino acids (Myashiro et al., 1980), and on the identification of appropriate cell cultivation conditions (Wight et al., 1988). Even though such approaches have been successful in the production of more than 10 g/L of S-layer proteins in the growth medium, the medium employed was complex, rich, and expensive in costly nutrients when viewed in commercial terms and considering larger scale production. Efforts at stimulating S-layer secretion during the course of batch fermentation through the addition of nutrient stimulations were shown to be relatively successful (Wight et al., 1989), although such stimulations tended to be administered at arbitrary times without full awareness of the metabolic state of the organism.

Recent work on simplifying (and reducing the cost of) the growth medium, and on identifying a "real time" metabolic marker which can be used to provide optimal nutrient stimulation at the appropriate time, has been quite successful (Wight et al.,1992). Consistent and reproducible production of 15 g/L of S-layer proteins has been possible in a simple fructose-based medium, using dissolved oxygen levels as indicators of the time at which stimulation by either fructose or polypeptone are required. The current work was intended to demonstrate that this approach could be scaled-up by one and two orders of magnitude (up to 150 L pilot plant capacity) and that expensive polypeptone could be replaced by a cheaper nutrient source.

MATERIALS AND METHODS

B. brevis 47 (strain 6285 from the Japan Collection of Microorganisms) was cultivated in the fructose and salts medium previously described (Wight et al., 1992) supplemented with either 20 g/L BBL polypeptone (Becton Dickinson, Cockeysville, MD) or 20 g/L Pancase S (Champlain Industries, Cornwall, Ontario). 5% (v/v) inocula were prepared in shake flasks in the case of the 2 L and 14 L fermentors, and in a 7.5 L MBR seed fermentor in the case of the 150 L fermentor, with inoculum ages of 12 to 18 hours and a cultivation temperature of 34 °C being employed.

The 2 L Bioflo fermentors were from New Brunswick Scientific (Edison, N.J, USA) with 1.2 L working volumes, the 14 L vessels were from CHEMAP (Volketswil, Switzerland) with 10 L working volumes, and the 150 pilot fermentor was an MBR (MBR BioReactor AG, Wetzikon, Switzerland) with a 100 L working volume. Cultivation temperatures were 34 °C, with an air flow of 1.7 vvm, and with Antifoam 204 organic (Sigma, St. Louis MO) used as required to control foam. Agitation rates for the 2 L, 14 L, and 150 L fermentors were 600, 400 and 300 rpm, respectively. The pH was not controlled, but the level of dissolved oxygen in each of the fermentors was measured by means of a dissolved oxygen (DO) electrode. Using the level of DO as a metabolic marker, the cultures in the fermentors were stimulated twice with nutrient supplements over a 48 hour period, first with a fructose (10 g/L), and second with a fructose (10 g/L) and 20 g/L polypeptone (for the 2 L and 14 L vessels) or Pancase S (for the 150 L fermentor) addition.

Analytical protocols for measuring fructose, cell density and protein were as described previously (Wight et al., 1992).

RESULTS AND DISCUSSION

When cultivated in a fructose medium with no post-inoculation stimulation with nutrients, *B. brevis* 47 secretes only a small amount (2-3 g/L) of S-layer protein into the medium (Fig. 1), with the majority of the protein appearing in the stationary phase of the culture. Fructose does appear to be used for cell growth, and this is in contrast to the glucose in glucose medium in which the sugar is used slowly and appears to contribute little carbon to growth (Wight et al., 1989). A second distinguishing feature of the cultivation of *B. brevis* on fructose is the rapid increase in DO which corresponds to the point of fructose depletion. This is again in contrast to cultivation of this organism in glucose medium where there appears to be very little relationship between glucose consumption and oxygen demand (Wight et al., 1989; Wight et al., 1992).

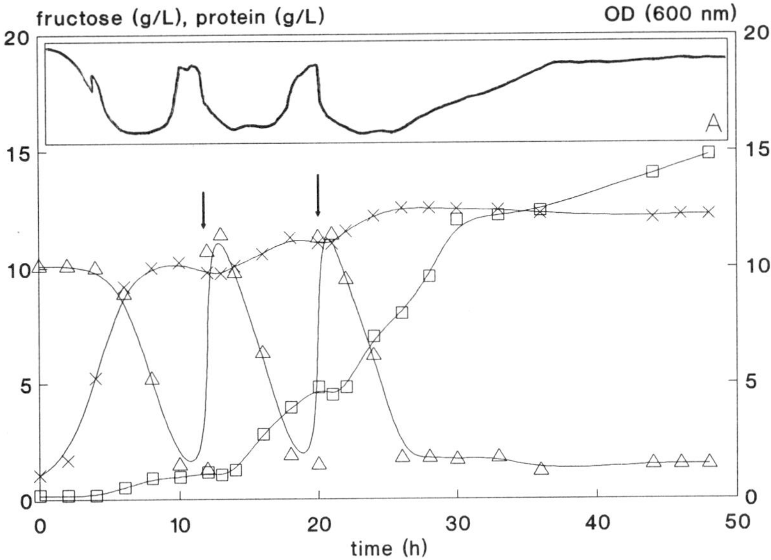

Figure 1. Cultivation of *B. brevis* 47 in a 2 L fermentor on fructose medium. Dissolved oxygen is shown in panel A with the scale from 0 to 100 % saturation. Cell density as expressed by optical density, X; fructose concentration, ▵; and protein concentration, ▢. (From Wight et al., 1992, by permission of John Wiley and Sons, Inc.).

It was this very dramatic rise in the DO level that was thought to indicate a shift in metabolism by the organism into a stationary growth phase in which S-layer proteins would be rapidly produced. Accordingly, a strategy was developed in which a carbon and energy source (fructose), plus a source of protein building blocks (polypeptone) were added at appropriate times during growth to stimulate S-layer protein production. Various combinations of the supplements were investigated, all in response to the DO level which indicated fructose depletion, and an optimum feeding strategy was developed in which fructose was first added and then fructose + polypeptone. This resulted in a significant increase in protein production (up to 15 g/L) as well as discernable inflections of protein production corresponding to the culture's response to stimulation (Fig. 2). This significant degree of stimulation was found to be highly reproducible when undertaken in response to the "real time" DO metabolic marker and was therefore felt to be of practical value. However, in order to examine parameters which would be relevant for using this approach in commercial applications with a genetically modified *B. brevis*, scale-up and culture medium costs had to be addressed.

The previous supplement regimen was scaled-up by a factor of approximately ten (from a 1.2 L working volume to 10.0 L) with additions of fructose and fructose + polypeptone occurring in response to the DO markers as before. From the data in Fig. 3 it is clear that very similar results were obtained at this larger scale. Total S-layer protein and the cell densities were almost exactly the same as in the smaller fermentation. The times of stimulation, corresponding to the times at which the culture consumed all of the fructose, were within one hour of one another. Inflections in the protein accumulation curves were similar in both cases, although it appears that some cell lysis did occur in the larger scale fermentation towards the end of the cultivation period. The remarkable similarities in the dynamic performance of the cultures shown in Figs. 2 and 3 suggest that the scale of operation did not effect the metabolic activity of the organism nor the resulting yield of protein.

In light of the above encouraging results, a further scale-up in volume to 150 L fermentor capacity was undertaken with the nutrient supplementation in both the original batch medium and in the stimulation consisting of Pancase S instead of polypeptone. Preliminary shake flask experiments had shown that there was only a slight reduction in the total amount of protein produced by *B. brevis* when the much cheaper Pancase S was employed. The strategy in this experiment was identical to that used in the two smaller scale fermentations, with DO being used as an indicator of when to add fructose and fructose+ Pancase S. The results, shown in Fig. 4, indicate several important differences between this scale of operation and the smaller ones. First, a considerable growth lag of about 12 hours was found in the pilot scale fermentation. This lag is primarily attributed to the use of Pancase S in the medium and the prior subculturing of the organism on polypeptone. Second, there was a complete depletion of DO in the pilot fermentor which was unlike the situation in the two smaller (2 L and 14 L) systems in which the DO never dropped to zero. Finally, both the cells produced and the protein secreted were both at levels which were only about one-half of those achieved at the smaller scale.

We feel that it was the complete depletion of DO which caused the amount of cells produced to reach only one-half of the anticipated level. Oxygen transfer becomes a significant challenge when fermentation processes are scaled up, and the present results are therefore not surprising since *B. brevis* is a highly aerobic organism. If oxygen depletion is to be avoided at this scale for this organism then some additional steps will need to be taken such as increased rotational speed of the impellors or the use of oxygen-enriched air at appropriate times in the cultivation

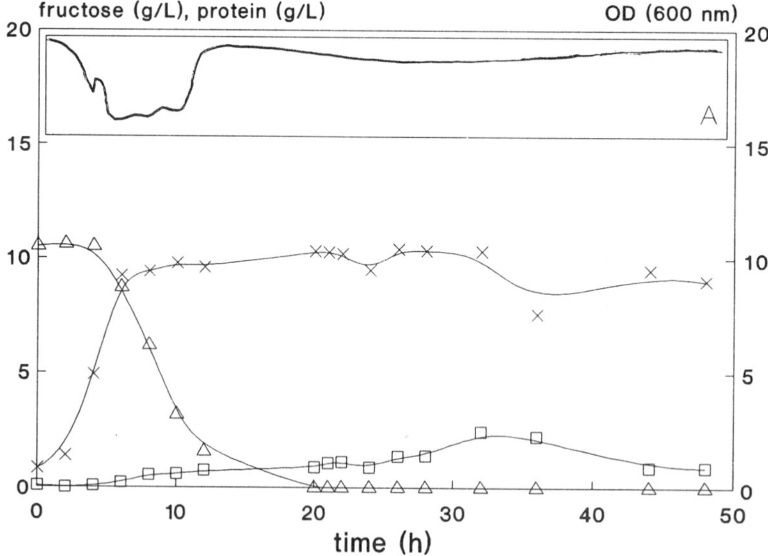

Figure 2. Cultivation of *B. brevis* 47 in a 2 L fermentor on fructose medium. DO is shown in panel A with the scale from 0 to 100 % saturation. Culture was stimulated with fructose and with fructose+ polypeptone as indicated by arrows. Cell density as expressed by optical density, X; fructose concentration, ▲; protein concentration, □. (From Wight et al., 1992, by permission of John Wiley and Sons, Inc.).

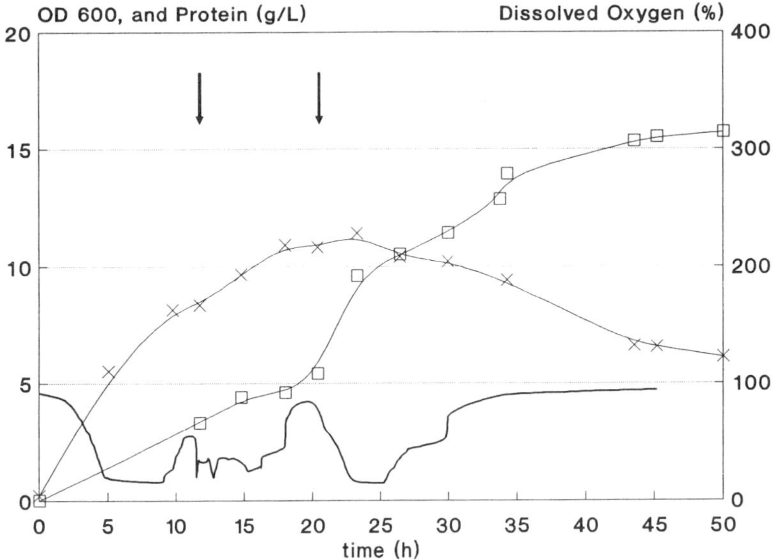

Figure 3. Cultivation of *B. brevis* 47 in a 14 L fermentor with stimulation by fructose and fructose+ polypeptone as indicated by arrows. DO is indicated in the lower panel. Cell density, X; and protein concentration, □.

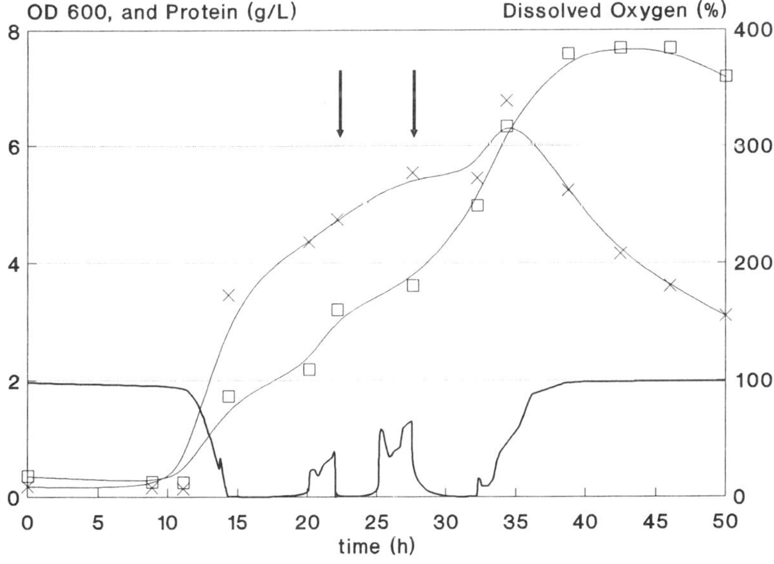

Figure 4. Cultivation of *B. brevis* 47 in a 150 L fermentor with stimulation by fructose and fructose+ Pancase S as indicated by arrows. DO is indicated in the lower panel. Cell density, X; and protein concentration, □.

process. Notwithstanding the periods of zero DO, the very sharp indicators of fructose depletion were still observed and used as markers for stimulation. The reduced protein level is also a potential concern. However, if the amount of produced protein is calculated on a <u>specific</u> basis (i.e., on a per unit cell mass basis) then the performance of the 150 L culture was very similar to the results at the smaller scales. There is therefore an expectation that the amount of extracellular protein produced at this scale may be comparable to that produced at smaller scales if the cell concentration could be increased to similar levels, perhaps by means of the techniques noted above.

It is important to note that the use of DO as a "real time" indicator for providing stimulation was particularly useful in this case because of the significant lag period observed in growth. If stimulation had been provided arbitrarily at about 12 and 21 hours, as was the case at the smaller scales, then it is unlikely that protein stimulation would have occurred. This conclusion is supported by Wight et al. (1989) who showed that timing of the nutrient additions was critical to the success of stimulation. Finally, the results in which Pancase S was used are encouraging because of the significantly reduced price of this material compared to polypeptone. It remains to be seen whether other even less expensive supplements (such as fish meal) could be used in a similar stimulatory fashion.

CONCLUSION

The use of DO as a measure of the extent of substrate depletion (and hence the state of the culture) has been shown to be a reliable means of determining when supplementation by either fructose alone or a mixture of fructose and protein precursors can be added to stimulate *B. brevis* 47 to produce large quantities of S-layer proteins. Such an approach has been successfully scaled-up from 2 L to 14 L to 150 L fermentor capacities. At least one less expensive alternative (Pancase S) to polypeptone has been found which could serve as the source of protein building blocks. It remains to be seen whether our molecular biological efforts at using the S-layer promoter of *B. brevis* to express a foreign gene will produce a similar response with these manipulations of cell culture conditions. Should this work be successful, it may be possible to secrete g/L quantities of heterologous proteins into the growth medium under controlled conditions at a reasonable scale.

ACKNOWLEDGEMENTS

This work was supported, in part, by the Natural Sciences and Engineering Research Council and Connaught Laboratories of Canada.

REFERENCES

Konishi, H., Sato, T., Yamagata, H., and Udaka, S., 1990, Efficient production of human α-amylase by a *Bacillus brevis* mutant, *Appl. Microbiol. Biotechnol.* 34:297.

Miyashiro, S., Enei, H., Hirose, Y., and Udaka, S., 1980, Extracellular production of proteins by microorganisms. Part III. Effect of glycine and L-isoleucine on protein production by *Bacillus brevis* 47. *Agric. Biol. Chem.* 44:105.

Takao, M., Morioko, T., Yamagata, H., Tsukagoshi, N., and Udaka, S., 1989,

Production of swine pepsinogen by protein-producing *Bacillus brevis* carrying swine pepsinogen cDNA, *Appl. Microbiol. Biotechnol.* 30:75.

Wight, C.P., Daugulis, A.J., Lau, R.H., and White, B.N., 1988, Conditions for production of extracellular protein by *Bacillus brevis* 47, *Biotechnol. Letters* 10:769.

Wight, C.P., Daugulis, A.J., Lau, R.H., and White, B.N., 1989, Stimulation of extracellular protein production in *Bacillus brevis* 47, *Appl. Microbiol. Biotechnol.* 31:338.

Wight, C.P., Whitney, G.K., Daugulis, A.J., and White, B.N., 1992, Enhancement and regulation of protein production by *Bacillus brevis* 47 through manipulation of cell culture conditions, *Biotechnol. Bioeng.* 40:46.

Yamagata, H., Nakahama, K., Suzuki, Y., Kakinuma, A., Tsukagoshi, N., and Udaka, S., 1989, Use of *Bacillus brevis* for efficient synthesis and secretion of human epidermal growth factor, *Proc. Natl. Acad. Sci. USA* 86:3589.

Chapter 23

INVESTIGATION OF LATTICE SURFACE LAYERS BY SCANNING PROBE MICROSCOPY

Max Firtel, Gordon Southam, and Terry J. Beveridge

Department of Microbiology, College of Biological Science
University of Guelph, Guelph, Ontario, Canada

Wei Xu, Manfred H. Jericho, Brad L. Blackford, and Peter J. Mulhern

Department of Physics, Dalhousie University
Halifax, Nova Scotia, Canada

INTRODUCTION

Scanning probe microscopy (SPM) is emerging as an important complementary technique to conventional microscopy for high-resolution surface investigation (Wickramasinghe, 1990). In SPM, no lenses are used and the image is formed by a finely tipped probe that scans the specimen surface (see Fig. 1). Using this technique, direct quantitative analysis of a specimen's surface topology, electronic structure, as well as other physical properties are possible in vacuum, gaseous, or liquid environments. Two types of SPM, scanning tunneling microscopy (STM; Binning et al., 1982), and atomic force microscopy, (AFM; Binning et al., 1986) image topographical details on biological surfaces to nanometer resolution. In STM, conductivity is an important factor in image formation since a tunneling current is established between the microscope tip and the surface under study. However, conduction through biological surfaces is a complex process (Spong et al., 1989; Travaglini et al., 1988), so these surfaces are often prepared for STM imaging by overlaying them with a conductive film. Such films tend to obscure structural detail and further progress is needed in the development of conductive fine-grained films suitable for high resolution work (Wepf et al., 1991; Amrein et al., 1991). The overall design of the AFM is based on that of the STM but the probe used is a force sensor which is mounted on a cantilever to measure interatomic attractive/repulsive forces (Burnham et al., 1991). This important operational difference enables the native topology of uncoated surfaces to be imaged. Biological applications of STM/AFM have been largely devoted to topographical analysis, but reports of chemical differentiation by tunneling spectroscopy (Spong et al., 1989), controlled 'nanodissection' of biological deposits by AFM (Hoh et al., 1991; Henderson, 1992), and AFM imaging of polymerizing macromolecules under water (Drake et al.,1989)

Advances in Bacterial Paracrystalline Surface Layers
Edited by T.J. Beveridge and S.F. Koval, Plenum Press, New York, 1993

have also been published. For recent reviews of biological applications of SPM see Engel, 1991; Edstrom et al., 1990; Blackford et al., 1991a; and Hansma et al., 1988.

SPM techniques complement those of transmission electron microscopy (TEM) in the investigation of biological surfaces. In terms of analyzing 3D space, vertical distances are directly measured and are better resolved than lateral distances when SPM is used. The electron dispersive quality of high atomic number metals is essential for contrast in TEM but not in SPM where such atoms are used instead to stabilize the specimen against tip forces or to produce a uniform conductive coating. Both SPM and TEM can be used in conjunction with powerful Fourier-based filter or correlation averaging techniques to remove aperiodic noise from the image. TEM structural analysis improves if the specimen is crystallizable into flat planar arrays, but SPM may be able to provide similar high resolution data on less ordered, or more deformable systems, such as lipid bilayers (Smith et al., 1987; Egger et al., 1990).

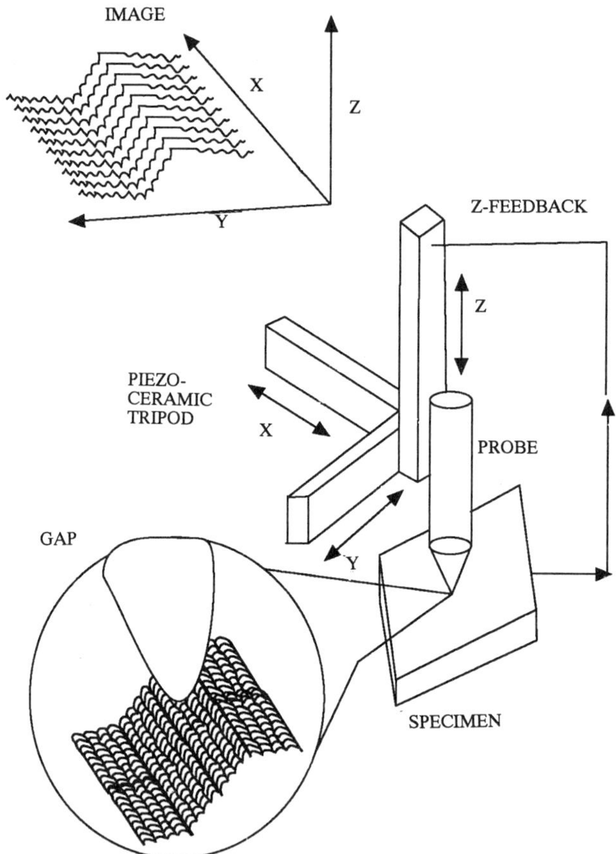

Figure 1. Simplified schematic of the operational design of a SPM (based on STM design). The probe is mounted on a piezoceramic tripod that controls the probe's raster scan over the x- and y- axes and its vertical (z-axis) movement. A gap of a few angstroms separates the penultimate atom of the probe's tip from the specimen surface. This gap distance is kept constant during scanning by feedback control which uses the signals generated by the tip-surface interaction as input. The resulting image is actually a record of the x-,y-, and z- movements of the probe during scanning. (Modified from Golovichenko, 1986).

Unquestionably, the most intriguing aspect of SPM for biology is the potential for high resolution imaging of specimens kept at normal pressures and in wet environments. However, two major factors unique to SPM appear as barriers to this goal; these are the macroscopic tip shape and tip compressive forces. The macroscopic dimensions of the tip effects the ability of the tip to physically trace the contours of irregular surfaces (Stemmer and Engel, 1990; Reiss et al., 1990). Edge broadening is a common artifact that is due to the use of insufficiently sharp tips (Keller et al., 1990; Butt et al., 1990). The second major factor is the compresive force of the tip during close contact surface scanning which can cause structural distortion (Dunlap and Bustamante, 1989; Burnham et al., 1991; Jericho et al., 1989; Stemmer and Engel, 1990; Stemmer et al., 1989) or even cause specimen displacement (Salmeron et al., 1990; Keller et al., 1990). High-resolution imaging of uncoated specimens is particularly limited by tip compression.

Considerable care is needed to interpret image detail since it is not yet possible to deconvolute these perturbing effects from the scanned image to produce a purely topographical representation (Salmeron et al., 1990; Stemmer and Engel, 1990). Furthermore, the substrate on which the biological specimen is deposited may also contain defects which may closely mimic biological structures, such as nucleic acid (Clemmer and Beebe, 1991). Highly-ordered surfaces, such as protein lattice layers, provide mechanically rigid surfaces to resist tip compression and they also contain structural landmarks for assessing image reproducibility and quality (Stemmer, et al., 1989; Amrein et al., 1991; Blackford et al., 1989; Stemmer and Engel, 1990). The usefulness of highly-ordered assemblies as test specimens for SPM has been demonstrated in a series of reports using bacteriophage supramolecular assemblies (Stemmer et al., 1988; Stemmer et al., 1989, Stemmer and Engel, 1990; see Fig. 2).

This chapter will examine the importance of bacterial S-layers to the development of SPM imaging techniques. At present, studies incorporating S-layers are mainly designed for comparative purposes and do not usually provide new

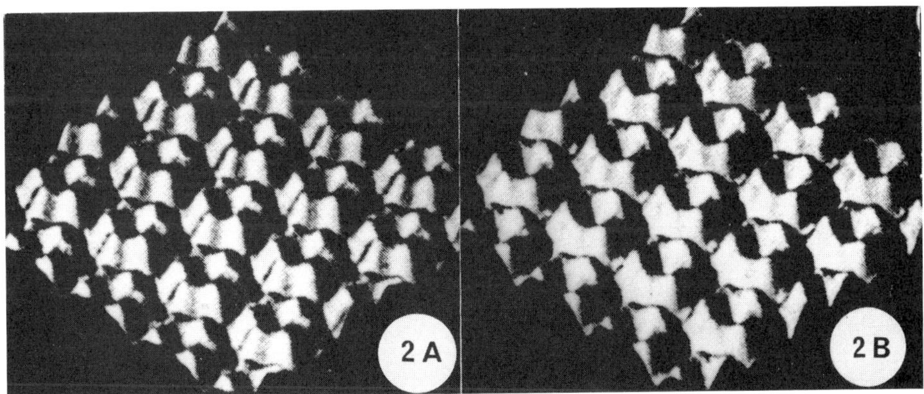

Figure 2. Reconstructed 3D surface relief maps of coated T4 polyhead capsomeres showing the close correspondence between (A) the TEM and (B) STM representations. The slight variation in the representations was thought to be due to the overlying carbon film, which is observed by STM but not by TEM. Height range is 2.3 nm. (Reprinted with permission from Stemmer et al., 1989).

structural information. However, some intriguing discoveries have been made using these bacterial layers which have broader implications for imaging three dimensional structure in biology. We will also mention some potential applications for SPM in the investigation of lattice layers.

SPM IMAGING OF BACTERIAL LATTICE LAYERS

Purple membrane of *Halobacterium halobium*

The purple membrane (PuM) of the extreme halophile *H. halobium* functions in photoreduction and contains a single protein, called bacteriorhodopsin (Br) (Oesterhelt and Stoeckenius, 1971; 1973). In the native membrane, Br is trimeric and is distributed as a planar hexagonal lattice with a center-to-center spacing of 6 nm between subunits (Blaurock and Stoeckenius, 1971). The 3D structure of Br is known to atomic dimensions based on TEM crystallographic evidence (Henderson et al., 1990), but less is known about the structure and packing of the lipid molecules in the membrane because of the technical difficulties associated with analyzing structure in lipid bilayers.

Worcester et al. (1988, 1990) used AFM to image PuM adsorbed to a quartz surface. Hexagonally-packed, raised regions extending 0.3 nm above the surrounding menstruum could be observed and likely corresponded to the Br trimer. An unusual attenuation of two of the three equivalent axes of the lattice, not seen by TEM techniques, was observed. It was not clear if the attenuation was a true representation of the membrane surface or was due to a tip effect.

Hansma and co-workers have designed instruments with submerged probes that allow scanning under water or buffer (Gould et al., 1990; Drake et al., 1989). Besides the obvious advantages, underwater scanning can be performed with smaller tip forces. Butt et al. (1990) prepared purified PuM for underwater scanning by allowing the membranes to passively adsorb to mica or silanized glass. The underwater signal-to-noise ratio was poor but visualization of the hexagonal lattice of the trimer units was achieved using Fourier-based filtering techniques. The reconstructed image indicated that the Br molecules were elevated above the lipid portion of membrane. However, a differential compressive effect of the tip on proteins and lipids could not be ruled out.

Fisher et al. (1990), have taken another approach to investigating PuM surfaces by using metal-coated replicas. They measured changes in thickness of papain-treated PuM adsorbed onto different substrates thereby taking advantage of the high vertical resolution capabilities of STM. These thickness measurements were in close agreement to those obtained by TEM methods.

The S-Layer of *Deinococcus radiodurans*

The S-layer (or HPI layer;Baumeister and Kubler, 1978) of *D. radiodurans* is a hexagonally-ordered lattice (center-to-center spacing = 18 nm) composed of a single protein species of M_r = 105000 (Thompson et al., 1982, Baumeister and Kubler, 1978). Using tomographic and relief reconstruction techniques, a 3D representation of the surface structure to a resolution of 1.5 nm has been proposed (Baumeister et al., 1986). The mass of the monomer protein is arranged in six-fold

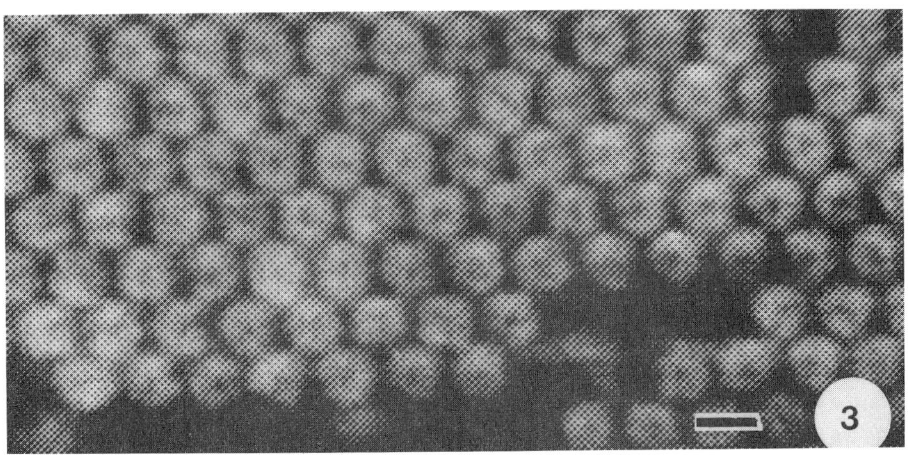

Figure 3. Unprocessed STM image of the uncoated S-layer. Pt/Ir tip operating at a high bias voltage (up to 10 V) and low current output (< 1pA). Details such as the central pore and interconnecting spokes between hexameric units are imaged. Grey scale indicative of height and range 8 nm from lowest (darkest regions) to highest (lightest regions). Bar = 15 nm. (Reprinted with permission from Guckenberger et al., 1989).

symmetry forming a dense core from which thin, interconnecting spokes emanate. Open spaces arranged in three-fold symmetry surround the core and the core, itself, is thought to enclose a central cavity or pore. The inner and outer faces are morphologically distinct.

The highly textured surface of this well characterized layer is suited to assessing SPM image quality and instrument performance (Michel and Travaglini, 1988; Guckenberger et al., 1989; Amrein et al., 1991). In an intriguing report which has broad implications for STM, Guckenberger et al. (1989), were able to reproducibily image the S-layer without the need of a conductive metal overlay (see Fig. 3). They found that imaging could occur under an unusual set of conditions including the use of blunted probes that produce low current with high applied voltages, and the use of relatively high humidities (35-45% vapour saturation). The reason for the good contrast in these images is not entirely known since the images of uncoated PuM obtained by the same authors were not as impressive (Guckenberger et al., 1991).

In a more conventional study, Amrien et al. (1991), coated the S-layer with a fine-grained film designed for high-resolution imaging. The S-layer was adsorbed from a suspension onto Pt/C-coated mica, freeze-dried and rotary shadowed at 65° with Pt/Ir/C (25-40% C content). The major structural features of the inner and outer surfaces were imaged, including the pores and interconnecting fibres, and were in good agreement with TEM. Due to the tip shape, elevations were measured with high accuracy, whereas pores and depressions did not appear as deep as expected from TEM. Asymmetrical cores were observed which were likely due to an anisometric tip. Interestingly, the signal-to-noise ratio was better in raw STM data than in raw TEM data. This indicates that STM may reveal more structural detail than TEM in specimens that are more disordered.

The intricate S-layer of the cyanobacterium *Synechococcus* GL24 has been examined by thin-sectioning, negative staining, and correlation averaging of images of negatively stained S-layer fragments. (see the Chapter by Schultze-Lam and Beveridge). The S-layer has hexagonal symmetry with a 22 nm repeat spacing and appears as a fine network (see Fig. 4). S-layer fragments tend to be small (i.e., a few hundred nm) and the layer is often found still attached to the underlying outer membrane.

Imaging the *Synechococcus* S-layer has proven to be a difficult undertaking because only a small proportion of the deposits were actually S-layer fragments. The time needed to search for specimens by SPM is greater than by lens-based microscopes since the image takes time to form. On the few S-layer fragments that we have been able to unequivocally identify by STM, we found that the repeat spacing and layer thickness measurements were consistent with results obtained by TEM methods. However, the intricate intermolecular network was likely obscured by the Pt/Ir coating so the STM imaged the layer as a lattice of circular, compact mounds without additional structural details (see Fig. 5). Imaging uncoated

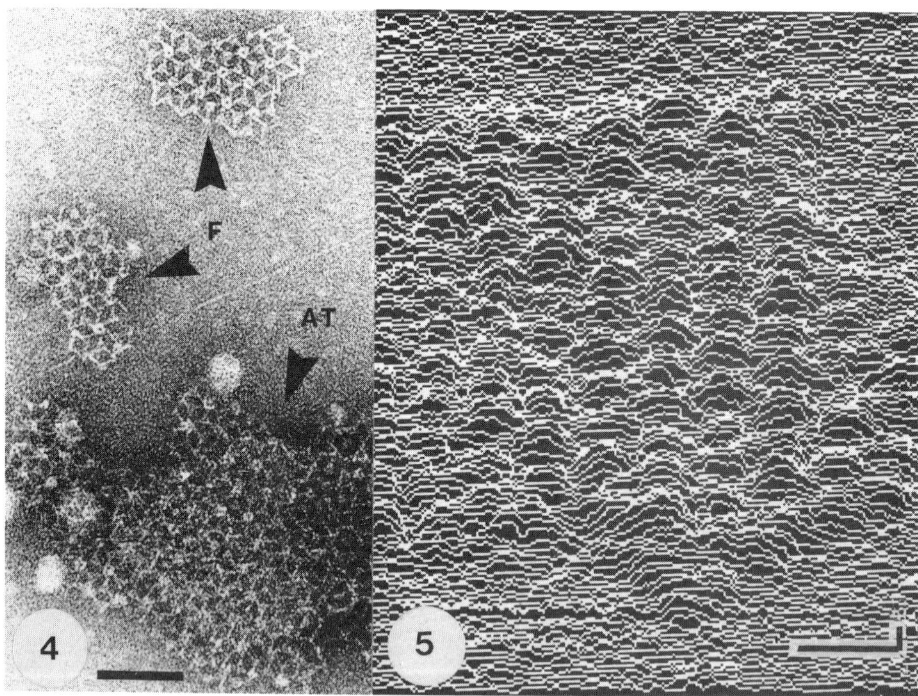

Figure 4. S-layer. TEM micrograph of negatively stained *Synechococcus* fragments showing the array still attached to the outer membrane (AT) and free fragments (F). The filament-like monomers coalesce around the 6-fold axis (enclosing a central pore) and interconnect with neighbouring units to produce large diamond-shaped openings exhibiting a definite handedness. Bar = 50 nm. **Figure 5.** A STM image of a Pt/Ir-coated S-layer fragment. The thin, interconnecting network of monomers was likely obscured by the coating. Bars: x,y = 60 nm; z = 10 nm. (TEM micrograph with kind permission of S. Schultze-Lam, Univ. of Guelph).

specimens by AFM was unsuccessful and was likely due to the high compression forces needed for imaging. At present we are experimenting with fine-grain coatings to reveal molecular details while at the same time maintaining sufficient structural rigidity to resist tip compression forces.

Lattice-bearing layers of *Methanospirillum hungatei*

This methanogenic filamentous archaeobacterium has a complex envelope. The rod-shaped cells are enclosed within separate compartments formed by a sheath and spacer plugs (Beveridge et al., 1987).

Sheath and its constituent hoops. *M. hungatei* sheath is a highly resilient, chemically stable structure capable of withstanding harsh denaturing conditions (Beveridge et al., 1985). The outer surface possesses a fine lattice with a 2.8 nm repeat arranged with oblique symmetry (Stewart et al., 1985). The sheath is composed of numerous discrete hoops that stack in register to form the sheath cylinder (Sprott et al., 1986). Detailed descriptions of purification methods and structural analysis are presented in the Chapter by Southam and Beveridge.

The sheath is an ideal specimen for SPM because it is a covalently bonded structure which can be prepared as large rectangular fragments (0.5 μm by several μm) free of contaminating debris. The characteristic broad corrugation of the sheath surface is a useful feature for positive identification in long-range searches and the finer lattice repeat structure (2.8 nm) is a suitable detail for assessing the quality of high resolution images. Furthermore, the height of the collapsed sheath is constant (18 nm) and can act as a calibration standard for height measurement by SPM.

A typical collapsed sheath and its component hoops are shown in Figs. 6 and 7, respectively. In our previous studies, long-range scanning and "hopping" mode capabilities were incorporated into our STM design based on information gathered from imaging sheath fragments (Blackford et al., 1989; Jericho et al., 1989). The "hopping" mode technique was designed to offset stress buildup during STM imaging of uncoated samples and has been successful in revealing new structural details on the sheath to a nominal vertical resolution of 0.4 nm (Beveridge et al., 1990). The sheath surface has also been used to test a variety of overlays such as pure carbon, Pt, and Au/Pd (Blackford et al., 1991b). To date the 2.8 nm repeat on the sheath surface has yet to unambiguously identified since the grain of the overlay material approaches the size of the lattice constant (Blackford et al., 1989; Southam et al., 1992).

The hoops could be reproducibly imaged by STM using carbon films thick enough to allow stable conduction (approximately 4 nm thick) but few structural details were observed (Blackford et al., 1991b). Because of their uniformity hoops may be useful test specimens for assessing the quality of the tip for imaging biological structures (Southam et al., 1992).

Recently, AFM has been applied to uncoated sheath and hoops. The broad corrugation pattern on the sheath has been observed by AFM but, to date, AFM images have not been as good as STM images. This was probably due to differences in tip sharpness and compression loads between AFM and STM. AFM images of sheath layers did demonstrate the ability of sheath to conform exactly to the topology of underlying substrate faults or objects (Southam et al., 1992). In a preliminary study, the strength of attachment of antibody to antigens present on the sheath was directly measured by AFM/STM techniques (Mulhern et al., 1992).

Multilaminar plugs. Cells lay down circular, complex, multilaminar "spacer" plugs during division which, along with the sheath, physically enclose each cell within the filament (Beveridge et al., 1987) Intact plugs were gently extracted from the filament using a freeze-thaw technique followed by 1.2 µm membrane filtration and sucrose gradient sedimentation. Pt-shadowed intact plugs were observed by TEM as thick stacks covered on opposite faces with amorphous material (Fig. 8). Exposure to alkali or even mild heat will cause the plug to dissociate into its individual layers. Plugs were considered excellent candidates for SPM studies because they have a characteristic circular shape with complex surface features.

Samples were diluted in ultrapure water, deposited on freshly cleaved graphite or TaSe$_2$ substrates for STM imaging, or mica substrates for AFM imaging, air dried, and mildly heat-fixed prior to shadowing with Pt. Thick, multilayered plugs were relatively easy to find but the thinner individual layers were obscured by the background debris. A 33 nm thick and 400 nm wide intact plug, is shown in Fig. 9. The outer layer was found to be about 40 nm smaller in diameter and about 8 nm thick. Both the "particulate" and "holey" layers previously observed by TEM (Beveridge et al., 1991) were imaged by STM (Figs. 10 and 11, respectively). These layers were imaged on plug specimens ranging from 5 - 24 nm thick. The images confirmed the existance of a central cavity in each of the subunits of the "holey" layer but we could not conclude that this cavity was continuous through the thickness of the layer. The center-to-center spacing of both types of lattice was ca. 17 nm, slightly less than expected.   Occasionally, single layer specimens still associated with amorphous material were found (Fig. 12). The amorphous material associated with the "holey" layer was marked with small depressions which were not observed in the

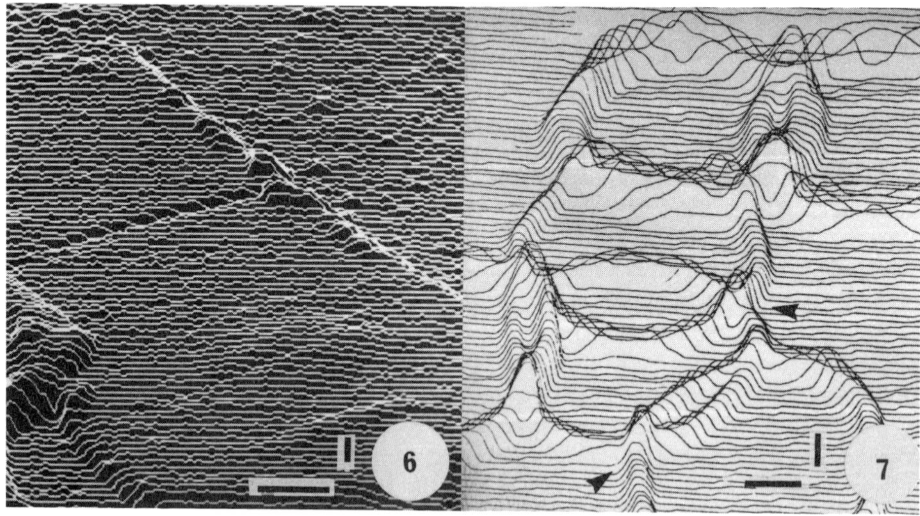

Figure 6. STM image of a single collapsed sheath (Sh) which has been Pt- shadowed at a 90° angle, Bars: x,y = 100 nm; z = 20 nm. Figure 7. C-coated hoops. The flexibility of the hoop structure is evident at overlapping regions (arrows). Bars: x,y = 100 nm; z = 10 nm.

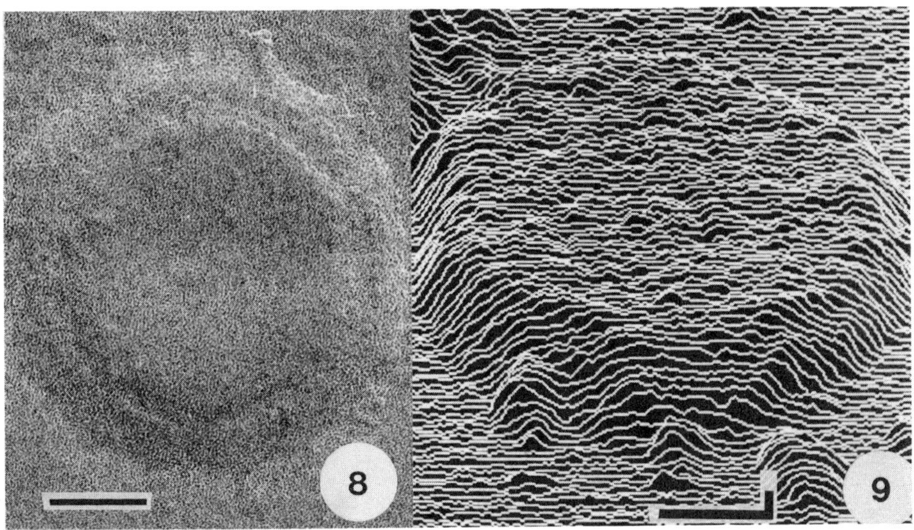

Figure 8. TEM shadowgraph of an intact plug with the outer face exposed. The amorphous covered upper lattice layer of the outer face is refered as the "cap" layer. Bar = 100 nm. **Figure 9.** Line-scan STM image of an intact plug with the outer face exposed for comparison to Fig. 8. Bars: x, y = 100 nm; z = 8 nm.

amorphous material associated with the "particulate" layer. Currently, we are attempting to reconstruct the 3D architecture of the intact plug by identifying the relationship between the thickness of the specimen with the topology of its exposed surface.

An interesting phenomenon of SPM imaging was observed on one Pt/Ir plug specimen with an exposed "particulate" array (Fig. 12). On repeated scanning of this specimen it appeared that the outermost layer(s) was physically moved to the side to expose an "underlying layer".However, we cannot rule out the possibility that this "underlyins layer" is in fact a virtual image due to multiple imaging points on the tip (i.e., a double-tip effect). Deliberate dissociation of the plug stack into component layers by force dissection is currently being attempted.

FUTURE CONSIDERATIONS

Layers produced by *in vitro* self-assembly using purified monomers are a logical choice for future studies, and such layers have recently been used (Ohnesorge et al., 1992) Self-assembled layers would provide large highly-ordered biolayers for testing film overlays or instrument performance. A range of systems from single layers to complex lipid bilayer/lattice layers could feasibly be developed as test specimens. As preliminary studies (Drake et al., 1989; Ohnesorge et al. 1992) indicate, the high-resolution imaging of the self-assembly process in real time may soon become an eventuality.

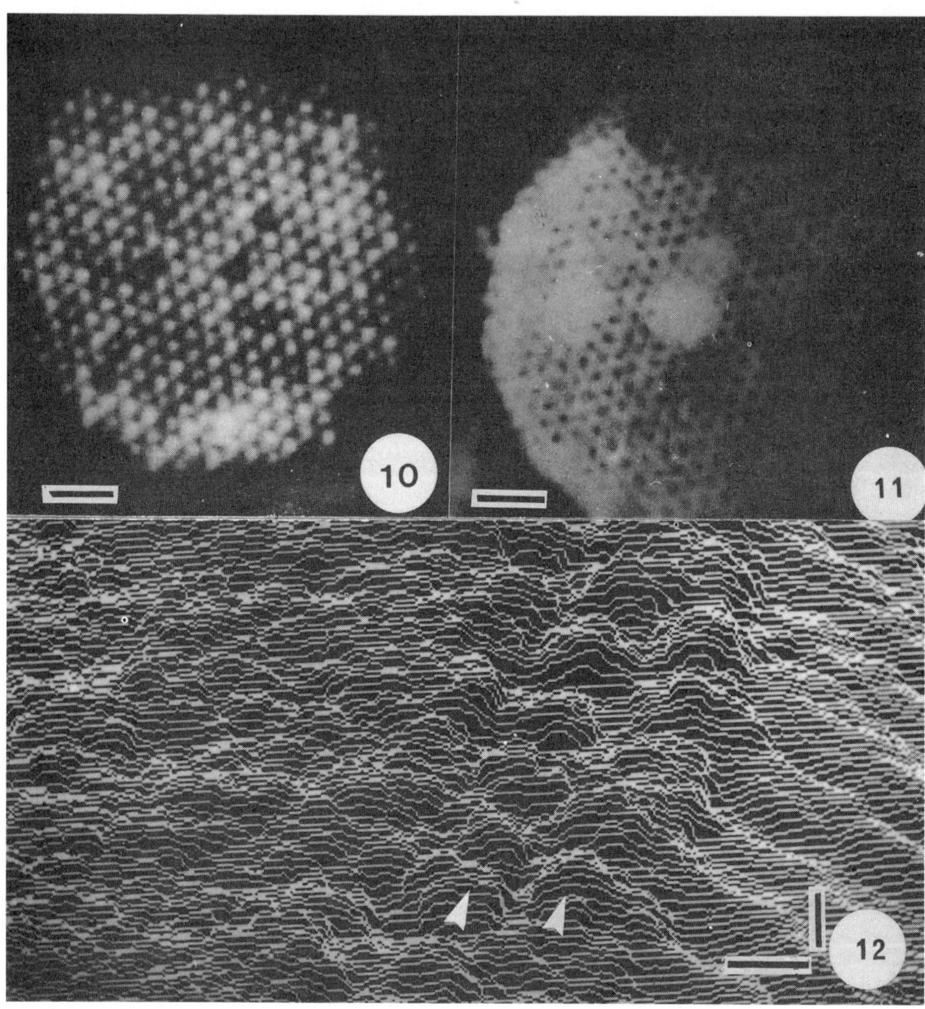

Figure 10. Processed STM image of the "particulate" layer. Height contrast is given by the grey scale; the lowest regions are dark. The layer is imaged as rows of compact, rounded masses spaced at 17 nm intervals. Bar = 50 nm. **Figure 11**. Processed STM image of the 'holey' layer. The subunits are flat circular masses containing a central cavity or pore (dark regions). It is not known if the pore is continuous since the probe was unable to transverse the entire thickness of the layer. Bar = 50 nm. **Figure 12**. Line-scan STM image of the edge of a plug specimen showing in detail the association of an amorphous layer with rows of compact masses (indicated by light arrows). It is most likely part of a 'particulate' layer. Bars: x,y = 20 nm; z = 5 nm.

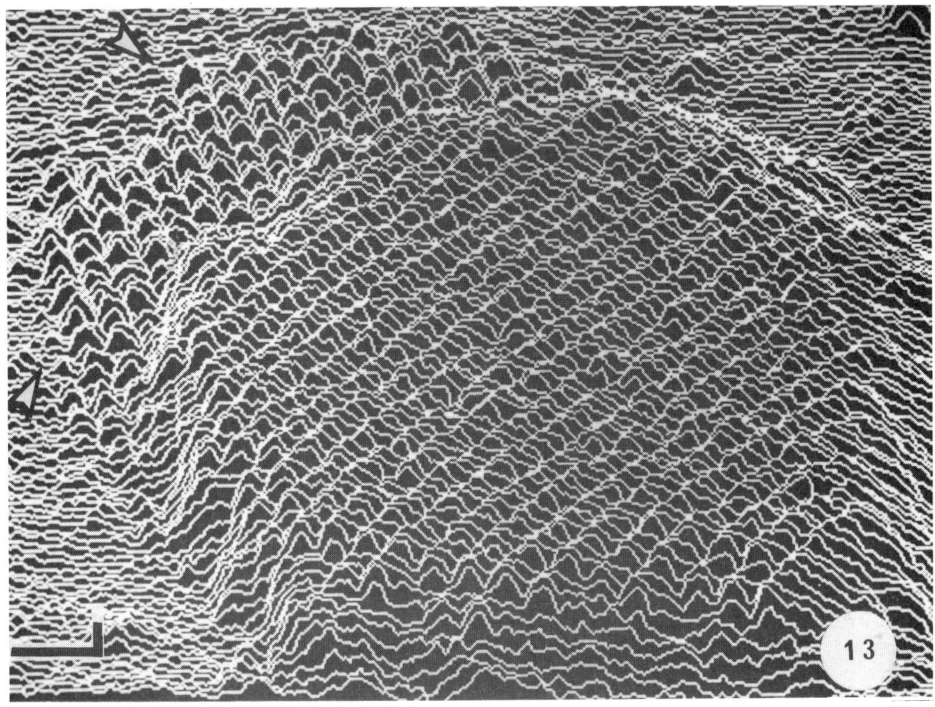

Figure 13. Demonstration of a likely double-tip effect on a multilayered plug with the "particulate" layer exposed. The area indicated by dark arrows corresponds to a possible virtual image. Bars: x,y = 50 nm; z = 7 nm.

ACKNOWLEDGEMENTS

We gratefully acknowledge support from the Natural Sciences and Engineering Research Council of Canada (NSERC) to MHJ and the Medical Research Council of Canada (MRC) to TJB. The TEM was performed in the NSERC Guelph Regional STEM Facility which is partially supported by a NSERC infrastructure grant. The STM and AFM is located in the Department of Physics, Dalhousie University. We also thank B. Harris for technical assistance on TEM.

REFERENCES

Amrein, M., Wang, Z., and Guckenberger, R., 1991, Comparative study of a regular protein layer by scanning tunneling microscopy and transmission electron microscopy, *J. Vac. Sci. Technol.* 9B:1276.

Baumeister, W., and Kubler, O., 1978, Topographical study of the cell surface of *Micrococcus radiodurans, Proc. Natl. Acad. Sci. USA* 75:5525.

Baumeister, W., Barth, M., Hegerl, R., Guckenberger, R., Hahn, M., and Saxton, W.O., 1986, Three-dimensional structure of the regular surface layer (HPI layer) of *Deinococcus radiodurans, J. Mol. Biol.* 187:241.

Beveridge, T.J., Stewart, M., Doyle, R.J., and Sprott, G.D., 1985, Unusual stability of the *Methanospirillum hungatei* sheath, *J. Bacteriol.* 162:728.

Beveridge, T.J., Southam, G., Jericho, M.H., and Blackford, B.L., 1990, High resolution topography of the S-layer sheath of the archaebacterium *Methanospirillum hungatei* provided by scanning tunneling microscopy, *J. Bacteriol.* 172:6589.

Beveridge, T.J., Sprott, G.D., and Whippey, P., 1991, Ultrastructure, inferred porosity, and Gram-staining character of *Methanospirillum hungatei* filament termini describe a unique cell permeability for this archaeobacterium, *J. Bacteriol.* 173:130.

Beveridge, T.J., Harris, B.J., and Sprott, G.D., 1987, Septation and filament splitting in *Methanospirillum hungatei, Can. J. Microbiol.* 33:725.

Binning, G., Rohrer, H., Gerber, Ch., and Weibel, E., 1982, Surface studies by scanning tunneling microscopy, *Phys. Rev. Lett.* 49:57.

Binnig, G., Quate, C.F., and Gerber, Ch., 1986, Atomic force microscope, *Phys. Rev. Lett.* 56:930.

Blackford, B.L., Jericho, M.H., and Mulhern, P.J., 1991a, A review of scanning tunneling microscope and atomic force microscope imaging of large biological structures:Problems and prospects, *Scanning Microsc.* 5:907.

Blackford, B.L., Jericho, M.H., Mulhern, P.J., Frame, C., Southam, G., and Beveridge, T.J., 1991b, Scanning tunneling microscope imaging of hoops from the cell sheath of the bacteria *Methanospirillum hungatei* and atomic force microscope imaging of complete sheaths, *J. Vac. Sci. Technol.* B9:1242.

Blackford, B.L., Watanabe, M.O., Dahn, D.C., Jericho, M.H., Southam, G. and Beveridge, T.J., 1989, The imaging of the complete biological structure with the scanning tunneling microscope, *Ultramicroscopy* 27:427.

Blaurock, A.E., and Stoeckenius, W., 1971, Structure of the purple membrane, *Nature New Biology* 233:152.

Burnham, N.A., Colton, R.J., and Pollock, H.M., 1991, Interpretation issues in force microscopy, *J. Vac. Sci. Technol.* A9:2548.

Butt, H.J., Downing, K.H., and Hansma, P.K., 1990, Imaging the membrane protein bacteriorhodopsin with the atomic force microscope, *Biophys. J.* 58:1473.

Clemmer, C., and Beebe, T., 1991, Graphite: a mimic for DNA and other biomolecules in scanning tunneling microscope studies, *Science* 251:640.

Dunlap, D.D. and Bustamante, C., 1989, Images of single stranded nucleic acids by scanning tunnelling microscopy, *Nature* 342:204.

Drake, B., Prater, C.B., Weisenhorn, A.L., Gould, S.A.C., Albrecht, T.R., Quate, C.F., Cannell, D.S., Hansma, H.G., and Hansma, P.K., 1989, Imaging crystals, polymers and processes in water with the atomic force microscope, *Science* 243:1586.

Edstrom, R.D., Yang, X., Lee, G., Evans, D.F., 1990, Viewing molecules with scanning tunneling microscopy and atomic force microscopy, *Federation of American Societies for Experimental Biology (FASEB) J.* 4:3144.

Egger, M., Ohnesorge, F., Weisenhorn, A.L., Heyn, S.P., Drake, B., Prater, C.B., Gould, S.A.C., Hansma, P.K., and Gaub, H.E., 1990, Wet lipid-protein membranes imaged at submolecular resolution by atomic force microscopy, *J. Struct. Biol.* 103:89.

Engel, A., 1991, Biological applications of scanning probe microscopes, *Annu. Rev. Biophys. Biophys. Chem.* 20:79.

Fisher, K.A., Whitfield, S.L., Thomson, R.E., Yanagimoto, K.-C., Gustafsson, M.G.L., and Clarke, J., 1990, Measuring changes in membrane thickness by scanning tunneling microscopy, *Biochim. Biophys. Acta.* 1023:325.

Golovichenko, J.A., 1986, The tunneling microscope: a new look at the atomic world, *Science* 232:48.

Gould, S.A.C., Drake, B., Prater, C.B., Weisenhorn, A.L., Manne, S., Hansma, H.G., Hansma, P.K., Massie, J., Longmire, M., Elings, V., Dixon Northern, B., Mukergee, B., Peterson, C. M., Stoeckenius, W., Albrecht, T.R., and Quate, C.F., 1990, From atoms to integrated circuit chips, blood cells and bacteria with the atomic force microscope, *J. Vac. Sci. Technol.* A8:369.

Guckenberger, R., Wiegrabe, W., Hillebrand, A., Hartmann, T., Wang, Z., and Baumeister, W., 1989, Scanning tunneling microscopy of a hydrated bacterial surface protein, *Ultramicroscopy* 31:327.

Guckenberger, R., Hacker, B., Hartmann, T., Scheybani, T., Wang, Z., Wiegrabe, W., and Baumeister, W., 1991, Imaging of uncoated purple membrane by scanning tunneling microscopy, *J. Vac. Sci. Technol.* B9:1227.

Hansma, P. K., Elings, V.B., Marti, O., and Bracker, C.E., 1988, Scanning tunneling microscopy and atomic force microscopy: application to biology and technology, *Science* 242:209.

Henderson, R., Baldwin, J.M., Ceska, T.A., Zemlin, F., Beckmann, E., and Downing, K.H., 1990, A model for the structure of bacteriorhodopsin based on high resolution electron cryo-microscopy, *J. Mol. Biol.* 213:899.

Henderson, E., 1992, Imaging and nanodissection of individual supercoiled plasmids by atomic force microscopy, *Nucleic Acids Res.* 20:445.

Hoh, J.H., Lal, R., John, S.A., Revel, J.-P., and Arnsdorf, M.F., 1991, Atomic force microscopy and dissection of gap junctions, *Science* 253:1405.

Jericho, M.H., Blackford, B.L., and Dahn, D.C., 1989, Scanning tunneling microscope imaging technique for weakly bonded surface deposits, *J. Appl. Phys.* 65:5237.

Keller, R.W., Dunlap, D.D., Bustamante, C., Keller, K.J., Garcia, R.G., Gray, C., and Maestre, M.F., 1990, Scanning tunneling microscopy images of metal-coated bacteriophages and uncoated, double stranded DNA, *J. Vac. Sci. Technol.* A8:706.

Michel, B. and Travaglini, G., 1988, An STM for biological applications:bioscope, *J.Microsc.* 152:681.

Mulhern, P.J., Blackford, B.L., Jericho, M.H., Southam, G., and Beveridge, T.J., 1992, AFM and STM studies of the interaction of antibody with the S-layer sheath of the archaeobacterium *Methanospirillum hungatei,* *Ultramicroscopy* 42-44:1214.

Oesterhelt, D., and Stoeckenius, W., 1971, Rhodopsin-like protein from the purple membrane of *Halobacterium halobium, Nature New Biology* 233:149.

Oesterhelt, D., and Stoeckenius, W., 1973, Functions of a new membrane photoreceptor, *Proc. Nat. Acad. Sci. USA* 70:2853.

Ohnesorge, F., Heckl, W.M., Haberle, W., Pum, D., Sara, M., Schindler, H., Schilcher, K., Kiener, A., Smith, D.P.E., Sleytr, U.B., and Binnig, G., 1992, Scanning force microscopy studies of the S-layers from *Bacillus coagulans* E38-66, *Bacillus sphaericus* CCM2177 and of an antibody binding process, *Ultramicroscopy* 42-44:1236.

Reiss, G., Vancea, J., Wittmann, H., Zweck, J., and Hoffmann, H., 1990, Scanning tunneling microscopy on rough surfaces: Tip-shape-limited resolution, *J. Appl. Phys.* 67:1156.

Salmeron, M., Beebe, T., Odriozola, J., Wilson, T., Ogletree, D.F., and Siekhaus, W., 1990, Imaging of biomolecules with the scanning tunneling microscope:problems and prospects, *J. Vac. Sci. Technol.* A8:635.

Salmeron, M., Ogletree, D.F., Ocal, C., Wang, H.-C, Neubauer, G., Kolbe, W., and Meyers, G., 1991, Tip-surface forces during imaging by scanning tunneling microscopy, *J. Vac. Sci. Technol.* B9:1347.

Smith, D.P.E., Bryant, A., Quate, C.F., Rabe, J.P., Gerber, Ch., and Swalen, J.D., 1987, Images of a lipid bilayer at molecular resolution by scanning tunneling microscopy, *Proc. Natl. Acad. Sci. USA* 84:969.

Southam, G., Firtel, M., Blackford, B.L., Jericho, M.H., Xu, W., Mulhern, P.J., and Beveridge, T. J., 1992, Transmission electron microscopy, scanning tunneling microscopy and atomic force micoscopy of the cell envelope layers of the archaeobacterium *Methanospirillum hungatei* GP1, *J. Bacteriol.*, in press.

Spong, J.K., Mizes, H.A., LaComb Jr., L.J., Dovek, M.M., Frommer, J.E., and Foster, J.S., 1989, Contrast mechanism for resolving organic molecules with tunnelling microscopy, *Nature* 338:137.

Sprott, G.D., Beveridge, T.J., Patel, G.B., and Ferrante, G., 1986, Sheath disassembly in *Methanospirillum hungatei* GP1. *Can. J. Microbiol.* 32:847.

Stemmer, A., Engel, A., Haring, R., Reichelt, R., and Aebi, U., 1988, Scanning tunneling microscope with integrated 2-axes heterodyne interferometer and light-microscope, *Ultramicroscopy* 25:171.

Stemmer, A., Hefti, A., Aebi, U., and Engel, A., 1989, Scanning tunneling and transmission electron microscopy on identical areas of biological specimens, *Ultramicroscopy* 30:263.

Stemmer, A., and Engel, A., 1990, Imaging biological macromolecules by STM: quantitative interpretation of topographs, *Ultramicroscopy* 34:129.

Stewart, M., Beveridge, T.J., and Sprott, G.D., 1985, Crystalline order to high resolution in the sheath of *Methanospirillum hungatei*:a cross-beta structure, *J. Mol. Biol.* 183:509.

Thompson, B.G., Murray, R.G.E., and Boyce, J.F., 1982, The association of the surface array and the outer membrane of *Deinococcus radiodurans, Can. J. Microbiol.* 28:1081.

Travaglini, G., Rohrer, H., Stoll, E., Amrein, M., Stasiak, A., Sogo, J., and Gross, H., 1988, Scanning tunneling microscopy of recA-DNA complexes, *Physica Scripta.* 38:309.

Wepf, R., Amrein, M., Burkli, U., and Gross, H., 1991, Platinum Iridium Carbon -a high-resolution shadowing material for TEM, STM and SEM of biological macromolecular structures, *J. Microsc.* 163:51.

Wickramasinghe, H.K., 1990, Scanning probe microscopy: Current status and future trends, *J. Vac. Sci. Technol.* A8:363.

Worchester, D.L., Miller, R.G., and Bryant, P.J., 1988, Atomic force microscopy of purple membranes, *J. Microsc.* 152:817.

Worchester, D.L., Kim, H.S., Miller, R.G., and Bryant, P.J., 1990, Imaging bacteriorhodopsin lattices in purple membranes with atomic force microscopy, *J. Vac. Sci. Technol.* A8:403.

Chapter 24

STABLE LIPOSOMES FORMED FROM ARCHAEAL ETHER LIPIDS

Christian G. Choquet, Girishchandra B. Patel, and G. Dennis Sprott[*]

Institute for Biological Sciences, National Research Council of Canada
Ottawa, Ontario, Canada

Terry J. Beveridge

Department of Microbiology, College of Biological Science
University of Guelph
Guelph, Ontario, Canada

INTRODUCTION

Liposomes have application in serving as a support, or matrix, to form two dimensional crystals from those proteins which normally interact with the cytoplasmic membrane (Wingfield et al., 1979). This interaction includes membrane proteins that are largely embedded within the lipid bilayer, others with relatively more surface area exposed, and surface proteins which interact either via a single transmembrane strand or via a covalently linked lipid (Andreas et al., 1992). Recently, Andreas et al. demonstrated the two-dimensional crystallization of a solubilized surface protein from *Comamonas acidovorans* on lipid vesicles.

Archaeobacterial (Archaeal; Woese et al., 1990) membrane lipids are structurally different than those from other sources. Fatty acyl chains, which are often unsaturated and are esterified to sn-1,2 carbons of glycerol, are replaced in archaeobacteria by fully saturated phytanyl chains in ether linkage to glycerol carbons with opposite sn-2,3 configuration (Kates, 1990).

Liposomes are known to form spontaneously from the isolated polar lipids of only a few archaeobacteria, including the ether lipids of *Halobacterium cutirubrum* (as multilamellar structures; Chen et al., 1974), the major tetraether lipid from *Thermoplasma acidophilum* (Ring et al., 1986), and a tetraether lipid subfraction from *Sulfolobus acidocaldarius* (Lo and Chang, 1990). Here we explore the possibility of forming liposomes from other archaeobacterial lipids by detergent dialysis and pressure extrusion, and characterize the liposomes formed. A simple method of

---

[*]Author presenting the paper as NRC Publication No. 34280.

Advances in Bacterial Paracrystalline Surface Layers
Edited by T.J. Beveridge and S.F. Koval, Plenum Press, New York, 1993

forming archaeobacterial liposomes should find application in reassembly of those surface layer proteins which normally interact with the cytoplasmic membrane.

POLAR LIPID STRUCTURES

Generally polar lipids (ether core lipid plus headgroup) constitute from 80-95% of the total lipids extracted. Since the neutral lipid fraction is not used in this report on liposome construction, only the polar lipid structures will be reviewed.

Various archaeobacteria have one or more of the lipid cores shown in Fig. 1. The standard diether (D_S) was first discovered by Kates and associates (reviewed by Kates, 1990) in the extreme halophile *Halobacterium cutirubrum*. The structure was a 2,3-diphytanyl-_sn_-glycerol, of opposite stereochemistry to the glycerolipids of non-archaeobacteria. Head group variations bonded to the _sn_-1 carbon of glycerol resulted in a series of diether polar lipids which accounted for the bulk of the total lipid, and all of the ether lipids found in this extreme halophile (Kates, 1990).

Variations to the D_S structure occur. For example, certain alkalophilic extreme halophiles add an extra five carbon unit on one or both of the C_{20} chains (Morth and Tindall, 1985). Also, the deep-sea methanogen *Methanococcus jannaschii* has D_S macrocyclic diether (Comita et al., 1984), and tetraether lipid cores which vary in relative amounts as a response to the temperature of growth (Sprott et al., 1991).

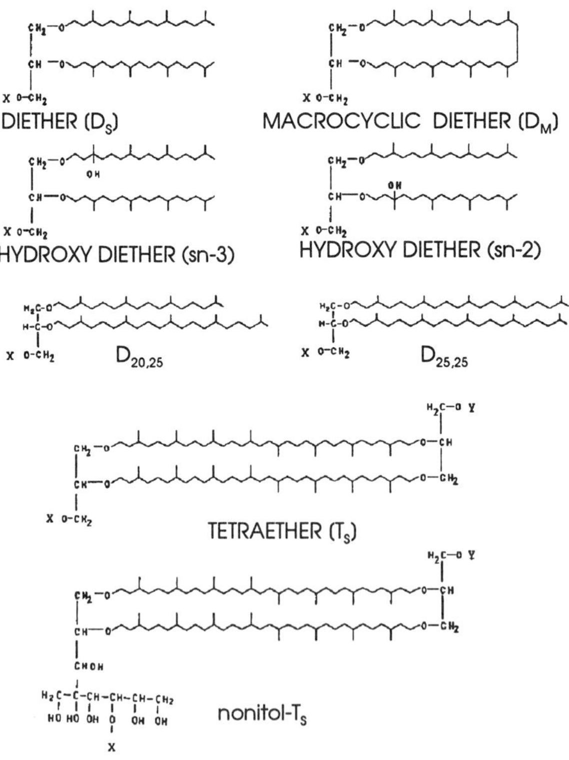

Figure 1. The most commonly found lipid moieties of the ether lipids in archaeobacteria (Archaea). X and Y represent the wide variety of head groups in the polar ether lipids. When hydrolyzed to remove the head groups, X and Y are replaced by a hydrogen atom and the resulting structures are referred to as the core lipids or lipid cores.

Another variation encountered in certain methanogens is the presence of a single -OH group at position 3 of one of the phytanyl chains (Ferrante et al., 1988; Sprott et al., 1990). Finally, certain thermoacidophiles, such as *Sulfolobus sulfataricus*, have about 10% D_S lipid core and 90% as standard tetraether (T_S) plus nonitol-tetraether (De Rosa and Gambacorta, 1988). This latter component is especially relevant in liposome studies, since it is a non-bilayer forming lipid (Lelkes et al., 1983). In Table 1, we summarize the core lipids and their abundance in the methanogen lipid extracts used here. The readers are referred to recent reviews for the structures of the numerous head groups found in these glycolipids and phospholipids (De Rosa and Gambacorta, 1988; Kates, 1990; Sprott, 1992).

Table 1. Known distribution of the different ether core lipids present in the methanogens used in our studies

Methanogen	Distribution (%) of:				References
	Diether	Hydroxy-diether	Tetra-ether	Macrocyclic Diether	
M. voltae	>90	<10			Sprott et al., 1990 Choquet et al., 1992
M. concilii	70	30 (sn-3)			Ferrante et al., 1988
M. mazei	43	57 (sn-2)			Sprott et al., 1990
M. hungatei	50		50		Choquet et al., 1992
M. jannaschii					
65° C[a]	15		42	43	Sprott et al., 1991
50° C[a]	60		21	19	Sprott et al., 1991

[a] Growth temperature of *M. jannaschii*; other methanogens were grown at 35°C.

LIPID PREPARATION

Methanogens were grown in a 75 L fermentor in 55 L of cysteine-sodium sulfide reduced medium (Choquet et al., 1992). Cells were harvested in the late exponential growth phase, and frozen as a paste at -20°C. Lipids were extracted from thawed cells and the polar lipids recovered by precipitation with cold acetone, as described. Following the third precipitation from acetone, the polar lipids were loaded onto a silica G column for removal of non-lipid contaminants (Choquet et al., 1992).

LIPOSOME FORMATION

Liposome formation from the total polar lipid extracts was attempted by the following two methods.

For the detergent dialysis technique (Choquet et al., 1992), polar lipids and n-octyl-β-D-glucopyranoside were dissolved in $CHCl_3$ in 1:20 molar ratio. Mixed micelles were formed from the dried mixture by dissolving in dialysis buffer (10 mM potassium phosphate, pH 7.14) containing 160 mM NaCl. Liposomes were formed

Table 2. Size characteristics of liposomes prepared by detergent dialysis*

Origin of total polar lipid extract	Result of:				Trapped vol (μl/mg of lipid[a])
	DLS		Electron microscopy		
	Mean diam. ± SD (nm)	Coefficient of variation[b]	Mean diam. ± SD	Coefficient of variation	
M. jannaschii					
65°C[c]	129 ± 31	0.24	113 ± 28 (282)[d]	0.24	4.5
50°C[c]	81 ± 22	0.27	Not determined		4.0
M. voltae	54 ±19	0.35	56 ± 23 (298)	0.41	4.5
M. mazei	45 ± 19	0.42	49 ± 21 (201)	0.43	1.8
M. concilii[e]	69 ± 30	0.43	70 ± 38 (212)	0.54	3.4

a Total polar lipid.
b Coefficient of variation of the size distribution = standard deviation/mean diameter.
c Growth temperature.
d Number of liposomes measured.
e Liposomes were filtered through a 0.22 μm pore-size nylon filter to remove the large aggregates.
* Reproduced from Choquet et al. (1992) with permission.

at ambient temperature with a Liposomat (Avestin Inc., Ottawa) operating for 4 h at a flow rate of 0.5 ml/min for the mixed micelles and 2.5 ml/min for the dialysis buffer.

A pressure extrusion method was conducted according to MacDonald et al. (1991) with a LiposoFast (Avestin Inc., Ottawa). Polar lipids (20 mg/ml) in 10 mM potassim phosphate, pH 7.14 plus 160 mM NaCl were homogenized in a 2 ml Potter-Elvehjem tissue grinder. The resulting multilamellar structures were passed 21 times through two (stacked) polycarbonate filters of either 100 or 200 nm porosity.

In some cases 5(6)-carboxyfluorescein or [^{14}C]sucrose was entrapped by adding 100 mM or 150 mM (33 nCi/μmol), respectively, to the extrusion buffer.

LIPOSOME ANALYSIS

Mean diameters and number-weighted size distributions of the vesicle preparations were determined by dynamic light scattering (DLS) in the vesicle mode using the NICOMP model 370 submicron particle sizer.

In some cases size was determined by transmission electron microscopy (TEM) analysis of liposomes negatively stained with a 1% solution of sodium phosphotungstate (pH 7.2). Grids were treated with bacitracin prior to sample application (Choquet et al., 1992).

For freeze-fracturing, liposomes were centrifuged (200,000 x g max for 5 h). Pellets were frozen, fractured, and etched according to Choquet et al. (1992) and Beveridge et al. (1993).

LIPOSOME MORPHOLOGY AND SIZE

The natural mix of polar lipids extracted from *Methanococcus voltae, Methanococcus jannaschii, Methanosarcina mazei* and *Methanosaeta concilii* (Table 1) produced liposomes upon detergent dialysis (Fig. 2). In the case of the *M. concilii* liposomes, large particles (>1 μm) were seen by phase contrast microscopy which accounted for a small fraction of the population. By electron microscopy they appeared to be aggregates of variously sized liposomes, and were easily removed by filtration through a 0.22 μm nylon filter. Once removed, the remaining liposome population was fairly homogeneous (Fig. 2D), but the most homogeneous population was obtained from the deep-sea thermophile, *M. jannaschii* (Fig. 2B).

Liposomes from *Methanospirillum hungatei* showed severe size and shape heterogeneity. Some vesicles were large, multilamellar (in freeze-fractures) and connected by fine tubules (Fig. 2E). These large aggregates were quite frequent, so that the population could not accurately be measured by dynamic light scattering (DLS). Sizing by electron microscopy revealed that 40% of the liposomes were 20-100 nm in diameter, 28% varied between 300 and 1000 nm, and the remainder were intermediate in size.

Liposomes prepared by detergent dialysis varied in size depending on the source of the lipids (Table 2). Sizing by electron microscopy or DLS yielded similar results, and showed that exclusively diether lipid extracts (*M. voltae, M. mazei* and *M. concilii*) resulted in smaller liposomes than diether/tetraether mixtures (*M. jannaschii*). Further, increasing the proportion of tetraether lipids by growth of *M. jannaschii* at a higher temperature resulted in larger liposomes. These liposomes were sealed vesicles, as seen from [^{14}C]sucrose entrapment studies (Table 2). Volume estimates ranged from 1.8 to 4.5 μl/mg lipid.

Freeze-fracturing was performed to determine whether the ether liposomes were unilamellar or multilamellar. Freeze-fractured liposomes of *M. voltae* had internal bilayer fractures (arrows, Fig. 3 panel 1) and all faces had smooth surfaces. *M. jannaschii* liposomes (65°C growth, Fig. 3 panel 2) cleaved along their outer surfaces and only a few cross-fractured to reveal a smooth inner liposome face (arrow). No fractures through the bilayer were seen. Freeze-fractured liposomes of *M. jannaschii* (50°C growth) had small internal bilayer fractures to reveal intramembrane particles (arrows, Fig. 3 panel 3) in these proteinless structures. In all cases the vesicles were unilamellar.

Liposomes were produced by a second technique (not employing a detergent), whereby a homogenate of hydrated polar lipids were extruded through a constant pore-sized filter. An obvious advantage with this technique compared to detergent dialysis was to avoid the presence of the large aggregates seen with *M. concilii* and *M. hungatei* polar lipids (Fig. 4). Another advantage was the ease of varying size (see *M. jannaschii*, Fig. 4) through the use of defined pore-sized filters. The size characteristics by DLS are shown in Table 3 for various archaeobacterial polar lipids. Size was affected not only by filter pore-size, but also by the type of lipids present and in the case of *H. cutirubrum* by the NaCl content of the extrusion buffer.

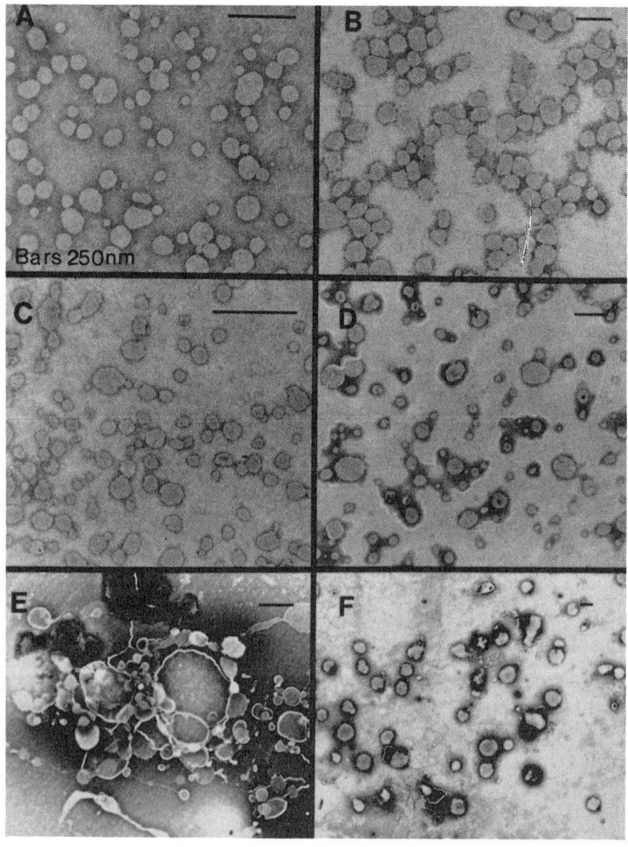

Figure 2. Transmission electron micrographs of liposomes prepared by detergent dialysis and negatively stained with phosphotungstate. The liposomes were prepared from the natural mix of total polar lipids obtained from *M. voltae* (A), *M. jannaschii* grown at 65°C (B), *M. mazei* (C), *M. concilii* (D), and *M. hungatei* (E and F). Reproduced with permission from Choquet et al., 1992.

Figure 3. Freeze-fractured liposomes prepared by detergent dialysis. *M. voltae* liposomes (panel 1) consisting of diether lipids. Arrows point to internal bilayer fractures. Liposomes prepared from the total polar lipids of *M. jannaschii* grown at 65°C (panel 2) cleaved primarily along their outer surfaces, although a few cross-fractured (small arrowhead) to reveal the inner surface after etching. *M. jannaschii* total polar lipids extracted following growth at 50°C formed liposomes (panel 3) which produced small internal bilayer fractures (arrows). Large arrowheads indicate the direction of shadowing.

Table 3. Size characteristics (liposome mean diameter in nm ± standard deviation of the mean) of the different liposome suspensions.

Origin of total polar lipid extract	100 nm pore filter	200 nm pore filter	400 nm pore filter
M. jannaschii	115 ± 26[1] (0.23[2]) 97 ± 23[3] (0.24)	158 ± 53 (0.34) not determined	182 ± 71 (0.39)
M. smithii	111 ± 35 (0.31)	198 ± 54 (0.27)	226 ± 71 (0.31)
M. hungatei	114 ± 27 (0.24)	182 ± 46 (0.25)	189 ± 63 (0.33)
M. mazei	79 ± 23 (0.29)	not determined	
M. voltae	93 ± 28 (0.30)	141 ± 43 (0.30)	174 ± 61 (0.35)
H. cutirubrum	78 ± 25 (0.32) 113 ± 42[4] (0.37)	139 ± 53 (0.38) 315 ± 44 (0.14)	160 ± 62 (0.38) 456 ± 57 (0.13)

[1] *M. jannaschii* grown at 65°C.

[2] coefficient of variation = standard deviation/mean diameter of the size distribution.

[3] *M. jannaschii* grown at 50°C.

[4] liposomes formed in extrusion buffer containing 4.0 M NaCl.

The coefficient of variation of the size distribution (between 0.20 and 0.40) obtained by DLS revealed that all the liposome suspensions had good size homogeneity. These findings supported the electron micrographs shown in Fig. 4.

The pH of the extrusion buffer can have an effect on the mean diameter of the resulting liposomes (Table 4). Of the lipid extracts used for this study, those of

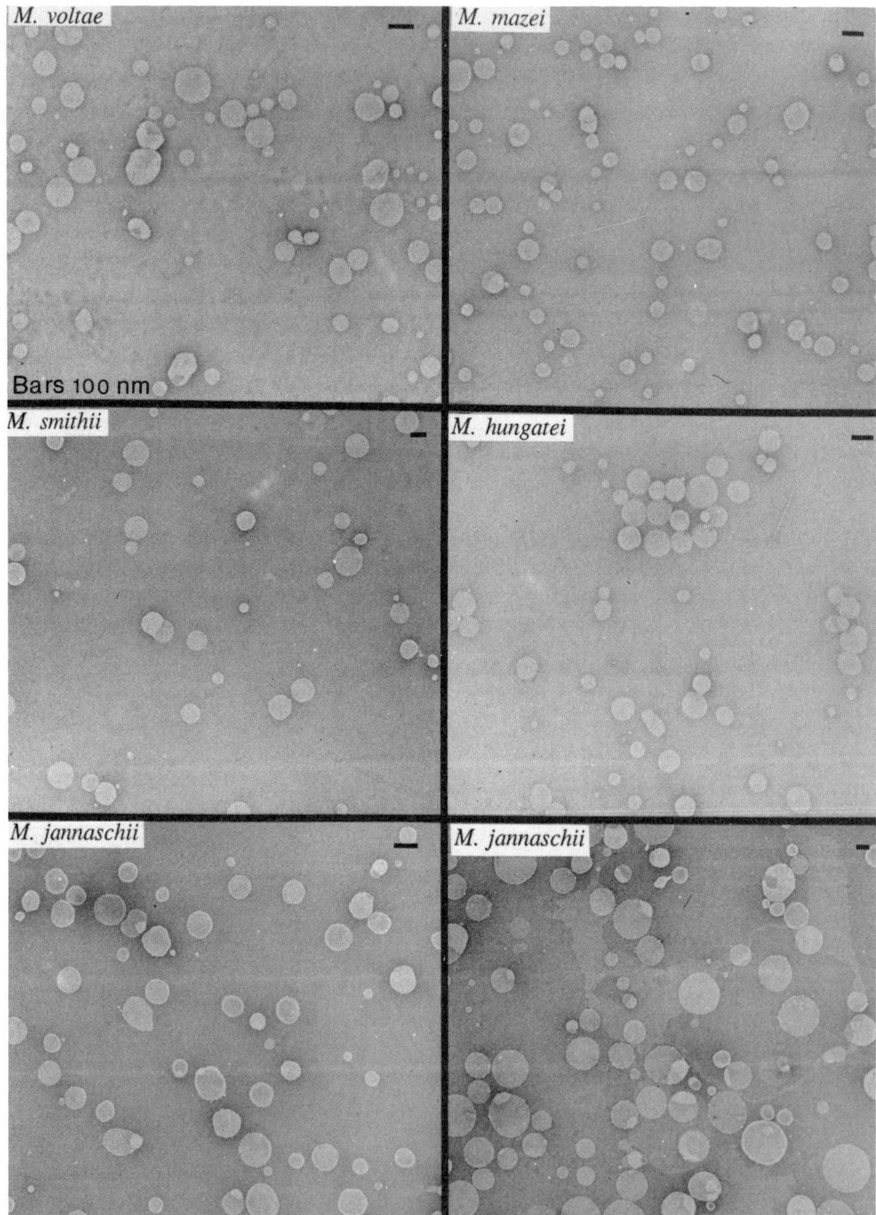

Figure 4. Transmission electron micrographs of liposomes prepared by pressure extrusion with 100 nm pore filters and negatively stained with phosphotungstate. For these experiments total polar lipid extracts were obtained from *M. voltae, Methanobrevibacter smithii, M. hungatei, M. mazei* and *M. jannaschii* grown at 65°C. Liposomes obtained with a 200 nm pore filter for total polar lipids of *M. jannaschii* grown at 65°C are shown on the right-hand side, lower panel.

M. mazei yielder larger liposomes with increasingly acidic conditions. Since this effect was not evident with polar lipids of *M. jannaschii* grown at 50°C, it must be dependent on the lipid composition of each extract.

Table 4. Effect of pH on the mean diameter (nm) of unilamellar liposomes during their formation and during their storage in aqueous media at 4°C

pH of liposome suspension[1]	storage time[2] (days)	Origin of total polar lipid extract		
		M. jannaschii[3]	*M. jannaschii*[4]	*M. mazei*
3.0	0	107 ± 28[5] (0.26[6])	130 ± 27 (0.21)	92 ± 24 (0.26)
	60	107 ± 27 (0.26)	134 ± 45 (0.33)	99 ± 40 (0.40)
7.1	0	96 ± 23 (0.24)	115 ± 27 (0.23)	79 ± 23 (0.29)
	60	99 ± 26 (0.26)	113 ± 33 (0.29)	72 ± 24 (0.33)
10.7	0	102 ± 23 (0.23)	97 ± 27 (0.28)	71 ± 22 (0.31)
	60	106 ± 26 (0.25)	95 ± 27 (0.28)	75 ± 20 (0.27)

[1] liposomes were generated with 100 nm pore filters and stored at the pHs indicated.
[2] liposomes were stored at 4°C.
[3] *M. jannaschii* grown at 50°C.
[4] *M. jannaschii* grown at 65°C.
[5] standard deviation of the mean.
[6] coefficient of variation of the size distribution.

Table 5. Entrapped [14C] sucrose remaining (% of initial entrapped concentration) after storage of unilamellar liposome suspensions at different pHs

pH[1] of liposome suspension	storage time (days)	egg lecithin	Origin of total polar lipid extract		
			M. jannaschii[2]	*M. jannaschii*[3]	*M. mazei*
3.0	1	78	75	72	98
	28	53	77	71	80
7.1	1	100	91	94	86
	26	98	94	98	64
10.7	2	78	89	89	83
	20	2	90	94	75

The header "[14C] Sucrose" spans the data columns above "Origin of total polar lipid extract".

[1] liposomes were generated with 100 nm pore filters at pH values shown and stored at 4°C. 0.2% sodium azide was added to prevent bacterial growth.
[2] *M. jannaschii* grown at 50°C.
[3] *M. jannaschii* grown at 65°C.

STABILITY

DLS analysis was used to detect physical instability (fusion or aggregation) of unilamellar ether liposomes during long storage periods. After 60 days of incubation

Table 6. Effect of temperature on leakiness of carboxyfluorescein

Incubation temperature (°C)	% leakage of carboxyfluorescein after 24 h storage			
	egg lecithin	DPPC	*M. jannaschii* (65°C)	*M. mazei*
65	100	100	17	74
50	83	100	11	56
35	49	4	3	29
25	24	1	6	6
4	11	1	0	0

at 4°C, the mean diameter of each liposome population had not changed (Table 4). In most cases the coefficients of variation of the size distributions were also unaffected, although a significant increase was observed at pH 3.0 with liposomes generated with lipids of *M. jannaschii* (65°C) and *M. mazei*.

Of the ether liposomes, those generated with lipids of *M. jannaschii* were the most stable. This is likely due to the presence of tetraether lipids in those extracts; *M. mazei* has no tetraether lipids. Sprott et al. (1991) have shown that *M. jannaschii* grown at 50°C contains higher proportions of standard diether lipids and less tetraether and macrocyclic diethers than in cells grown at 65°C. Therefore, the stability of liposomes made from lipids of cells grown at both temperatures was tested during storage at 35, 50 and 65°C (data not shown). The liposomes containing less tetraethers were consistently more leaky than those prepared from lipids of cells grown at 65°C.

The leakage of [^{14}C]sucrose from liposomes stored at 4°C was studied (Table 5). At pH values 3.0 and 7.1, ether liposomes retained the solute as effectively as the ester liposomes. However, at pH 10.7 ether liposomes were much more effective in solute retention than the lecithin liposomes.

Temperature is an important parameter to explore during attempts to develop two-dimensional crystals on lipid surfaces. Crystallization of the *C. acidovorans* surface array protein was greatly improved by elevating the crystallization temperature above 23°C (the phase transition temperature of the dimyristoylphosphatidylcholine vesicles; Andreas et al., 1992).

As a sensitive test, the stability of archaeobacterial liposomes was monitored by rates of leakage of a fluorescent entrapped dye (Table 6). Liposomes prepared from the polar lipids of *M. mazei* were considerably more leaky than those of *M. jannaschii*. In comparison, ester liposomes prepared from egg lecithin were very leaky at 23°C or higher, and those prepared from DPPC were leaky near the phase transition temperature for this lipid.

CONCLUDING REMARKS

Archaeobacterial ether lipids are likely to have advantages compared to ester lipids for the preparation of stable liposomes suitable for preparing planar crystal coatings of membrane-associated proteins. The natural mix of polar ether lipids from several archaeobacteria formed unilamellar liposomes by either detergent dialysis or pressure extrusion. These lipids have fully saturated phytanyl chains not sensitive to oxidation reactions, as well as phospholipase-insensitive ether bonds to the glycerol moiety. Ether liposomes were generally found to resist fusion/aggregation over long storage periods and were stable to pH and temperature extremes. In addition, their lipids appear not to undergo a phase transition over physiological temperatures. The variety of unique polar lipid structures found within the archaeobacteria and the possibility that a protein interaction may occur with a specific lipid, suggests that it may be prudent for crystallizations to use the lipids from the species from which the protein was solubilized. This may be especially relevant in reassembly of isolated S-layer proteins of archaeobacteria, where an intimate association with the cytoplasmic membrane is common (Baumeister et al., 1989).

REFERENCES

Andreas, P., Engelhardt, H., Jakubowski, U., and Baumeister, W., 1992, Two-dimensional crystallization of a bacterial surface protein on lipid vesicles under controlled conditions, *Biophys. J.* 61:172.

Baumeister, W., Wildhaber, I., and Phipps, B.M., 1989, Principles of organization in eubacterial and archaebacterial surface proteins, *Can. J. Microbiol.* 35:215.

Beveridge, T.J., Choquet, C.G., Patel, G.B., and Sprott, G.D., 1993, Freeze-fracture planes of methanogen membranes correlate with the content of tetraether lipids, *J. Bacteriol.* 175: in press.

Chen, J.S., Barton, P.G., Brown, D., and Kates, M., 1974, Osmometric and microscopic studies on bilayers of polar lipids from the extreme halophile, *Halobacterium cutirubrum, Biochim. Biophys. Acta* 352:202.

Choquet, C.G., Patel, G.B., Beveridge, T.J., and Sprott, G.D., 1992, Formation of unilamellar liposomes from total polar lipid extracts of methanogens, *Appl. Environ. Microbiol.* 58:2894.

Comita, P.B., Gagosian, R.B., Pang, H., and Costello, C.E., 1984, Structural elucidation of a unique, macrocyclic membrane lipid from a new, extremely thermophilic, deep-sea hydrothermal vent archaebacterium, *Methanococcus jannaschii, J. Biol. Chem.* 259:15234.

De Rosa, M., and Gambacorta, A., 1988, The lipids of archaebacteria, *Prog. Lipid Res.* 27:153.

Ferrante, G., Ekiel, I., Patel, G.B., and Sprott, G.D., 1988, A novel core lipid isolated from the aceticlastic methanogen, *Methanothrix concilii* GP6, *Biochim. Biophys. Acta* 963: 173.

Kates, M., 1990, Glyco-, phosphoglyco- and sulfoglycoglycerolipids of bacteria, *in*: "Glycolipids, Phosphoglycolipids, and Sulfoglycolipids", M. Kates, ed., Plenum Press, New York.

Lelkes, P.I., Goldenberg, D., Gliozzi, A., De Rosa, M., Gambacorta, A., and Miller, I.R., 1983, Vesicles from mixtures of bipolar archaebacterial lipids with egg phosphatidylcholine, *Biochim. Biophy. Acta* 33:337.

Lo, S.L., and Chang, E.L., 1990, Purification and characterization of a liposomal-forming tetraether lipid fraction, *Biochem. Biophys. Res. Commun.* 167:238.

MacDonald, R.C., MacDonald, R.I., Menco, B.P.M., Takeshita, K., Subbarao, N.K., and Hu, L., 1991, Small-volume extrusion apparatus for preparation of large, unilamellar vesicles, *Biochim. Biophys. Acta* 1061:297.

Morth, S., and Tindall, B.J., 1985, Variation of polar lipid composition within haloalkaliphilic archaebacteria, *System. Appl. Microbiol.* 6:247.

Ring, K., Henkel, B., Valenteijn, A., and Gutermann, R., 1986, Studies on the permeability and stability of liposomes derived from a membrane spanning bipolar archaebacterial tetraether lipid, *in*: "Liposomes as Drug Carriers", K.H. Schmidt, ed., Georg Thieme Verlag, Stuttgart.

Sprott, G.D., 1992, Structures of archaebacterial membrane lipids, *J. Bioenergetics Biomembr.* 24:555.

Sprott, G.D., Meloche, M., and Richards, J.C., 1991, Proportions of diether, macrocyclic diether, and tetraether lipids in *Methanococcus jannaschii* grown at different temperatures, *J. Bacteriol.* 173:3907.

Sprott, G.D., Ekiel, I., and Dicaire, C., 1990, Novel, acid-labile, hydroxydiether lipid cores in methanogenic bacteria, *J. Biol. Chem.* 265:13735.

Wingfield, P., Arad, T., Leonard, K., and Weiss, H., 1979, Membrane crystals of ubiquinone:cytochrome c reductase from *Neurospora* mitochondria, *Nature* 280:696.

Woese, C.A., Kandler, O., and Wheelis, M.L., 1990, Towards a natural system of organisms: Proposal for the domains Archaea, Bacteria, and Eucarya, *Proc. Natl. Acad. Sci. USA* 87:4576.

POSTER PRESENTATIONS

STRUCTURAL ANALYSIS OF THE PARACRYSTALLINE LAYER OF *CAMPYLOBACTER FETUS* SUBSP. *VENEREALIS* UA809

Lori L. Graham and Terry J. Beveridge

Department of Microbiology
College of Biological Science
University of Guelph
Guelph, Ontario, Canada

Campylobacter fetus subsp. *venerealis* (*C. venerealis*) is a major concern in the cattle and sheep industries since infections with this organism can result in infertility (Penner, 1988). The chronic inflammation associated with *C. venerealis* infections is likely due to the persistence of the whole organism or its constituents within the host. Recently, proteinaceous or glycoproteinaceous bacterial S-layers have been implicated as factors enhancing bacterial virulence. In some *Campylobacter fetus* strains the S-layer appears to protect the infecting bacterium from serum killing and phagocytosis by PMN leukocytes (Blaser et al., 1991), whereas in other strains, the S-layer undergoes antigenic variation (Dubreuil et al., 1990). These functional attributes would facilitate persistence of campylobacters within a host and in the case of *C. venerealis* infections, contribute to the chronic inflammation.

Thin section electron (EM) microscopy of *C. venerealis* cells grown 24 h in biphasic Mueller-Hinton medium at 37°C in an atmosphere of 5% O_2, 10% CO_2 and 85% N_2 and processed by freeze-substitution (Graham and Beveridge, 1990) revealed an electron dense layer, 10 nm in thickness, intimately associated with the external face of the outer membrane (Fig. 1).

No S-layer periodicity was apparent in most cells and was likely due to the super-imposition of individual protein subunits within the thin section or to the interdigitation of lipopolysaccharide O-antigen polymers between protein subunits (LPS can be well preserved by freeze-substitution; Lam et al., 1992). However, on a small subpopulation of cells periodicity was observed (inset, Fig. 1). Similarly, deep etching after freeze-fracture revealed two distinct subunit arrangements on the surfaces of cells from the same culture. The most common arrangement was a tetragonal or p4 pattern (Fig. 2). Some other cells were covered with a hexagonal or p6 array (Fig. 3). Although the deposition of platinum onto the cleaved surface makes accurate measurements of center-to-center spacing difficult, subunits of the p6 array were larger than those comprising the p4 array (Figs. 2 and 3). It is possible that these two patterns reflect differences in shadow angle, but the relative size difference of the individual subunits in each array and the two distinct surface structures observed in thin section (see Fig. 1) makes this unlikely. p4 and p6 arrays have been reported for other *Campylobacter spp.* isolates (Dubreuil et al., 1988; Fujimoto et al., 1989). In fact, both p4 and p6 arrays have been observed on a single *C. fetus* cell (Fujimoto et al., 1991), although at no time were two different arrays visualized on a single *C. venerealis* cell in this study. From thin section images, it seems unlikely that this strain of *C. venerealis* is synthesizing two arrays, one overlying

Figure 1. Thin section of a *C. venerealis* cell processed by freeze-substitution. The S-layer is immediately adjacent to the outer membrane (arrow). **Inset:** This section reveals the periodicity observed in the S-layer of some cells. Bars represent 100 nm.

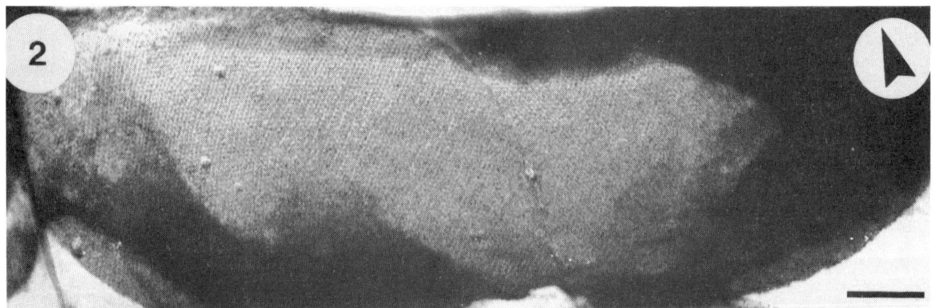

Figure 2. Freeze-etch image revealing the p4 array on whole cells. Arrow indicates direction of shadowing. Bar represents 100 nm.

Figure 3: Other cells were covered with a p6 array. Note the larger size of the subunits in this array compared with those of the p4 array. Arrow indicates direction of shadowing. Bar represents 200 nm.

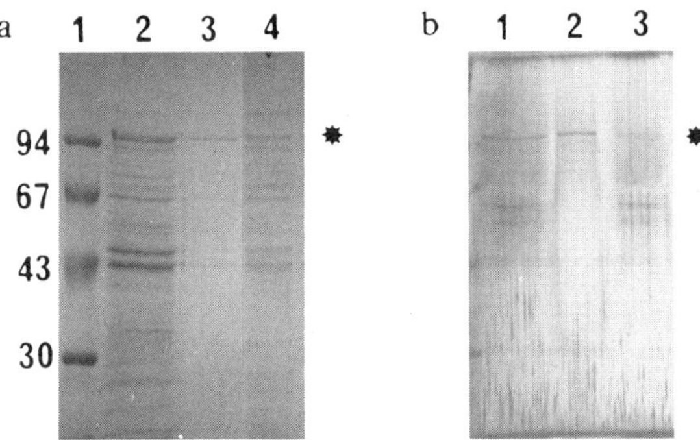

Figure 4. SDS-PAGE analysis. a) Coomassie stained gel, Lane 1, molecular weight markers; Lane 2, whole cells; Lane 3, culture supernatant; Lane 4, shear preparation; b) Silver stained lyophilized culture supernatant; Lane 1, complete lysis buffer; Lane 2, lysis buffer lacking β-mercaptoethanol; Lane 3, lysis buffer lacking SDS. * indicates S-layer protein. Molecular weights are indicated at left.

the other, and since mixed or "patchy" arrays of p4 and p6 symmetry were never observed on the same cell, it is also unlikely that a second array could be reassembling over an underlying pre-existing array. Although only one array is actually expressed per cell, it appears that the information necessary to create different arrays is contained within individual *C. venerealis* cells.

No array was ever visualized on negatively stained whole cells except along the periphery of a few cells. In an attempt to improve these images, numerous chemical treatments were employed to specifically remove the surface protein layer. SDS-PAGE analysis of whole cells revealed a dominant band at M_r 103.5 kDa (Fig. 4). All of the chemical agents listed in Table 1 non-specifically removed this protein as well as numerous other proteins. No planar arrays were ever observed by EM of these preparations, even though the 103.5 kDa protein was always detected by SDS-PAGE. Shearing of whole cells in water was found to release the surface array together with the envelope sacculi (Fig. 5). These sacculi, which retained cell shape, showed an S-layer only along their periphery when negatively stained. However, arrays identical to those shown in Fig. 2 and 3 were observed when these sacculi were freeze-etched.

Culture supernatants were found to contain large quantities of the 103.5 kDa protein (Fig. 4). This material was easily concentrated by freeze-drying or by ammonium sulfate precipitation and its banding pattern by SDS-PAGE was not altered when run under non-reducing conditions. In the absence of SDS, however, no protein was detected suggesting that the protein may still be associated with outer membrane material in aggregates too large to enter the gel (Fig. 4). No array was detected by EM examination of this material.

Sloughing of S-layer proteins could be an additional reason for maintenance of chronic inflammation. This, coupled with variability of S-layer protein arrangement would facilitate persistence of the bacterium within the host and further contribute to the chronic inflammation associated with *C. venerealis* infections. Studies are currently underway to determine the role of the S-layer in *C. venerealis* infections and to better define the mechanisms of switching between the two arrays expressed by this isolate.

Table 1. Chemical treatments used to extract S-layer protein from *C. venerealis* UA809 whole cells

Deionized water[a]		Detergent 1% SDS[c] pH 7.06

Chelators
  EDTA[b] 1 mM - 100 mM pH 6.7
  EGTA[c] 1 mM pH 6.5
  citrate[c] 10 mM
		pH
		  0.05 M HEPES[c] pH 2.0
		  0.2 M glycine buffer[c] pH 1.95, 2.2
		  Universal buffer[b] pH 4.0-12.0

Denaturants
  urea[b] 0.5 M - 6.0 M
  guanidine hydrochloride[c]
    1.0 M, 2.0 M pH 6.7
		Salts[d]
		  NaCl, LiCl, KCl 10 mM pH 6.2
		  $BaCl_2$, $CaCl_2$, $MgCl_2$ 10 mM pH 6.2

[a]Cells were resuspended in 1.5 mL deionized water and heated to 60°C for 15, 30 or 60 min. Control cells were incubated at 21°C. Cells were pelleted and the supernatants and resulting pellets analyzed by SDS-PAGE.

[b]Cells from a biphasic culture were resuspended in deionized water to OD_{600} = 0.19. 50 µl of this suspension was mixed with 1.2 ml of the reagent and incubated at 21°C for 0, 1, 5, 10, 15, 30, and 60 min. Cells were pelleted and the supernatant and the remaining cell pellet were analyzed by SDS-PAGE.

[c]1.4 ml of a biphasic culture was pelleted in microfuge tubes. The pellet was resuspended in 1.0 ml of the reagent, incubated 15 min at 21°C and repelleted. The second pellet was resuspended in 1.0 ml of the reagent, incubated 5 min at 21°C and pelleted a total of 5 times. The remaining cell pellet and each supernatant were analyzed separately by SDS-PAGE.

[d]1.5 ml of a 36 h biphasic culture was pelleted in microfuge tubes. The pellet was resuspended in 1.0 ml of salt solution and incubated 15 min at 21°C. After pelleting, the supernatant was removed and the pellet resuspended in 1.0 ml of salt solution, incubated and repelleted as described. Both supernatants and resulting cell pellets were analyzed by SDS-PAGE.

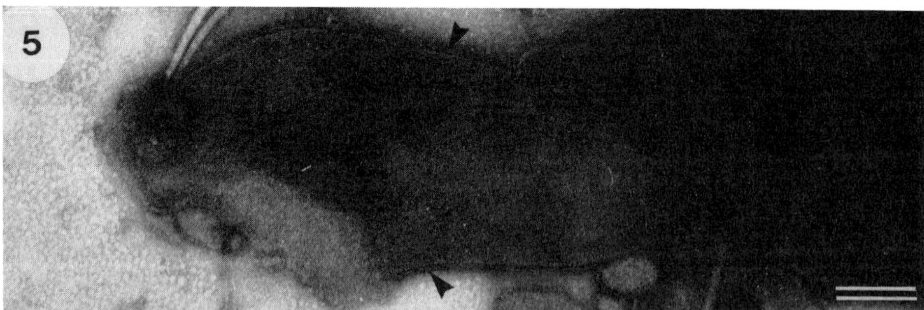

Figure 5. Negative stain (2% uranyl acetate) of sacculi sheared from whole cells.   Note the array present around the periphery (arrows). Bar represents 200 nm.

ACKNOWLEDGEMENTS

C. venerealis UA809 was supplied by D. Taylor, University of Alberta, Edmonton, Alberta, Canada. All microscopy was performed at the NSERC Guelph Regional STEM Facility which is partially supported by a Natural Sciences and Engineering Research Council infrastructure grant. The research was made possible by a Medical Research Council grant to TJB.

REFERENCES

Blaser, M.J., Smith, P.F., Repine, J.E., and Joiner, K.A., 1991, Pathogenesis of *Campylobacter fetus* infections. Failure of encapsulated *Campylobacter fetus* to bind C3b explains serum and phagocytosis resistance, *J. Clin. Invest.* 81:1434.

Dubreuil, J.D., Kostrzynska, M., Austin J.W., and Trust, T.J., 1990, Antigenic differences among *Campylobacter fetus* S layer proteins, *J. Bacteriol.* 172:5035.

Dubreuil, J.D., Logan, S.M., Cubbage, S., Eidhin, D.N., McCubbin, W.D., Kay, C.M., Beveridge, T.J., Ferris, F.G., and Trust, T.J., 1988, Structural and biochemical analysis of a surface array protein of *Campylobacter fetus*, *J. Bacteriol.* 170:4165.

Graham, L.L. and Beveridge, T.J., 1990, Evaluation of freeze-substitution and conventional embedding protocols for routine electron microscopic processing of eubacteria, *J. Bacteriol.* 172:2141.

Fujimoto, S., Takade, A., Amako, K., and Blaser, M.J., 1991. Correlation between molecular size of the surface array protein and morphology and antigenicity of the *Campylobacter fetus* S layer, *Infect. Immun.* 59:2017.

Fujimoto, S., Umeda, A., Takade, A., Murata K., and Amako, K., 1989, Hexagonal surface layer of *Campylobacter fetus* isolated from humans, *Infect. Immun.* 57:2563.

Lam, J.S., Graham, L.L., Lightfoot, J., Dasgupta T., and Beveridge, T.J., 1992, Ultrastructural examination of the lipopolysaccharide of *Pseudomonas aeruginosa* strains and their isogenic rough mutants by freeze substitution, *J. Bacteriol.* 174:7159.

Penner, J.L., 1988, The genus *Campylobacter*: a decade of progress, *Clin. Microbiol. Rev.* 1:157.

Chapter 26

EFFECT OF TRITON X-100 ON THE S-LAYER OF METHANOCULLEUS MARISNIGRI

Douglas P. Bayley* and Susan F. Koval

Department of Microbiology and Immunology
University of Western Ontario
London, Ontario, Canada

INTRODUCTION

Methanoculleus marisnigri is an irregularly-shaped coccus originally isolated by Romesser et al. (1979) from the anoxic sediment of the Black Sea. Cells are peritrichously flagellated, and grow optimally at 37°C. The cell wall is composed exclusively of a glycosylated S-layer with hexagonal symmetry. The S-layer protein contains (based on dry weight) 0.8% galactose, 0.3% glucose, and 0.1% mannose (Romesser et al., 1979) and has a M_r = 138 kDa (Zabel et al., 1984). We have isolated the S-layer glycoprotein of this organism by utilizing its sensitivity to Triton X-100 at elevated temperatures.

RESULTS

Fig. 1 shows a freeze-etched preparation of a *M. marisnigri* cell. Note the hexagonal pattern of the S-layer on the surface and the absence of a fracture plane through the plasma membrane. Cell envelopes were assayed for sensitivity of the S-layer to Triton X-100 at varying temperatures from room temperature up to 80°C. Although an incubation temperature of 60°C solubilized much of the S-layer protein, 80°C was found to be the most effective (Fig. 2). The S-layer protein was solubilized (lane 4) whereas many other proteins remained insoluble (lane 3). Electron microscopy of thin sections showed that the plasma membrane bilayer structure remained resistant to this treatment while the S-layer was removed (Fig. 3b). After phase-partitioning of the detergent soluble extract at 75°C (above the cloud point of 64°C), the S-layer glycoprotein was detected by SDS-PAGE in the aqueous phase

---

*Present address: Department of Microbiology and Immunology
                           Queen's University
                           Kingston, Ontario, Canada

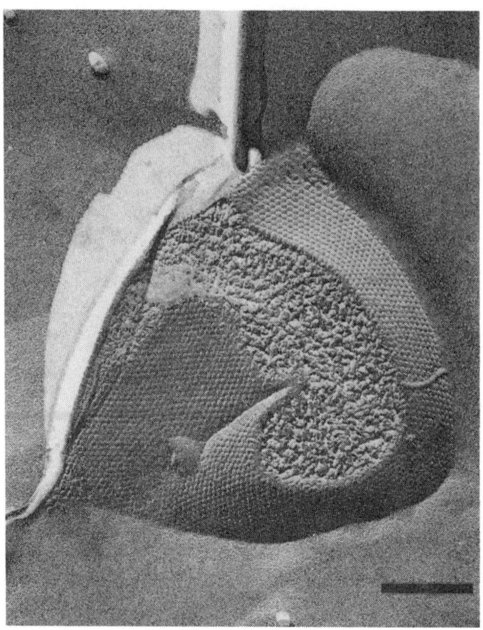

Figure 1. Freeze-etched preparation of *M. marisnigri*. Bar=200 nm.

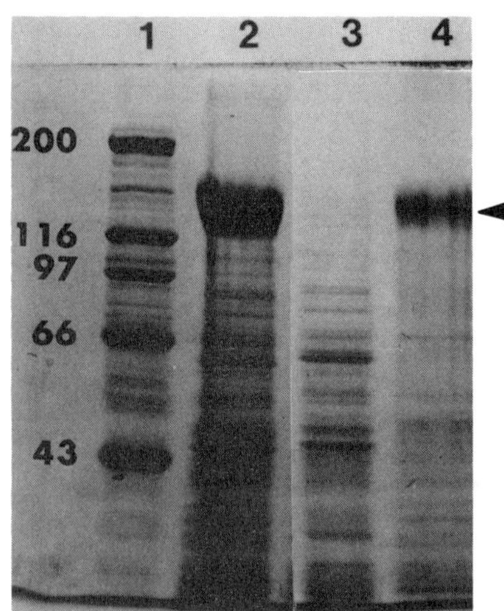

Figure 2. SDS-PAGE of envelopes treated in 1% Triton X-100 for 1h at 80C. Lane 1, molecular weight markers in kDa are along the left side of the figure; lane 2, control-untreated envelopes; lane 3, Triton X-100 insoluble fraction; lane 4, Triton X-100 soluble fraction. Arrow head indicates position of S-layer protein.

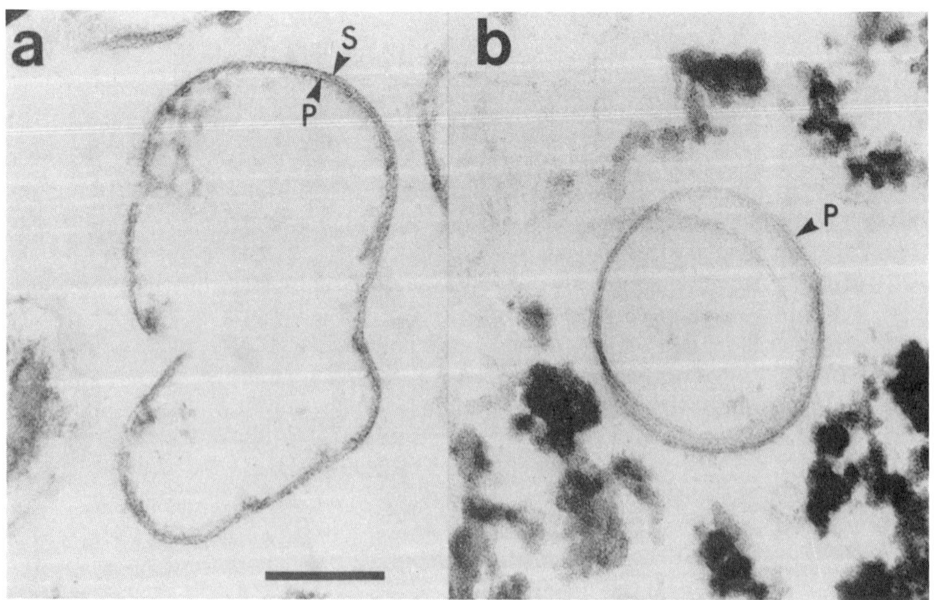

Figure 3. Thin-section preparation of *M. marisnigri* envelope before (A) and after (B) treatment in 1% Triton X-100 at 80°C. S denotes the S-layer, and P the plasma membrane. Bar=200 nm.

(data not shown). The S-layer was not removed from envelopes by incubation in distilled water alone at temperatures as high as 80°C as was shown for *Methanococcus voltae* (Koval and Jarrell, 1987). Thus, temperature alone was not responsible for the solubilization.

DISCUSSION

These studies indicate that the S-layer of *M. marisnigri* is sensitive to Triton X-100 at 80°C. Many S-layers composed of non-glycosylated proteins are not sensitive to Triton X-100. The detergent can be used to solubilize adherent membranes (Nußer and König, 1987) prior to isolation of the S-layer proteins. Thus, the sensitivity of the S-layer of *M. marisnigri* to Triton X-100 was unexpected. The glycosylation of the S-layer protein may be responsible in part for this result, since Faguy et al. (1992) have shown that glycosylated archaeobacterial flagellar filaments, such as those of *Methanospirillum hungatei*, are also sensitive to Triton X-100.

The fact that freeze-etched preparations of cells grown at 37°C failed to yield a fracture plane within the hydrophobic region of the plasma membrane might suggest the presence of C_{40} tetraether lipids. The plasma membrane of *M. marisnigri* grown at 25°C has been shown to possess both diether and tetraether lipids, but the relative amounts were not reported (Grant et al., 1985). Tetraether lipids add vertical stability to membranes and at concentrations near 45% make the lipid bilayers practically impossible to fracture (T.J. Beveridge, personal communication). The plasma membrane of *M. hungatei* contains 50% tetraether lipids (Choquet et al., 1992) and does not yield a fracture plane (T.J. Beveridge, personal communication). The resistance of the plasma membrane of the mesophilic *M. marisnigri* to Triton X-100, even at 80°C, is not understood. However, the membrane of *Thermoplasma*, a thermophilic archaeobacterium which does not possess a cell wall, but has a high proportion of tetraether lipids, is also resistant to this detergent (Langworthy, 1985).

REFERENCES

Choquet, C.G., Patel, G.B., Beveridge, T.J., and Sprott, G.D., 1992, Formation of unilamellar liposomes from total polar extracts of methanogens, *Appl. Environ. Microbiol.* 58: 2894.

Faguy, D.M., Koval, S.F., and Jarrell, K.F., 1992, Correlation between glycosylation of flagellin proteins and sensitivity of flagellar filaments to Triton X-100 in methanogens, *FEMS Microbiol. Lett.* 90:129.

Grant, W.D., Pinch, G., Harris, J.E., DeRosa, M., and Gambacorta, A., 1985, Polar lipids in methanogen taxonomy, *J. Gen. Microbiol.* 131:3277.

Koval, S.F., and Jarrell, K.F., 1987, Ultrastructure and biochemistry of the cell wall of *Methanococcus voltae*, *J. Bacteriol.* 169:1298.

Langworthy, T.A., 1985, Lipids of archaebacteria, *in*: "The Bacteria", vol. 8, C.R. Woese and R.S. Wolfe, eds., Academic Press, Inc., New York, NY.

Nußer, E., and König, H., 1987, S-layer studies on three species of *Methanococcus* living at different temperatures, *Can. J. Microbiol.* **33**: 256.

Romesser, J.A., Wolfe, R.S., Mayer, F., Spiess, E., and Walther-Mauruchat, A., 1979, *Methanogenium*, a new genus of marine methanogenic bacteria, and characterization of *Methanogenium cariaci* sp. nov. and *Methanogenium marisnigri* sp. nov., *Arch. Microbiol.* 121:147.

Zabel, H.P., König, H., and Winter, J., 1984, Isolation and characterization of a new coccoid methanogen, *Methanogenium tatii* spec. nov. from a solfataric field on Mount Tatio, *Arch. Microbiol.* 137:308.

Chapter 27

CHARACTERIZATION OF THE S-LAYER GLYCOPROTEINS OF TWO LACTOBACILLI

Alexander Möschl, Christina Schäffer, Uwe B. Sleytr,
and Paul Messner
Center for Ultrastructure Research and
Ludwig Boltzman Institute for Molecular Nanotechnology
University of Agriculture, Vienna, Austria,

Rudolf Christian,
Scientific Software Company, Vienna, Austria

Gerhard Schulz,
Sandoz Forschungsinstitut, Vienna, Austria

Surface layer (S-layer) glycoproteins have been described on many archaeobacteria. Among eubacteria their presence has only been reported for the Bacillaceae (for a review see Messner and Sleytr, 1992). Lactic bacteria often produce extracellular polymers such as slimes and capsules (for example those described by Nakajima et al., 1990 and Racine et al., 1991). Most of these polymers are polysaccharides but a few reports claim the presence of glycoproteins in lactic streptococci (Macura and Townsley, 1984) and lactobacilli (Garcia-Garibay and Marshall, 1991). These observations have encouraged us to extend our survey on glycosylated S-layer proteins to lactobacilli.

We screened different strains from the culture collection of the Institut für Milchforschung und Bakteriologie, Universität für Bodenkultur, Wien, Austria. Using SDS-PAGE, two strains out of approximately 40 have revealed a positive carbohydrate staining reaction (Figure 1a, lane 3). One strain was *Lactobacillus buchneri* 41021/251, the other strain was *L. plantarum* 41021/252. Electron microscopy of freeze-etched preparations (Fig. 1b) and thin sections (Fig. 1c) of both bacteria indicated the presence of oblique S-layers on intact cells. The morphological and chemical characterization of the glycosylated S-layers of both strains is summarized in Table 1. *L. buchneri* was chosen for subsequent more detailed characterization of the S-layer glycoprotein.

Several attempts were made to isolate the S-layer. This included extraction of the crystalline array from intact bacteria either with 5 M lithium chloride (Lortal et al., 1992), or by standard methods using 5 M guanidine hydrochloride (see Messner and Sleytr, 1988). Due to the thick peptidoglycan layer (Fig. 1c) sonication of the

Table 1. Morphological and chemical characterization of S-layer glycoproteins of two *Lactobacillus* strains

Organism	Lattice type	Lattice spacing (nm)	SDS-PAGE (kDa)	Protein content (%)	Sugar content (%)	Phosphate content (%)
L. buchneri 41021/251	Oblique	6.1/5.4	53	90	2.0	1.9
L. plantarum 41021/252	Oblique	8.3/5.3	56	94	2.0	1.6

cells was not successful. Therefore, after freezing in dry ice, the cells were ground in a Bead Beater™ (Biospec Products, Bartesville, OK, U.S.A.) with glass beads. The disrupted cell walls were subsequently washed and extracted with 5 M guanidine hydrochloride. To prevent degradation by nonspecific proteases, the isolation procedure was performed in the presence of PMSF (phenylmethylsulfonylfluoride, 0.1 mM). As checked by SDS-PAGE, purified S-layer self-assembly products showed the same molecular weight as on intact cells (Fig. 1a, lane 2). The purified S-layer self-assembly products were digested with 6% (w/w) pronase according to standard procedures (Messner et al., 1992). Glycopeptides were purified by gel permeation chromatography through Bio-Gel P-4 and P-100 columns which yielded both high and low molecular mass glycopeptide fractions (data not shown). Since the high molecular mass fraction was only present in small amounts we focussed our efforts on the isolation and characterization of the low molecular mass glycopeptide fraction. Impurities were removed by ion exchange chromatography (Dowex 50W-X8,H$^+$,200-400 mesh; BioRad Labs., Richmond, CA, U.S.A.) and followed by reversed-phase chromatography through a RP18 column. A single glycoprotein fraction was collected showing approximately 86% carbohydrate content by weight. In the remaining peptide moiety serine and alanine were found in a ratio of approximately 3:2. By

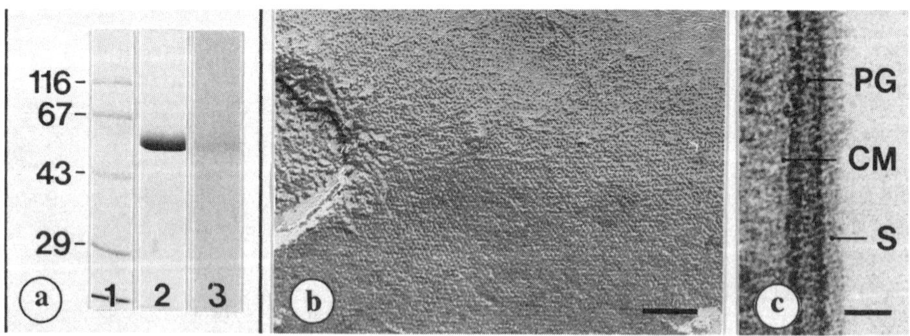

Figure 1. (a) SDS-PAGE of *L. buchneri*. Lane 1, Molecular mass standards (kDA), β-galactosidase (116), bovine serum albumin (67), ovalbumin (43), carbonic anhydrase (29); lane 2, whole cell extract, Coomassie Brilliant Blue staining; lane 3, periodic acid-Schiff staining. (b) Electron micrograph of a freeze-fractured cell of *L. buchneri*. (c) Electron micrograph of a thin sectioned cell of *L. buchneri*. CM, cytoplasmic membrane; PG, peptidoglycan; S, S-layer. Bars = 50 nm.

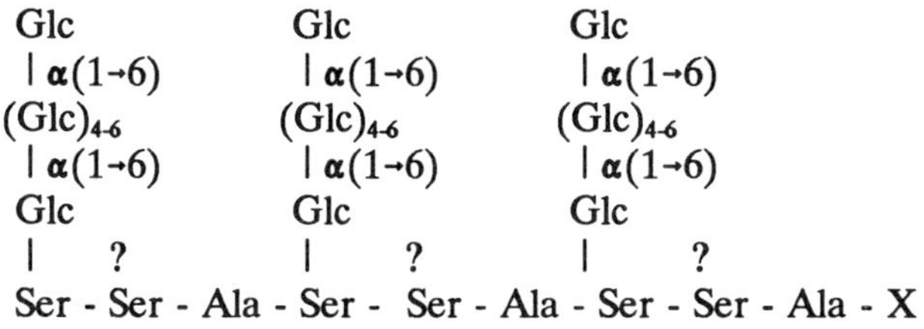

$$
\begin{array}{ccc}
\text{Glc} & \text{Glc} & \text{Glc} \\
|\ \alpha(1{\to}6) & |\ \alpha(1{\to}6) & |\ \alpha(1{\to}6) \\
(\text{Glc})_{4\text{-}6} & (\text{Glc})_{4\text{-}6} & (\text{Glc})_{4\text{-}6} \\
|\ \alpha(1{\to}6) & |\ \alpha(1{\to}6) & |\ \alpha(1{\to}6) \\
\text{Glc} & \text{Glc} & \text{Glc} \\
|\quad ? & |\quad ? & |\quad ?
\end{array}
$$

Ser - Ser - Ala - Ser - Ser - Ala - Ser - Ser - Ala - X

Figure 2. Schematic drawing of the proposed structure of the glycopeptide of *L. buchneri*.

high pressure liquid chromatography (HPLC) glucose was demonstrated to be the exclusive sugar component. This glycopeptide was subjected to N-terminal sequence analysis and the following amino acid sequence was found: SSASSASSA-X. ^{1}H and ^{13}C NMR measurements revealed the presence of 30 to 35 glucose residues, each $\alpha(1{\to}6)$ linked. From the NMR data and the sequence analysis it was obvious that the glycans in the decapeptide were linked via *O*-glycosidic linkages to the serine residues. At the moment it is not clear whether all serine residues are substituted by the glucose homooligomers (Fig. 2).

To our knowledge this is the first description of S-layer glycoproteins in the genus *Lactobacillus*. Masuda and Kawata (1983) have reported only non-glycosylated S-layers in their survey on lactobacilli. Interestingly, structurally comparable glucose-containing S-layer glycoproteins were also found in the archaeobacterium *Haloferax volcanii* (Mengele and Sumper, 1992). However, both the linkage between the glucose residues and the protein-carbohydrate linkage region were different. In *L. buchneri* we found $\alpha(1{\to}6)$ linked glucose residues whereas for *H. volcanii* a $\beta(1{\to}4)$ linkage was reported. Furthermore, the archaeobacterial glycans were bound to the polypeptide chain by *N*-glycosidic linkages.

ACKNOWLEDGMENTS

We thank S. Zayni, E. Mayerhofer, and A. Scheberl for excellent technical assistance and K. Vorauer for sequencing.

This work was supported in part by grants from the Hochschuljubiläumsstiftung der Stadt Wien, project H111/91, the Fonds zur Förderung der wissenschaftlichen Forschung in Österreich, project P8922-MOB, and the Österreichisches Bundesministerium für Wissenschaft und Forschung.

REFERENCES

Garcia-Garibay, M., and Marshall, V.M.E., 1991, Polymer production by *Lactobacillus delbrueckii* subsp. *bulgaricus, J. Appl. Bacteriol.* 70:325.

Lortal, S., van Heijenoort, J., Gruber, K., and Sleytr, U.B., 1992, S-layer of *Lactobacillus helveticus* ATCC 12046: Isolation, chemical characterization and re-formation after extraction with lithium chloride, *J. Gen. Microbiol.* 138:611.

Macura, D., and Townsley, P.M., 1984, Scandinavian ropy milk - identification and characterization of endogenous ropy lactic strepococci and their extra-cellular secretion, *J. Dairy Sci.* 67:735.

Masuda, K., and Kawata, T., 1983, Distribution and chemical characterization of regular arrays in the cell walls of strains of the genus *Lactobacillus, FEMS Microbiol Lett.* 20:145.

Mengele, R., and Sumper, M., 1992, Drastic differences in glycosylation of related S-layer glycoproteins from moderate and extreme halophiles, *J. Biol. Chem.* 267:8182.

Messner, P., and Sleytr, U.B., 1988, Separation and purification of S-layers from gram-positive and gram-negative bacteria, *in*: "Bacterial Cell Surface Techniques", I.C. Hancock and I.R. Poxton, eds., John Wiley and Sons, Chichester.

Messner, P., and Sleytr, U.B., 1992, Crystalline bacterial cell-surface layers, *Adv. Microbial Physiol.* 33:213.

Messner, P., Christian,R., Kolbe, J., Schulz, G., and Sleytr, U.B., 1992, Analysis of a novel linkage unit of *O*-linked carbohydrates from the crystalline surface layer glycoprotein of *Clostridium thermohydrosulfuricum* S102-70, *J. Bacteriol.* 174:2236.

Nakajima, H., Toyoda, S., Toba, T., Itoh, T., Mukai, T., Kitazawa, H., and Adachi, S., 1990, A novel phosphopolysaccharide from slime-forming *Lactococcus lactis* subsp. *cremoris* SBT 0495, *J. Dairy Sci.* 73:1472.

Racine, M., Dumont, J., Champagne, C.P., and Morin, A., 1991, Production and characterization of the polysaccharide from *Propionibacterium acidi-propionici* on whey-based media, *J. Appl. Bacteriol.* 71:233.

Chapter 28

DOES THE S-LAYER OF *AEROMONAS SALMONICIDA* EXIST IN MORE THAN ONE FUNCTIONAL ORGANIZATIONAL STATE?

Rafael A. Garduño, Julian C. Thornton, and William W. Kay

Department of Biochemistry and Microbiology and
The Canadian Bacterial Disease Network
University of Victoria
Victoria, British Columbia, Canada

INTRODUCTION

The S-layer of the fish pathogen *Aeromonas salmonicida*, the causative agent of furunculosis in salmonids, is probably the best characterized S-layer in regards to function. Not only does it constitute an important determinant in pathogenesis (Ishiguro et al., 1981), but it has also been assigned specific virulence functions (reviewed by Kay and Trust, 1991). We have recently demonstrated that the S-layer of *A. salmonicida* (or A-layer) possesses significant flexibility and plasticity. It normally exists in a single type of lattice capable of adopting different conformations (Garduño and Kay, 1992a), but under calcium depletion it forms two novel structural patterns as a consequence of structural rearrangements within the layer (Garduño et al., 1992). It is the objective of this present work to explore the functional relevance of one of these novel structural patterns, the "big squares" pattern.

METHODS

Three *A. salmonicida* strains were used: A450, a wild type strain; A450-1, a mutant that secretes S-layer sheets due to a deficiency in its lipopolysaccharide O-polysaccharide chain, and A450-3, an S-layer negative mutant. All strains were grown at 20°C. Fish peptone calcium deficient media (FPM) was prepared as previously reported (Garduño and Kay, 1992b). Macrophage association experiments were performed as reported previously (Garduño and Kay, 1992b). Survival to challenges with serum (1 part of a bacterial cell suspension (OD $_{650\,nm}$ = 1.0) mixed with 4 parts of fresh trout serum), or with reduced oxygen radicals (approx. 10^7 bacterial cells/ml challenged with either 1 mM hydrogen peroxide, or superoxide generated by the xanthine/xanthine oxidase system at a rate of 3 μmoles/min) was evaluated through viable cell counts by the dilution plate method. In vivo growth of *A. salmonicida* was

Advances in Bacterial Paracrystalline Surface Layers
Edited by T.J. Beveridge and S.F. Koval, Plenum Press, New York, 1993

achieved in intraperitoneal chambers surgically implanted in rainbow trout (*Oncorhynchus mykiss*). Electron microscopy of negatively stained specimens was performed by conventional methods.

RESULTS AND DISCUSSION

A. salmonicida cells grown in FPM possessed S-layers arranged in a novel pattern called "big squares" (BS). Another novel pattern obtained as a consequence of calcium limitation was the "white dots" pattern (WD). Normal and these altered patterns are shown in Fig. 1. Interestingly, in vivo grown cells of *A. salmonicida* produced A-layers displaying WD and BS patterns (Fig. 1).

We have initiated the characterization of the functional competence of the BS structural pattern in FPM grown cells. Control cells were grown in the calcium replete medium, trypticase soy broth (TSB). The S-layer negative (S⁻) strain A450-3, grown in TSB or FPM, was also included as control. Since a known function of normal A-layer is protection against serum-mediated bacteriolysis (Munn et al., 1982), we assayed the serum resistance of FPM grown bacteria. The results presented in Fig. 2a suggest that the BS pattern is as effective as the normal S-layer in protecting cells from complement-mediated bacteriolysis. S⁻ cells grown in both FPM or TSB had reduced serum resistance, indicating enhanced resistance depended on the presence of S-layer and not on the growth medium. Another known S-layer-dependent function is the specific association of *A. salmonicida* with macrophages (Garduño et al., 1992; Garduño and Kay, 1992b). Surprisingly, the BS structural pattern mediated a very efficient association of FPM grown cells with macrophages (Fig. 2b). Again, the FPM or TSB grown S⁻ mutant, A450-3, had similar low macrophage association levels, confirming that the enhanced association was an A-layer-mediated process and that the dramatic increase in macrophage association was due to the structural change produced in the A-layer and not to some other effect induced by growing cells in FPM. FPM grown cells were also competent in binding the hemin analog, Congo red (Kay et al., 1985) (not shown).

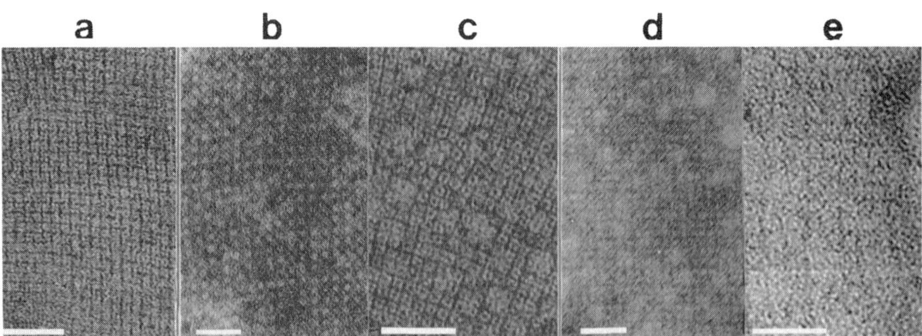

Figure 1. Electron micrographs showing a normal array of the *A. salmonicida* S-layer (a), as well as altered patterns obtained as a result of calcium limitation in vitro (b and c), or as a result of growth in vivo (d and e). WD patterns are shown in (b) and (d) and BS patterns in (c) and (e). Bar represents 50 nm for all micrographs.

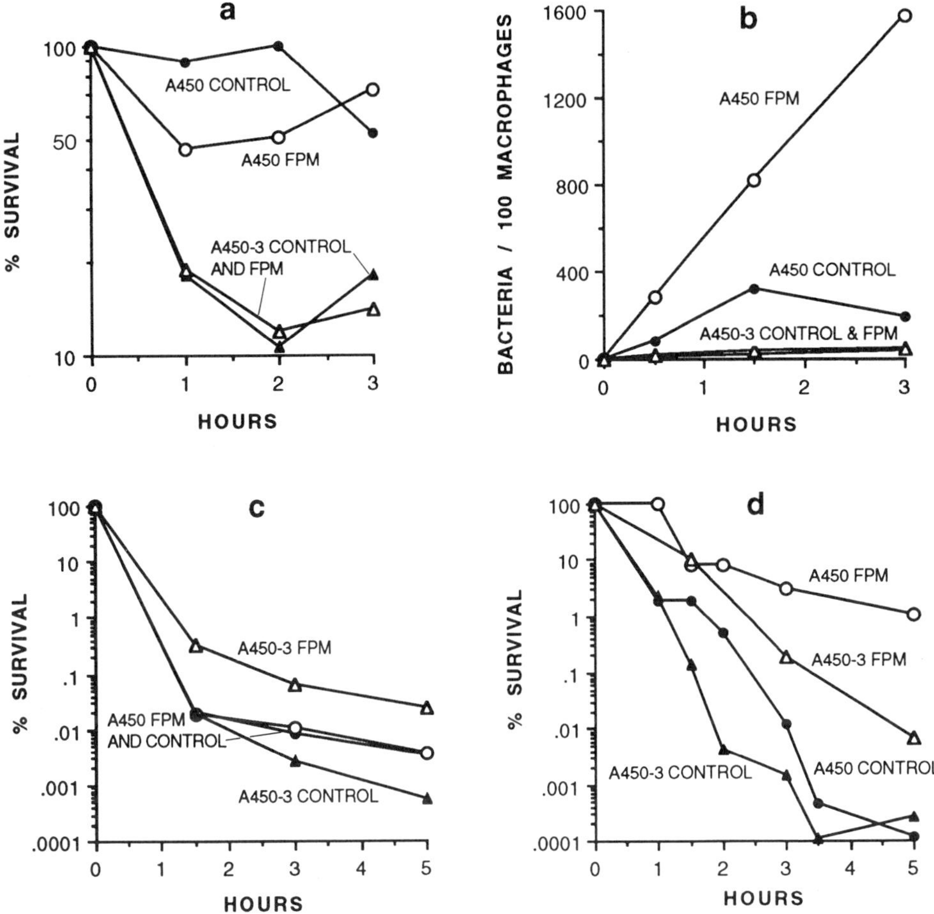

Figure 2. FPM grown bacterial cells (displaying the BS pattern) were compared with control cells grown in TSB (displaying normal patterns) to compare the functional competence of the BS pattern. The S⁻ mutant A450-3 was included as a control to define S-layer dependent processes. (a) Results from the serum resistance assay, (b) from the macrophage association assay, and from the challenge assays with peroxide (c) and superoxide (d).

We have been interested in the response of *A. salmonicida* to oxidative agents. Our own results, as well as results reported by others (Karczewski et al., 1991), indicate that the A-layer has some protective properties against reduced oxygen species. However, the main protective mechanism against oxidative killing is inducible and independent of the presence of A-layer (R.A. Garduño and W.W. Kay, unpublished results). We challenged FPM or TSB grown bacteria with peroxide (Fig. 2c) or superoxide (Fig. 2d) and observed that possession of the BS structural pattern (as well as the normal A-layer) was advantageous as compared to the S⁻ strain. However, the enhanced protection observed in FPM grown cells also appeared to involve A-layer independent mechanisms.

Although further experimentation is needed to evaluate the functional competence of the BS pattern, the results obtained so far are quite encouraging. *A. salmonicida* cells possessing BS or normal A-layer patterns are similarly protected against oxidative or non-oxidative killing, but the former pattern has the extra advantage with regards to adherence to host cells. Since novel structural patterns are present in vivo (Fig. 1) and in energy-starved cells (unpublished results), it is tempting to ask the question: does the A-layer possess more than one functional organizational state? Free cells in the environment, perhaps energy-limited, may produce altered A-layers with enhanced adhesiveness. Once in the host, changes in the levels of calcium and availability of nutrients may modulate the expression of different structural A-layer patterns on the surface of *A. salmonicida* with consequent implications in pathogenesis.

REFERENCES

Garduño, R.A., and Kay, W.W., 1992a, A single structural type in the regular surface layer of *Aeromonas salmonicida*, *J. Struct. Biol.* 108:202.

Garduño, R.A., and Kay, W.W., 1992b, Interaction of the fish pathogen *Aeromonas salmonicida* with rainbow trout macrophages, *Infect. Immun.* 60: 4612.

Garduño, T.A., Phipps, B.M., Baumeister, W., and Kay W.W., 1992, Novel structural patterns in divalent cation-depleted surface layers of *Aeromonas salmonicida*, *J. Struct. Biol.* 109:(in press).

Garduño, R.A., Lee, E.J.Y., and Kay, W.W., 1992, S-layer-mediated association of *Aeromonas salmonicida* with murine macrophages, *Infect. Immun.* 60:4373.

Ishiguro, E.E., Kay, W.W., Ainsworth, T., Chamberlain, J.B., Austen, R.A., Buckley, J.T., and Trust, T.J., 1981, Loss of virulence during culture of *Aeromonas salmonicida* at high temperature, *J. Bacteriol.* 148:333.

Karczewski, J.M., Sharp, G.J.E., and Secombes, C.J., 1991, Susceptibility of strains of *Aeromonas salmonicida* to killing by cell-free generated superoxide anion, *J. Fish Dis.* 14:367.

Kay, W.W., Phipps, B.M., Ishiguro, E.E., and Trust, T.J., 1985, Porphyrin binding by the surface array virulence protein of *Aeromonas salmonicida*, *J. Bacteriol.* 164:1332.

Kay, W.W., and Trust, T.J., 1991, Form and functions of the regular surface array (S-layer) of *Aeromonas salmonicida*, *Experientia* 47:412.

Munn, C.B., Ishiguro, E.E., Kay, W.W., and Trust, T.J., 1982, Role of surface components in serum resistance of virulent *Aeromonas salmonicida*, *Infect. Immun.* 36:1069.

Chapter 29

ATTACHMENT OF THE S-LAYER OF *CAULOBACTER CRESCENTUS* TO THE CELL SURFACE

Stephen G. Walker and John Smit

Department of Microbiology
University of British Columbia
Vancouver, British Columbia, Canada

In most cases S-layers attach to the outer membrane of gram-negative bacteria by non-covalent interactions (Koval and Murray, 1984; Sleytr and Messner, 1988). Lipopolysaccharide (LPS) (Dooley et al., 1985) or proteins (Thorne et al., 1975) have been implicated as the molecules to which the S-layer protein interacts to maintain cell association. This report suggests that another outer membrane component, termed the S-layer-associated oligosaccharide (SAO), is the molecule responsible for the attachment of RsaA, the S-layer protein of *Caulobacter crescentus*, to the cell surface.

C. crescentus produces a hexagonal S-layer (Smit et al., 1981) that requires calcium ions for subunit recrystallization in vitro (Walker et al., 1992). The nucleotide sequence of the *rsaA* gene indicates the protein contains four putative calcium binding sites (Gilchrist et al., 1992). Other gram-negative species which require calcium for S-layer assembly produce an unstructured layer on the cell surface or shed S-layer protein when calcium is absent or limited in the growth media (Beveridge and Murray, 1976; Koval and Murray, 1985; Doran et al., 1987). When *C. crescentus* NA1000 (formerly CB15A) is cultured in calcium-free media, growth does not occur (Edwards and Smit, 1991). This strain requires approximately 100 μM calcium for viability (S.G. Walker and J. Smit, unpublished observations). Spontaneous mutants, that do not require calcium for growth, appear at a frequency of 1 in 10^6 when wild-type cells are plated on calcium-free medium. These "calcium-independent" mutants share a second phenotype; when grown in liquid culture they shed a macroscopic precipitate. The precipitate consists largely of S-layer protein, as determined by Western blot analysis, and electron microscopy revealed that it is in an unstructured state. However, when the calcium-independent strains are grown on calcium-containing plates, negative-stain electron microscopy of carefully resuspended colonies revealed sheets of assembled S-layer that were not cell-associated (Smit et al., 1992). Therefore, the calcium-independent mutants are defective in S-layer attachment but not in subunit assembly.

The "rough" LPS (R-LPS) and extracellular polysaccharide (EPS) were purified from two calcium-independent mutants. Chemical analysis revealed no detectable alteration in these macromolecules (S.G. Walker, N. Ravenscroft and J. Smit, unpublished observations). Initial qualitative analysis of the LPS by the method of Hitchcock and Brown (1983) failed to show any difference in electrophoretic mobility of the R-LPS produced by the mutants compared to the parent strain (Ravenscroft et al., 1992). However, we found that extending the periodic oxidation step of the staining procedure (Walker et al., 1992) permitted the visualization of a polysaccharide species of slower electrophoretic mobility than the R-LPS in the parent strain. This molecule, subsequently termed the SAO, was absent in calcium-independent mutants.

The SAO and R-LPS are both extracted from the cell surface, along with 3 - 5 major proteins and EPS, by treatment with 0.77M NaCl and 0.12M EDTA (pH 7.2) (Walker et al., 1991). Small scale isolation and purification of the SAO was accomplished by preparative sodium dodecyl sulphate - polyacrylamide electrophoresis (SDS-PAGE) of proteinase K-treated extracts. When the purified SAO is solubilized in SDS-PAGE sample buffer and re-electrophoresed the apparent molecular weight did not change, indicating that the band did not result from R-LPS aggregation (Logan and Trust, 1984). The SAO also differs antigenically from the R-LPS. A polyclonal serum, raised in a failed attempt to produce antibody against the *Caulobacter* holdfast, fortuitously labels the SAO but not the R-LPS by Western blotting (Walker et al., 1992). Preliminary indirect immunofluorescent microscopy studies suggest that the anti-SAO serum labels the entire cell surface (but not the holdfast) of JS1003 (NA1000 with a chromosomal deletion of the S-layer gene) but does not label the surface of NA1000. This implies that the epitopes on the SAO recognized by the antiserum are masked by the S-layer protein.

Although the R-LPS and the SAO are antigenically, electrophoretically and, most likely, functionally distinct molecules, attempts to separate them by gel permeation chromatography were unsuccessful. Several gel types, buffer systems and detergents have been evaluated (Kasper et al., 1983; Peterson and McGroarty, 1985) but in all cases both species co-migrate during chromatography. This suggests that they are either similar in size (despite their different mobility in SDS-PAGE) or are associated with each other.

Initial chemical studies of isolated SAO were inconclusive, primarily due to limitations imposed by the small amount of purified material available through preparative SDS-PAGE. Future studies will concentrate on development of a large scale purification method. We are also attempting to genetically characterize the calcium-independent phenotype by screening a NA1000 Tn5 library. Using both biochemical and genetic analysis we hope to discover the molecular basis of the toxicity produced by growth in the absence of calcium for the wild-type strain. We hypothesize that calcium neutralizes a strong negative charge, located on the SAO, by creating a salt-bridge between the S-layer and the SAO. When calcium is no longer available, the negative charge on the SAO destabilizes the cell membranes causing cell death. Spontaneous mutants that no longer produce an SAO escape membrane destabilization and become "calcium-independent" but consequently are unable to attach the S-layer to the cell surface.

REFERENCES

Beveridge, T.J., and Murray, R.G.E., 1976, Dependence of the superficial layers of *Spirillum putridiconchylium* on Ca^{2+} or Sr^{2+}, *Can. J. Microbiol.* 22: 1233.
Dooley, J.S.G., Lallier, R., Shaw, D.H., and Trust, T.J., 1985, Electrophoretic and

immunochemical analyses of the lipopolysaccharides of various strains of *Aeromonas hydrophila, J. Bacteriol.* 164: 263.

Doran, J.L., Bingle, W.H., and Page, W.J., 1987, Role of calcium in assembly of the *Azotobacter vinelandii* surface array, *J. Gen. Microbiol.* 133: 399.

Edwards, P., and Smit, J., 1991, A transducing bacteriophage for *Caulobacter crescentus* uses the paracrystalline surface layer protein as a receptor, *J. Bacteriol.* 173: 5568.

Gilchrist, A., Fisher, J.A., and Smit, J., 1992, Nucleotide sequence analysis of the gene encoding the *Caulobacter crescentus* paracrystalline surface layer protein, *Can. J. Microbiol.* 38: 193.

Hitchcock, P.J., and Brown,T.M., 1983, Morphological heterogeneity among *Salmonella* lipopolysaccharide chemotypes in silver-stained polyacrylamide gels. *J. Bacteriol.* 154: 269.

Kasper, D.L., Weintraub, A., Lindberg, A.A., and Lonngren, J., 1983, Capsular polysaccharides and lipopolysaccharides from two *Bacteroides fragilis* strains: chemical and immunochemical characterization, *J. Bacteriol.* 153: 991.

Koval, S.F., and, Murray R.G.E., 1984, The isolation of surface array proteins from bacteria, *Can. J. Biochem. Cell Biol.* 62: 1181.

Koval, S.F., and, Murray R.G.E., 1985, Effect of calcium on the *in vivo* assembly of the surface protein of *Aquaspirillum serpens* VHA, *Can. J. Microbiol.* 31: 261.

Logan, S.M., and Trust, T.J., 1984, Structural and antigenic heterogeneity of lipopolysaccharides of *Campylobacter jejuni* and *Campylobacter coli*, *Infect. Immun.* 45: 210.

Peterson, A. A., and McGroarty, E.M., 1985, High molecular weight components in lipopolysaccharides of *Salmonella typhimurium*, *Salmonella minnesota*, and *Escherichia coli*, *J. Bacteriol.* 162: 738.

Ravenscroft, N., Walker, S.G., Dutton, G.G.S., and Smit, J., 1992, Identification, isolation and structural studies of the outer membrane lipopolysaccharide of *Caulobacter crescentus, J. Bacteriol.*, in press.

Sleytr, U. B., and Messner, P., 1988, Crystalline surface layers in procaryotes, *J. Bacteriol.* 170: 2891.

Smit, J., Grano, D.A., Glaser, R.M., and Agabian, N., 1981, Periodic surface array in *Caulobacter crescentus*: fine structure and chemical analysis, *J. Bacteriol.* 146: 1135.

Smit, J., Engelhardt, H., Volker, S., Smith, S.H., and Baumeister, W., 1992, The S-layer of *Caulobacter crescentus*: Three-dimensional image reconstruction and structural analysis by electron microscopy, *J. Bacteriol.* 174:6527.

Thorne, K.J.I., Thornley, M.J., Naisbitt, P., and Glauert, A.M., 1975, The nature of the attachment of a regularly arranged surface protein to the outer membrane of an *Acinetobacter sp.*, *Biochim. Biophys. Acta* 389: 97.

Walker, S.G., Ravenscroft, N., and Smit, J, 1991, Isolation of the cell surface molecules of *Caulobacter crescentus*, Proceed. Can. Soc. Microbiol., Western Branch Annu. Meet., Vol. 22 abstr. II - 41.

Walker, S.G., Smith, S.H., and Smit, J., 1992, Isolation and comparison of the paracrystalline surface layer proteins of freshwater *Caulobacters, J. Bacteriol.* 174: 1783.

Chapter 30

LINKER MUTAGENESIS OF THE *CAULOBACTER CRESCENTUS* S-LAYER PROTEIN

Wade H. Bingle, Peter Awram, and John Smit

Department of Microbiology
University of British Columbia
Vancouver, British Columbia, Canada

As a part of its life cycle, the gram-negative bacterium *Caulobacter crescentus* exhibits a characteristic morphological switch between a sessile stalked cell and a flagellated dispersive or swarmer cell. During this complex differentiation process however, both cell types continue to elaborate a paracrystalline S-layer of hexagonal organization. The S-layer is composed of a single 98 kDa secreted protein (RsaA) noncovalently attached to other protein monomers and to the surface of the outer membrane; the latter interactions may be mediated by calcium ions and a specific S-layer associated molecule present in the outer membrane. Because RsaA is a secreted protein which interacts with itself as well as other molecules present in the outer membrane, multiple functional regions are expected to exist within the protein including those involved in secretion, calcium binding, outer membrane attachment and formation of the core and connectivity regions of the S-layer. Despite this expectation, analysis of the translated nucleotide sequence of the *rsaA* gene has not revealed possible functional regions of the protein beyond the existence of a probable calcium binding region. Similarly, N-terminal amino acid sequencing and sequencing of C-terminal peptide fragments derived from the S-layer protein have shown that no N- or C-terminal processing of the protein occurs, providing few clues to the mechanism of secretion. However, gene fusion studies have shown that the first 35-52 amino acids of the RsaA N-terminus can direct reporter proteins to the periplasm. (For a review of the present state of knowledge surrounding the *C. crescentus* S-layer, see Bingle et al., this book).

In order to identify functional regions of the S-layer protein, we used linker insertion mutagenesis. This method produces local disruption of protein structure, and thus destruction of functional domains, without long-range polar effects. It also allows the positions of interesting mutations to be approximated without sequencing the entire gene. In the method, a double-stranded oligonucleotide coding for a unique restriction site is inserted into a gene at random sites created by partial endonuclease digestion. Because the inserted linker is in the order of 12 bp (coding for 4 amino acids) major perturbations to the resulting protein are avoided and stable species exhibiting mutant properties are formed.

To simplify the selection of linker insertions, we tagged a *Bam*HI linker with a kanamycin resistance gene and used this cassette (tagged linker *Bam*HI-1021K; Bingle and Smit, 1991) to mutagenize a promoterless version of the *rsaA* gene carried on a plasmid. Insertion sites in the *rsaA* gene were created by partial digestion with *Aci*I, *Hin*PI, *Taq*I and *Msp*I, all of which produce 5'-CG extensions. The mutagenesis cassette was first excised from its carrier plasmid with *Acc*I (to provide complimentary 5'-CG cohesive termini) and then ligated to the partially digested plasmid carrying the *rsaA* gene. After selecting for linker insertions using the antibiotic resistance marker, the kanamycin resistance gene was removed by cleavage with *Bam*HI leaving the *Bam*HI linker in the *rsaA* gene (Bingle and Smit, 1991).

The *rsaA* gene was routinely manipulated on a high copy number plasmid in *Escherichia coli*; in this host the *rsaA* gene is transcribed using the *lacZ* promoter present on the plasmid. It was found that certain strains of *E. coli* produced inclusion bodies when expressing the *rsaA* gene. Secondly, when linkers designed to produce out-of-frame insertions in *rsaA* (tagged linkers *Bam*HI-1065K and *Bam*HI-1071K) were used to mutagenize *rsaA*, the resulting proteins did not form inclusion bodies. Thus, inclusion body formation in *E. coli* could be used as an initial screen for in-frame linker insertions in *rsaA*. This proved to be a reliable screen: Western analysis of total cell protein prepared from *E. coli* cells expressing *rsaA* gene carrying linker insertions later indicated a 100% correlation between in-frame linker insertion in *rsaA* and subsequent RsaA inclusion body formation. Western analysis also revealed RsaA species containing small deletions; in these cases, it was presumed that the *Bam*HI linker had replaced the deleted portion of the *rsaA* gene and maintained the correct reading frame.

Versions of the promoterless *rsaA* gene carrying in-frame *Bam*HI linker insertions were subcloned into *C. crescentus* expression vectors (Bingle and Smit, 1990) and transferred to an S-layer deficient *C. crescentus* strain. All *rsaA* genes carrying linker insertions were tolerated by the bacterium, i.e., no insertions produced a clearly toxic phenotype. Cell surface localization of RsaA protein species carrying linker-encoded amino acids was evaluated by extracting whole cells washed from solid medium with Tris/EDTA. Of 240 linker insertions in *rsaA* examined so far, 55 permitted S-layer protein secretion to the exterior of the cell. These insertions were termed "permissive insertions."

Permissive insertions mapped throughout the length of the *rsaA* gene indicating that there appeared to be numerous regions of the S-layer protein which could accept additional amino acids without deleterious effects on secretion (Fig. 1).

In no case did a non-permissive insertion result in an obvious large scale cytoplasmic accumulation of full-length S-layer protein suggesting that non-secreted S-layer protein was either degraded or its production regulated in some manner. Seven permissive insertions were evaluated by electron microscopy for their effect on S-layer assembly; in all cases the S-layer attached to the cell surface and assembled correctly at a gross level.

The choice of restriction enzyme used for creating insertion sites in *rsaA* significantly affected the distribution of subsequent permissive insertions recovered (Fig. 1). This may be due to the location of the restriction site with respect to the *rsaA* reading frame which in turn determines which of the three possible linker encoded peptides will be incorporated into the S-layer protein. For example, *Aci*I sites occur almost exclusively in one location with respect to the *rsaA* reading frame except in the region of the gene encoding amino acids 650 to 750. This may explain the lack of permissive insertions in this region (see over).

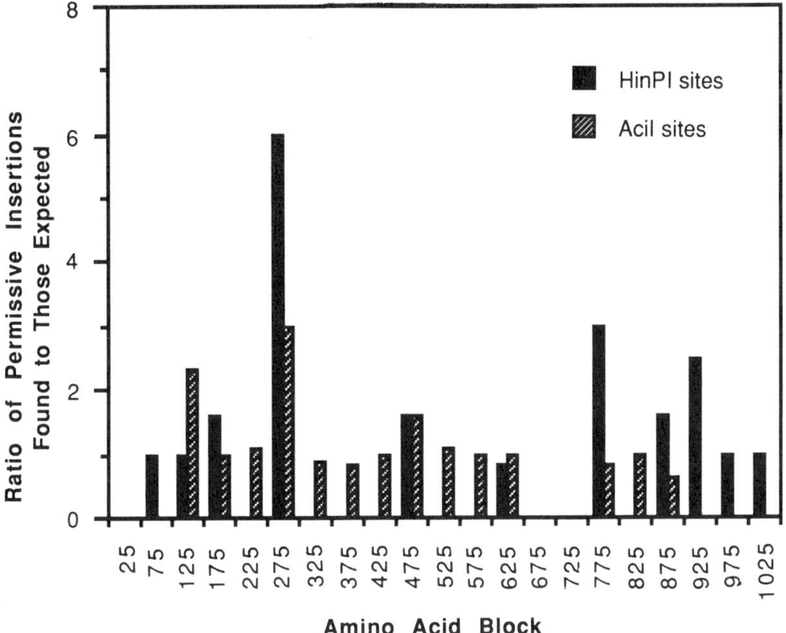

Figure 1. Permissive linker insertions at sites in RsaA corresponding to *Aci*I and *Hin*PI sites in the *C. crescentus rsaA* gene. The RsaA protein has been arbitrarily divided into blocks of 50 amino acids and the midpoint of each block is indicated on the X-axis. The ratio plotted on the Y-axis was calculated by dividing the number permissive insertions found in an amino acid block (expressed as percentage of the total number in the protein) by the number of theoretical insertion sites in that block (expressed as a percentage of the total number in the protein). A ratio of 1 indicated that insertions were found in a particular amino acid block in the proportion expected from the proportion of restriction sites corresponding to that area of the *rsaA* gene. If no theoretical insertion sites were present in a block of amino acids because no restrictions sites occurred in the corresponding area of the gene, a value of 1 was plotted.

Although the locations of nonpermissive insertions have yet to be mapped, we analyzed the locations of the permissive insertions to determine whether there were any regions of the RsaA protein expected to contain permissive insertions yet did not exhibit any of them (Fig. 1). This analysis indicated that no permissive insertions were found in the first 50 amino acids from the N-terminus or in a region covering amino acids 650-750. The lack of permissive insertions at the N-terminus supports secretion reporter studies (see above) that this region is important for S-layer protein secretion. We are currently mapping the locations of the nonpermissive insertions in RsaA to confirm this hypothesis.

We are also continuing to evaluate the effect of permissive linker insertions on S-layer assembly, i.e., those causing the S-layer to be released from the cell surface and shed into the culture medium and those causing the S-layer to remain cell-associated in an unorganized form. However, many assembly defective phenotypes are expected to be subtle and may require computer analysis of electron microscopic images. Finally, those permissive insertions not causing gross defects to S-layer attachment and assembly may serve as useful sites for the insertion of larger peptides;

such additional mass may be a useful tag to allow regions of S-layer to be visible in image-processed electron micrographs and to be correlated with the primary amino acid sequence of the S-layer protein.

REFERENCES

Bingle, W.H., and Smit, J., 1990, High-level expression vectors for *Caulobacter crescentus* incorporating the transcription/translation initiation regions of the paracrystalline surface-layer-protein gene, *Plasmid* 24: 143.
Bingle, W.H., and Smit, J., 1991, A method of tagging specific-purpose linkers with an antibiotic-resistance gene for linker mutagenesis using a selectable marker, *Biotechniques* 10: 150.
Gilchrist, A., and Smit, J., 1992, Nucleotide sequence analysis of the gene encoding the *Caulobacter crescentus* paracrystalline surface layer, *Can. J. Microbiol.* 38: 193.

Chapter 31

CAN S-LAYERS OF BACILLACEAE CONTROL THE RELEASE OF THEIR OWN EXOPROTEINS ?

Elke Sturm, Eva Egelseer, Margit Sára, and Uwe B. Sleytr

Center for Ultrastructure Research and
Ludwig Boltzmann Institute for Molecular Nanotechnology
Vienna, Austria

INTRODUCTION

Crystalline bacterial cell surface layers (S-layers) represent the outermost cell wall component in many strains of gram-positive and gram-negative eubacteria and have been observed during all states of growth and division (for recent review see Messner and Sleytr, 1992). High-resolution electron microscopy and permeability studies on isolated S-layers revealed a pore size in the range of 4 nm for most of the *Bacillus* species, which corresponds to a molecular weight cut-off of approximately 45000 Da (Sára and Sleytr, 1987). Many gram-positive eubacteria are able to produce enzymes like amylases, proteases or cellulases which are released into the culture fluid. Such bacteria as those of the genus *Bacillus* have frequently been the organisms of choice with regard to the commercial production of large quantities of extracellular enzymes (Priest, 1989). The amylases secreted by different *Bacillus* species have been shown to have M_rs in the range of 55000 to 60000, but these can be synthesized as larger precursors with M_r up to 130000 (Uozumi et al., 1989). Because the molecular weights of the majority of extracellular proteins are just within the range of the molecular weight cut-off of the S-layers of the *Bacillaceae*, the question arises how these proteins can be efficiently exported by S-layer carrying strains. In the present study, the size of the exported amylases, as well as the molecular weight cut-off of the S-layer was determined for *Bacillus polymyxa* CCM 1459.

CHARACTERISATION OF THE S-LAYER LATTICE OF *BACILLUS POLYMYXA* CCM 1459

As shown by thin-sectioning and freeze-etching of whole cells (Fig. 1), the S-layer of *B. polymyxa* CCM 1459 completely covered the cell surface and exhibited

Advances in Bacterial Paracrystalline Surface Layers
Edited by T.J. Beveridge and S.F. Koval, Plenum Press, New York, 1993

square symmetry with a center-to-center spacing of the morphological units of 14 nm (Sára et al., 1990). The M_r of the S-layer protein by SDS-PAGE was 170000.

DETERMINATION OF AMYLASE ACTIVITY IN THE CELL CULTURE SUPERNATANT

For determination of the amylase activity, cells were grown on LS-medium (10 g tryptone, 5 g yeast extract, 5 g NaCl and 10 g potato starch of per litre, pH 7.3) at 33°C for up to 20 hours. The optical density of the cell suspension was determined at 600 nm; the amylase activity was measured according to the method of Bernfeld (1955) which is based on the reaction of reducing sugars liberated by amylase activity with 3,5- dinitrosalicylic acid. Instead of the 0.02 M phosphate buffer described by Bernfeld (1955), in our experiments a solution of 20 mM NaCl and 15 mM $CaCl_2$ was used. Bands with enzymatic activity were detected in situ after electrophoresis as described by Lacks and Springhorn (1980). The integrity of whole cells and of the S-layer was ascertained by light microscopy and thin-sectioning; pellets of whole cells were also investigated by SDS-PAGE (see Fig. 2).

As shown by thin-sectioning, whole cells showed the typical cell envelope profile, consisting of a plasma membrane, a 5-10 nm thick peptidoglycan-containing layer and the S-layer lattice. This wall profile was identical in cells harvested from either early exponential growth phase or after 20 hours of growth where the highest amylase activity could be detected (Fig. 2). This showed that the enzymes are released through a coherent envelope layer and not through openings present in autolysed cells.

AMMONIUM SULFATE PRECIPITATION OF TOTAL PROTEIN IN THE CELL CULTURE SUPERNATANT

The protein in the cell culture supernatant was precipitated by addition of ammonium sulfate to a final concentration of 70% (w/v). Purification of the dialysed

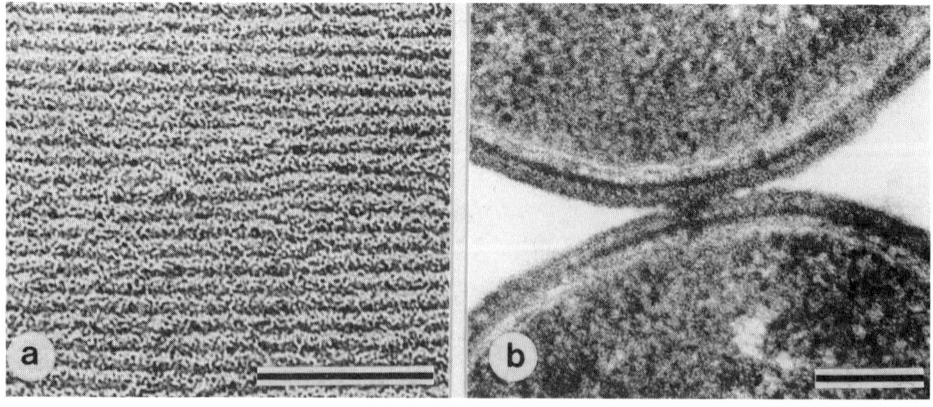

Figure 1. a) Freeze-etched preparation and b) thin-section of whole cells from *B. polymyxa* CCM 1459. Bar = 100 nm.

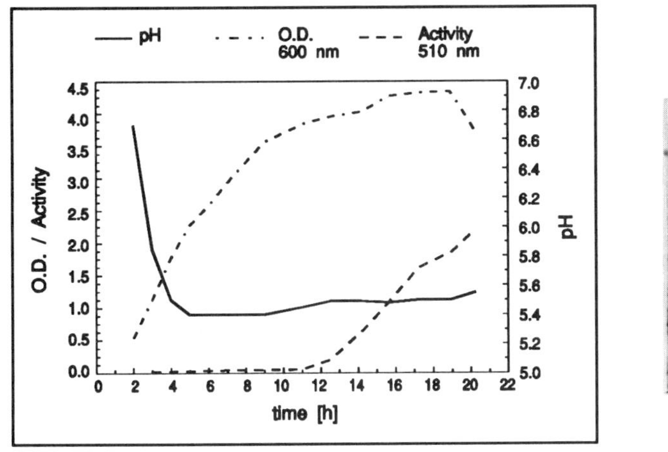

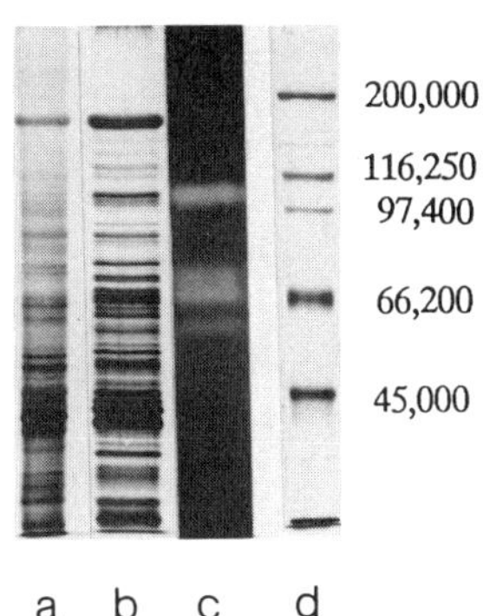

Figure 2. The optical density, pH and amylase activity of the culture fluid are plotted against the time of cultivation. **a)** SDS-extract of whole cells; **b)** SDS-extract of the culture supernatant; **c)** detection of the enzyme activity bands on SDS-polyacrylamide gels **d)** molecular weight standard.

precipitate was carried out by means of chromatofocusing using a Mono PHR 5/20 column (Pharmacia Inc.). Bands of the fractions eluted from the chromatofocusing column exhibiting amylase activity are shown in Fig. 3.

PERMEABILITY PROPERTIES OF THE S-LAYER LATTICE OF *B. POLYMYXA* CCM 1459

To define the molecular weight cut-off of the S-layer lattice, permeability studies were carried out on S-layer ultrafiltration membranes (SUM's) which were produced by depositing *B. polymyxa* S-layers on microfiltration membranes as described by Sára and Sleytr (1987). By using test proteins with well defined molecular weights it could be demonstrated that the exclusion limit of the S-layer lattice is in the range of $M_r = 45000$.

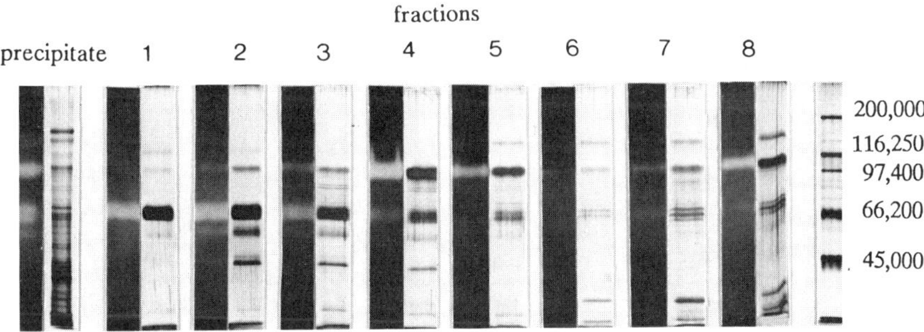

Figure 3. SDS-PAGE and detection of amylase activity in the precipitate and in fractions #1 to #8 after chromatofocusing. In all fractions the amylases were visualized by applying the iodine-starch reagent to the proteins separated by SDS-PAGE. Left lanes show the amylase activities whereas the right lanes show the proteins present in the respective fractions. In the last lane the molecular weight standards are shown.

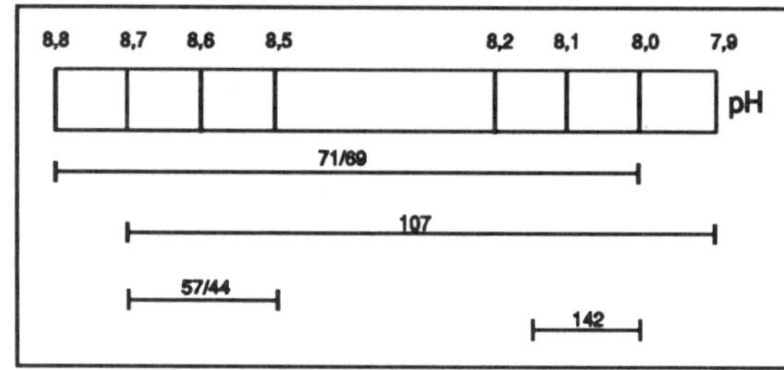

Figure 4. M_rs and pH values at which the fractions were eluted from the chromatofocusing column. The M_rs are given in kDa above the lines denoting their pH range.

In the supernatant of the bacterial suspension six protein bands with amylase activity could be detected. The M_rs and the pH values at which the proteins were eluted from the column are given in Fig. 4.

CONCLUSION

In the present study it was shown that the S-layer-carrying strain, *B. polymyxa* CCM 1459, produces large amounts of amylases with a broad spectrum of molecular weights (i.e., M_rs are 142000, 107000, 71000, 69000, 57000, and 44000).

From permeability studies, the S-layer of *B. polymyxa* CCM 1459 had revealed a molecular weight cut-off of approximately 45000. Because of this we suggest that the larger amylases with $M_r = 142000$ and 107000 cannot pass through the S-layer lattice once they have folded into their final conformation. Therefore one can speculate, that if these amylases are released through the pores of the S-layer lattice, they pass either as linear polypeptide chains or in a partially prefolded conformation. In the case of the smaller amylases (i.e., those with $M_r = 71000$, 69000, 57000, and 44000), the shape and molecular size of the protein molecules will also determine whether they can still pass through the pores of the lattice after folding into their three dimensional structure. Assuming that the size of the enzyme molecules or their prefolded domains correspond to the size of the pores in the S-layer lattice, the crystalline arrays could define a periplasmic space and consequently delay or "control" the release of the exoenzymes. Further investigations need to be done to elucidate the complex mechanism(s) of secretion by S-layer-carrying Bacillaceae.

REFERENCES

Bernfeld, P., 1955, Amylases, α and β, *Methods Enzymol.* 1:149.

Lacks, S.A., and Springhorn, S.S., 1980, Renaturation of enzymes after polyacrylamide gel electrophoresis in the presence of sodium dodecyl sulfate, *J. Biol. Chem.* 255:7467.

Messner, P., and Sleytr, U.B., 1992, Crystalline bacterial cell-surface layers, *Adv. Microbial Physiol.* 33:213.

Priest, F.G., 1989, Products and applications, *in:* "Bacillus Biotechnology Handbook 2," C.R. Harwood, ed., Plenum Press, New York.

Sára, M., and Sleytr, U.B., 1987, Production and characteristics of ultrafiltration membranes with uniform pores from two-dimensional arrays of proteins, *J. Membrane Sci.* 33:27.

Sára, M., Moser-Thier, K., Kainz, U., and Sleytr, U.B., 1990, Characterisation of S-layers from mesophilic Bacillaceae and studies on their protective role towards muramidases, *Arch. Microbiol.* 153:209.
Uozumi, N., Sakurai, K., Sasaki, T., Takekawa, S., Yamagata, H., Tsukagoshi, N., and Udaka, S., 1989, A single gene directs synthesis of a precursor protein with β- and α-amylase activities in *Bacillus polymyxa*, *J.Bacteriol.* 171:375.

Chapter 32

S-LAYER OF *BACILLUS STEAROTHERMOPHILUS* PV72

Bea Kuen and Werner Lubitz

Institute of Microbiology and Genetics
University of Vienna
Vienna, Austria

Margit Sára and Uwe B. Sleytr

Center for Ultrastructure Research
University of Agriculture
Vienna, Austria

GENERAL PROPERTIES OF *BACILLUS STEAROTHERMOPHILUS*

B. stearothermophilus PV72 is a thermophilic gram-positive bacterium, which is surrounded by a hexagonally-arranged (p6) S-layer (Sleytr et al., 1986). As estimated by SDS-PAGE, the S-layer protein of *B. stearothermophilus* has a $M_r = 130$ kDa (Messner et al., 1984). The S-layer can be dissociated into subunits which have the ability to self-assemble into lattices identical to those observed on intact cells (Sleytr, 1976).

CLONING AND SEQUENCING OF THE S-LAYER GENE

A purified S-layer protein preparation was separated by two dimensional gel electrophoresis and tryptic peptides of the S-layer protein were analysed by amino acid micro-sequencing (Frank and Ashman, 1986). The N-terminal sequence of 15 amino acids and 8 internal peptides of 15 to 20 amino acids were identified. Oligonucleotides which were deduced from the peptide sequences and an anti-S-layer antiserum were used to screen a gene library which was established in *Escherichia coli*. Only one out of three positive clones which reacted with the antiserum gave a positive hybridisation signal with 5 out of 8 oligonucleotides (clone 24). This clone directed the synthesis of a polypeptide with a $M_r = 70$ kDa which is much lower than the M_r of the authentic S-layer protein (data not shown). DNA sequencing revealed that clone 24 contained an open reading frame (ORF) of 760 codons, including 5 of the 8 amino acid microsequences and the 3'-end of the S-layer gene.

BamHI and PstI chromosomal DNA digests each gave one hybridisation signal with the cloned S-layer fragment in a size range of 4.6 kb and 5.8 kb, respectively. Despite the use of a variety of cloning vectors and strategies we were unable to clone the corresponding chromosomal fragments. These findings lead to the speculation that the S-layer gene or other sequences carried on the chromosomal DNA are unstable or toxic in *E. coli*. For this reason the missing 5' sequences of the S-layer gene were transcribed from the chromosome using PCR and oligonucleotide primers which corresponded to the 5' end of the gene (N-terminal amino acid sequence) and to an S-layer gene-specific sequence of clone 24. From the sequence of the 1.3 kb PCR fragment the missing 5' ORF of the mature S-layer protein could be identified by the 3 amino acid microsequences which were not included in clone 24.

The complete amino acid sequence of the mature S-layer protein of *B. stearothermophilus*, obtained from clone 24 and the PCR fragment, is given in Fig. 1. The mature S-layer protein consists of 1199 amino acids. As the mature protein sequence starts with alanine we conclude that the S-layer protein is synthesized as a precursor.

The amino acid composition (Table 1) of the S-layer protein shows that the most abundant amino acids are threonine, alanine and valine. As in most other S-layer proteins, cysteine is missing.

```
ATDVATVVSQAKAQFKKAYYTYSHTVTETGEFPNINDVYAEYNKAKKRYR        50
DAVALVNKAGGAKKDAYLADLQKEYETYVFKANPKSGEARVATYIDAYNY        100
ATKLDEMRQELEAAVQAKDLEKAEQYYHKIPYEIKTRTVILDRVYGKTTR        150
DLLRSTFKAKAQELRDSLIYDITVAMKAREVQDAVKAGNLDKAKAAVDQI        200
NQYLPKVTDAFKTELTEVAKKALDADEAALTPKVESVSAINTQNKAVELT        250
AVPVNGTLKLQLSAAANEDTVNVNTVRIYKVDGNIPFALNTADVSLSTDG        300
KTITVDASTPFENNTEYKVVVKGIKDKNGKEFKEDAFTFKLRNDAVVTQV        350
FGTNVTNNTSVNLAAGTFDTDDTLTVVFDKLLAPETVNSSNVTITDVETG        400
KRIPVIASTSGSTITITLKEALVTGKQYKLAINNVKTLTGYNAEAYELVF        450
TANASAPTVATAPTTLGGTTLSTGSLTTNVWGKLAGGVNEAGTYYPGLQF        500
TTTFATKLDESTLADNFVLVEKESGTVVASELKYNADAKMVTLVPKADLK        550
ENTIYQIKIKKGLKSDKGIELGTVNEKTYEFKTQDLTAPTVISVTSKNGD        600
AGLKVTEAQEFTVKFSENLNTFNATTVSGSTITYGQVAVVKAGANLSALT        650
ASDIIPASVEAVTGQDGTYKVKVAANQLERNQGYKLVVFGKGATAPVKDA        700
ANANTLATNYIYTFTTEGQDVTAPTVTKVFKGDSLKDADAVTTLTNVDAG        750
QKFTIQFSEELKTSSGSLVGGKVTVEKLTNNGWVDAGTGTTVSVAPKTDA        800
NGKVTAAVVTLTGLDNNDKDAKLRLVVDKSSTDGIADVAGNVIKEKDILI        850
RYNSWRHTVASVKAAADKDGQNASAAFPTSAIDTTKSLLVEFNETDLAE        900
VKPENIVVKDAAGNAVAGTVTALDGSTNKFVFTPSQELKAGTVYSVTIDG        950
VRDKVGNTISKYITSFKTVSANPTLSSISIADGAVNVDRSKTITIEFSDS        1000
VPNPTITLKKADGTSFTNYTLVNVNNENKTYKIVFHKGVTLDEFTQYELA        1050
VSKDFQTGTDIDSKVTFITGSVATDEVKPALVGVGSWNGTSYTQDAAATR        1100
LRSVADFVAEPVALQFSEGIDLTNATVTVTNITDDKTVEVISKESVDADH        1150
DAGATKETLVINTVTPLVLDNSKTYKIVVSGVKDAAGNVADTITFYIKZ        1199
```

Figure 1. Amino acid sequence of the S-layer protein of *B. stearothermophilus*. Microsequences identified from tryptic digests of the mature S-layer protein are underlined.

Table 1. Amino acid composition of the S-layer protein of *B. stearothermophilus*

Amino acid	Number		Percentage
Ala	135		11.3
Arg	21		1.2
Asn	74		6.2
Asp	83		6.9
Cys	-		-
Gln	31		2.6
Glu	63		5.3
Gly	71		5.9
His	5		0.4
Ile	53		4.4
Leu	80		6.7
Lys	109		9.1
Met	3		0.3
Phe	40		3.3
Pro	28	5.8	2.3
Ser	69		13.0
Thr	156		0.3
Trp	4		3.5
Tyr	42		10.9
Val	131		

Sequencing of the signal sequence and further upstream regulatory sequences are in progress.

ACKNOWLEDGEMENT

We thank R. Frank, EMBL, Heidelberg and C. Forauer, Institut für angewandte Mikrobiologie, Universität für Bodenkultur, Wien for amino acid sequencing.

REFERENCES

Frank, R., and Ashman, K., 1986, A new covalently modified support for gas-liquid phase sequencing, *Biol. Chem. Hoppe-Seyler* 367:573.

Messner, P., Hollaus, F., and Sleytr, U.B., 1984, Paracrystalline cell wall surface layers of different *Bacillus stearothermophilus* strains, *Int. J. Syst. Bacteriol.* 34:202.

Sleytr, U.B., 1976, Self-assembly of the hexagonally and tetragonally arranged subunits of bacterial surface layers and their reattachment to cell walls, *J. Ultrastruct. Res.* 55:360.

Sleytr, U.B., Sara, M., Küpcü, Z., and Messner, P., 1986, Structural and chemical characterization of S-layers of selected strains of *Bacillus stearothermophilus* and *Desulfotomaculum nigrificans*, *Arch. Microbiol.* 146:19.

Chapter 33

ROLE OF A C-TERMINAL DOMAIN IN THE STRUCTURE AND SURFACE ANCHORING OF THE TETRAGONAL S-LAYER OF *AEROMONAS HYDROPHILA*

Stephen R. Thomas, John W. Austin, and Trevor J. Trust

Department of Biochemistry and Microbiology
and Canadian Bacterial Diseases Network
University of Victoria
Victoria, British Columbia, Canada

INTRODUCTION

A number of species of *Aeromonas* are capable of producing S-layers which exhibit p4 symmetry (Stewart et al., 1986; Murray et al., 1988; Al-Karadaghi et al., 1988; Dooley et al., 1989; Kokka et al., 1990; Kokka et al., 1992). In wild type strains, these tetragonal S-layers are anchored to the surface of the *Aeromonas* cell and little, if any, assembled layer is released from the cell surface. However, *A. salmonicida* mutants which produce a lipopolysaccharide (LPS) deficient in O-antigen chains (rough LPS), while still capable of assembling the S-layer protein subunits into an array, are unable to anchor this assembled array to their cell surface (Chart et al., 1984; Dooley et al., 1989). Indeed the released S-layer forms a stable sheet consisting of two exactly superimposed layers interacting through their inner surfaces (Dooley et al., 1989). In the case of *A. hydrophila*, an intact core LPS appears to be required to anchor the array to the cell surface because while rough strains can maintain S-layer on their surface, deep rough LPS mutants shed their S-layer (Dooley et al., 1988). The regions of the S-layer proteins participating in this anchoring of S-layer to the surface of the *Aeromonas* cells have not been identified.

The p4-arrayed S-layers of *Aeromonas* contain a major tetragon at one four-fold axis of symmetry, and a minor tetragon at the second four-fold axis of symmetry. Each tetrameric morphological unit is formed by four S-layer protein domains, the major tetragonal core being formed by the larger mass domain, and the second center of symmetry formed by a lesser linker domain. Protease digestion studies with both of the *A. salmonicida* and *A. hydrophila* S-layer proteins have provided evidence for the possible structural identity of these two morphological domains (Chu et al., 1991; Kostrzynska et al., 1992). Treatment of $M_r = 50000 - 52000$ S-layer proteins with trypsin resulted in a $M_r = 37000 - 39000$ major peptide which was resistant to trypsin cleavage. In both cases automated Edman degradation mapped this trypsin resistant fragment to the N-terminal end of the S-layer protein. In the case of the *A. hydrophila* protein, treatment with chymotrypsin or endoproteinase Glu-C also resulted in a 38 kDa protease-resistant peptide. This data suggested that the major trypsin-resistant structural domain could be that segment of the S-layer protein which forms the larger mass or major morphological core of the S-layer, while the lesser C-terminal domain of the S-protein which is far more susceptible to proteases could contribute to the lesser linker domain.

We have now isolated a single insertion transposon mutant of *A. hydrophila* which produces a truncated S-layer protein. Ultrastructural and biochemical analysis of this $M_r = 38000$ protein has allowed us to confirm this structural prediction, and has also provided information on the role of the N- and C-terminal domains of the *A. hydrophila* S-layer protein in anchoring the S-layer to the surface of the *Aeromonas* cell.

RESULTS and DISCUSSION

Western immunoblot analysis of the outer membrane and culture supernatant fractions of liquid grown cells of *A. hydrophila* Tn5 insertion mutant TF7-ST1 showed the presence of a 38 kDa immunoreactive polypeptide in the culture supernatant (Fig. 1). In the wild type the 52 kDa S-layer protein was only detected in the outer membrane fraction (Fig. 1). In the case of *A. hydrophila* TF7-ST1 cells grown on solid media, any truncated S-protein which accumulated on the cell surface was readily removed by washing, indicating that it was not anchored to the surface. N-terminal sequence analysis of the 38 and 52 kDa proteins showed that they shared identical N-terminal amino acid sequences (Kostrzynska et al., 1992). When the 52 kDa native S-protein was digested with trypsin and analysed by SDS-PAGE (Fig. 1), the N-terminal peptide was of similar subunit molecular weight to the truncated product.

Culture supernatants of the mutant TF7-ST1 containing the truncated 38 kDa polypeptide were studied by electron microscopy and were found to contain two predominant macromolecular assemblies which, based on previous computer reconstructions of both the *A. hydrophila* (Murray et al., 1988; Al-Karadaghi et al., 1988) and *A. salmonicida* S-layers (Stewart et al., 1986; Dooley et al., 1988), would appear to consist of four copies of the truncated protein: 1) a ring-like structure with an external diameter of 6.5 nm and an internal diameter of 3.0 nm, probably representing the view down the four fold axis of the major S-layer tetramer, and 2),

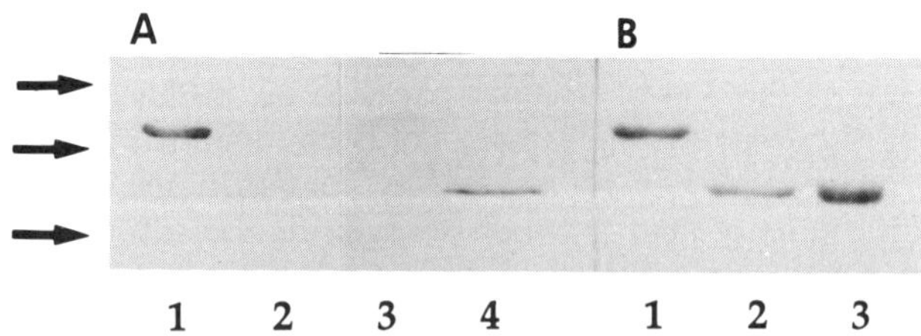

Figure 1. Western immunoblot of SDS-PAGE of outer membrane and supernatant fractions (A) of native TF7 S-protein, and truncated S-layer protein of *A. hydrophila* Tn5 insertion mutant TF7-ST1. Lane 1 and 3 are outer membrane fractions of TF7 and lane 2 and 4, supernatant fractions, respectively. (B) Coomassie blue stain showing a comparison of truncated S-protein produced by mutant TF7-ST1 (lane 3), with NH2-terminal peptide of purified wild type S-protein produced by trypsin cleavage (lane 2), and native S-protein (lane 1). Arrows at left indicate (top to bottom) $M_r = 66200$, 45000, and 31000.

a cup-like assembly consisting of two parallel 12.0 nm long lines joined at one end and 6.5 nm apart. This is consistent with the width of the major tetramer of the native *A. hydrophila* S-layer, likely lying on its side (Murray et al., 1988). Side views of single major tetramers and triple major tetramers joined at their bases were most common, although side views of double major tetramers were also seen. The truncated protein was never seen to form a tetragonal array, either after natural excretion from the cell surface, or following reassembly studies using low pH extraction and gel filtration purified product.

The truncated S-layer protein was analysed biochemically. Sedimentation analysis placed the subunit M_r of the truncated S-layer protein at 38650, and circular dichroism studies showed that it contained approximately 42% β-sheet, 10% α-helix, and 19% β-turn. Amino acid composition analysis showed that it contained 387 residues per molecule. The truncated protein also had an increased relative hydrophobic content compared to the native S-protein. This is consistent with the functional requirements of a structural domain which probably associates hydrophobically with the equivalent domains of other monomers to produce the major protein mass of the tetragonal surface array.

We conclude therefore that the N-terminal, $M_r = 38650$, protease resistant domain of the *A. hydrophila* S-layer protein is responsible for the formation of the primary morphological unit of the S-layer. In the native layer this unit probably interacts hydrophobically with the underlying outer membrane, however this interaction by itself is not adequate to anchor the array to the cell surface as evidenced by the loss of the truncated S-protein into the growth media. The minor C-terminal domain of the S-protein contributes to the minor linker morphological unit of the S-layer. This domain is essential for array assembly. In addition, the C-terminal domain appears to participate in anchoring of the S-layer to the cell surface, presumably via interaction with LPS (Dooley et al., 1988).

REFERENCES

Al-Karadaghi, S., Wang, D. N., and Hövmoller, S., 1988, Three-dimensional structure of the crystalline layer from *Aeromonas hydrophila, J. Ultrastruct. and Mol. Struct. Res.* 101:92.

Chart, H., Shaw, D. H., Ishiguro, E. E., and Trust, T. J., 1984, Structural and immunochemical homogeneity of *Aeromonas salmonicida* lipopolysaccharide, *J. Bacteriol.* 158:16.

Chu, S., Cavaignac, S., Feutrier, J., Phipps, B. M., Kostrzynska, M., Kay, W. W., and Trust, T. J., 1991, Structure of the tetragonal surface virulence array protein and gene of *Aeromonas salmonicida, J. Biol. Chem.* 266:15258.

Dooley, J. S. G., and Trust, T. J., 1988, Surface protein composition of *Aeromonas hydrophila* virulent for fish: Identification of an S-layer protein, *J. Bacteriol.* 170:499.

Dooley, J. S. G., McCubbin, W. D. M., Kay, C. M., and Trust, T. J., 1988, Isolation and biochemical characterization of the S-layer protein from a pathogenic strain of *Aeromonas hydrophila, J. Bacteriol.* 170:2631.

Dooley, J. S. G., Engelhardt, H., Baumeister, W., Kay, W. W., and Trust, T. J., 1989, Three dimensional structure of the surface layer from the fish pathogen *Aeromonas hydrophila, J. Bacteriol.* 171:190.

Kokka, R. P., Vedros, N. A., and Janda, J. M., 1990, Electrophoretic analysis of the surface components of autoagglutinating surface array protein positive and surface array protein negative *Aeromonas hydrophila* and *Aeromonas sobria*, *J. Clin. Microbiol.* 28:2240.

Kokka, R. P., Lindquist, D., Abbott, S. L., and Janda, J. M., 1992, Structural and pathogenic properties of *Aeromonas schubertii*, *Infect. Immun.* 60:2075.

Kostrzynska, M., Dooley, J. S. G., Shimojo, T., Sakata, T., and Trust, T. J., 1992, Antigenic diversity of the S-layer from a pathogenic strain of *Aeromonas hydrophila* and *Aeromonas veronii* biotype *sobria*, *J. Bacteriol.* 174:40.

Murray, R. G. E., Dooley, J. S. G., Whippey, P. W., and Trust, T. J., 1988, Structure of an S-layer on a pathogenic strain of *Aeromonas hydrophila*, *J. Bacteriol.* 170:2625.

Stewart, M., Beveridge, T. J., and Trust, T. J., 1986, Two patterns in the *Aeromonas salmonicida* A-layer may reflect a structural transformation that alters permeability, *J. Bacteriol.* 166:120.

Chapter 34

IDENTIFICATION AND CHARACTERIZATION OF AN *AEROMONAS SALMONICIDA* GENE WHICH AFFECTS A-PROTEIN EXPRESSION IN *ESCHERICHIA COLI*

Shijian Chu and Trevor J. Trust

Department of Biochemistry and Microbiology
and Canadian Bacterial Disease Network
University of Victoria
Victoria, British Columbia, Canada

The S-layer of *Aeromonas salmonicida* (A-layer) is composed of a protein (A-protein) of subunit $M_r = 50800$. The gene (*vapA*) coding for A-protein has been cloned, sequenced and expressed in *Escherichia coli*. Because deletions of flanking DNA $3'$ to *vapA* resulted in decreased A-protein expression in *E. coli*, we determined the down stream DNA sequence of the 1200 bp of DNA downstream of *vap A*. This analysis revealed the presence of a gene which codes for a protein with three potential structural domains, i.e., a nucleotide-binding domain, a membrane-associated region, and a leucine zipper-basic region sequence.

ANALYSIS OF *vapA* GENE EXPRESSION IN *E. COLI*

The *vapA* gene from *A. salmonicida* strain A450 was previously subcloned as a 3.4 Kb insert in pSC150 (Fig. 1) and expressed (Chu et al., 1991). Removal of 0.55 or 0.85 Kb of DNA sequences from the $3'$ end of the insert (pSC151 and pSC152, Fig. 1) resulted in a decrease in A-protein production of at least 16-fold. When the deleted sequence was ligated into a second plasmid pSC163 (Fig. 1) and introduced into the DH5α cells containing either pSC151 or pSC152, A-protein production was restored to the level of pSC150/DH5α.

DNA SEQUENCE ANALYSIS

Sequencing of this down stream DNA revealed an open reading frame (ORF) of 924 bp which started 205 bp after the *vapA* gene (Fig. 1). In pSC151 and pSC152 which produced decreased amounts of A-protein, the *vapA* gene was still intact, but this down stream ORF was truncated at the $3'$ end by 541 bases and 846 bases respectively. Southern blot and PCR analysis of a variety of *Aeromonas* strains showed that this ORF was present in all the *A. salmonicida* strains tested, but was not present in S-layer producing strains of *A. hydrophila* and *A. veronii* biotype *sobria*.

IDENTIFICATION OF THE GENE PRODUCT AND ITS ATP BINDING ACTIVITY

The ORF encodes a predicted protein of 308 amino acid residues with $M_r =$ 34015 and a predicted pI of 5.7. The deduced protein lacks a consensus leader sequence, but a membrane-associated region was predicted between residues 196-209 (Fig. 2). According to the secondary structure prediction, the longest α-helix region in the molecule lies between residues 211 and 251. In this region, a heptad repeat of leucine and valine residues (leucine zipper) was observed starting at residue 211 (Fig. 2). Similar to the eukaryotic basic region-leucine zipper (bZIP) proteins (Vinson et al., 1989), this potential leucine zipper sequence is flanked at the $3'$ end by a highly positively charged sequence about 20 amino acid residues long (Fig. 2) with a predicted pI of 11.29.

Sequence similarity searches showed that the ORF contained regions of sequence with high similarities to a number of proteins belonging to a transport protein super family (Blight and Holland, 1990). The most recognizable sequence was a nucleotide binding sequence (residues 58-66 in the ORF, Fig. 2) known as a P-loop (phosphate-binding loop), GXXXXGKST (Saraste et al., 1990).

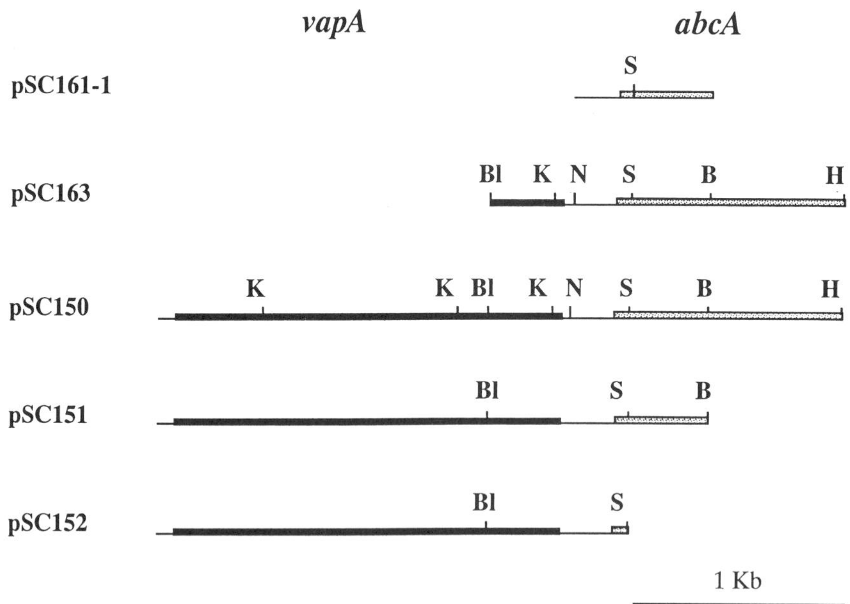

Figure 1. Restriction maps of inserts in each subclone used in this study. Names of the subclones are marked on the left. All subclones are vertically aligned to each other. Thicker lines indicate *vapA* coding regions. Dotted boxes represent the *abcA* gene. B, *Bam*HI; Bl, *Bgl*II; H, *Hind*III; K, *Kpn*I; N, *Nar*I; S, *Sph*I.

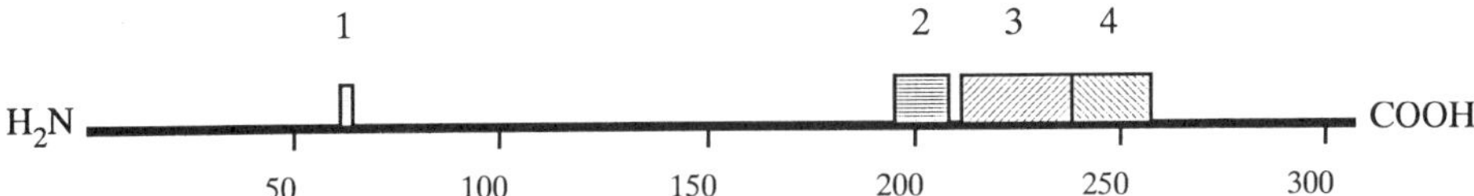

Figure 2. Schematic map of deduced sequence of AbcA. Numbers below the long line represent amino acid residues. Shaded areas 1 to 4 above the long line indicate the P-loop, the putative membrane-associated sequence, the leucine zipper region, and the basic region, respectively.

EXPRESSION AND ATP BINDING ACTIVITY

In both in vitro transcription-translation and in vivo T7 polymerase expression studies, the protein showed an apparent $M_r = 43000$, suggesting that the protein displays aberrant migration during SDS-PAGE analysis. A *lacZ* fusion was also constructed. This clone, pSC161-1 (Fig. 1), without the membrane-associated sequence, produced a fusion protein of apparent $M_r = 130000$ which was confined to the cell cytosol in *E. coli*. The *lacZ* fusion gene was also introduced into *A. salmonicida* A450. The β-galactosidase activity was shown to be similar in *A. salmonicida* and *E. coli*. The fusion protein was purified using an affinity column with anti-β-galactosidase monoclonal antibody.

Because the P-loop was included in the β-galactosidase fusion protein, the ATP binding ability of the purified fusion protein was tested using an ATP affinity column. The fusion protein was shown to bind to the column and was washed off with an elution buffer containing ATP confirming the ability of the protein to bind ATP. Because of this ATP binding ability and the sequence similarity with the ATP-binding-cassette (ABC)-transport proteins, the gene is named *abcA* (= <u>A</u>TP <u>b</u>inding <u>c</u>assette protein <u>A</u>).

The precise function of the *abcA* gene in *A. salmonicida* has yet to be determined. However, the predicted protein has the characteristics of a protein which may play a role in transport, signaling, or gene regulation. Also, the gene maps immediately downstream of *vapA*, and its presence affects the amount of VapA produced in *E. coli*. Taken together, this suggests that *abcA* may play a role in A-layer production in *A. salmonicida*.

REFERENCES

Blight, M.A., and Holland, I.B., 1990, Structure and function of haemolysin B, P-glycoprotein and other members of a novel family of membrane translocators, *Mol. Microbiol.* 4:873.

Chu, S., Cavaignac, S., Feutrier, J. Phipps, B.M., Kostrzynska, M., Kay, W.W., and Trust, T., 1991, Structure of the tetragonal surface virulence array protein and gene of *Aeromonas salmonicida*, *J. Biol. Chem.* 266:15258.

Saraste, M., Sibbald, P.R., and Wittinghofer, A., 1990, The P-loop - a common motif in ATP- and GTP-binding proteins, *Trends Biochem. Sci.* 15:430.

Vinson, C.R., Sigler, P.J.B., and McKnight, S.L., 1989, Scissors-grip model for DNA recognition by a family of leucine zipper proteins, *Science* 246:911.

Chapter 35

LOCALISATION AND CLONING OF GENES INVOLVED IN THE EXPORT OF THE A-PROTEIN OF *AEROMONAS SALMONICIDA*

Brian Noonan, Sonia Cavaignac and Trevor J. Trust

Department of Biochemistry and Microbiology
and Canadian Bacterial Diseases Network
University of Victoria
Victoria, British Columbia, Canada

INTRODUCTION

The paracrystalline surface protein array of *Aeromonas salmonicida* is composed of tetragonally arranged subunits of $M_r = 50800$ (Chu et al., 1991). The gene encoding the A-protein has recently been cloned (Belland and Trust, 1987) and sequenced (Chu et al., 1991), and studies on its 3 dimensional structure have been performed (Dooley et al., 1989). The A-layer is involved in the ability of *A. salmonicida* to produce disease in fish and it appears to protect the bacteria from the bacteriocidal effect of immune and non-immune serum (Ishiguro et al., 1981; Munn et al., 1982) as well as from phagocytic cells (Trust et al., 1983).

In order to form a layer on the *A. salmonicida* cell surface, the A-protein subunits must be transported across both the cytoplasmic and outer membranes. Previous studies have determined that the A-protein possesses a leader peptide which allows it to be transported across the cytoplasmic membranes of both *A. salmonicida* and *Escherichia coli* (Chu et al., 1991).

There are two distinct mechanisms by which proteins are exported from gram-negative cells (reviewed by Lory, 1992). In the first case, the proteins are transported across the cytoplasmic membrane to the periplasm and from there across the outer membrane. The *Pseudomonas aeruginosa* alkaline phosphatase (Poole and Hancock, 1983) and the haemolysin of certain *E. coli* strains (Wagner et al., 1983) are examples of this mechanism. The other observed mechanism of protein secretion bypasses the periplasm as the proteins to be exported pass through the cytoplasmic and outer membranes by means of the Bayer junctions (Lory, 1992).

The evidence to date suggests that the export of the *A. salmonicida* A-protein is via the periplasmic space as discussed in the first mechanism above. Transposon Tn5 mutagenesis of *A. salmonicida* A449 led to the isolation of a number of mutants with altered surface morphologies (Belland and Trust, 1985). Two mutants from this study, TM1 and TM2 were found to actively synthesize A-protein but did not form an S-layer. It was demonstrated that these mutants are unable to transport the A-protein across the outer membrane. In the case of both mutants, considerable quantities of A-protein were localised in the periplasm (Belland and Trust, 1985).

RESULTS AND DISCUSSION

The ability of the mutants TM1 and TM2 to export other proteins was examined (Table 1). It was found that TM1 and TM2 were indistinguishable from the wild type when tested for haemolytic activity and casein degradation. This would indicate that the general protein export mechanisms in these mutants are unaffected by the Tn5 insertions.

The lipopolysaccharide (LPS) of TM1 and TM2 has previously been studied and silver staining showed that TM2 was unable to produce the O-polysaccharide chains observed in the wild type. This result was confirmed by western blotting with antisera raised against *A. salmonicida* LPS. This result raises the possibility that the mutated gene in TM2 is required for O-polysaccharide production as well as A-protein transport.

Table 1. Phenotypes of Tn5 mutants of *A. salmonicida*

Strain	A-Protein location	LPS O-chains	Protease[1] activity	Haemolysis[2]
A449 (wild type)	Cell Surface	+	+	+
TM1	Periplasm	+	+	+
TM2	Periplasm	-	+	+

[1]casein degradation; [2]blood agar

In order to investigate the export of A-protein in *A. salmonicida*, total cellular DNA from TM1 and TM2 was isolated, partially restricted with *Sau*3A, and ligated into the cosmid cloning vector, pHC79 (Hohn and Collins, 1980). Tn5-containing clones from both TM1 and TM2 were isolated by selection with kanamycin. Two of the clones obtained, pHC79-11 (TM1) and pHC79-20 (TM2) were selected for further study.

Restriction analysis of the two clones showed them to be completely different indicating that there must be at least two distinct regions involved in the export of the A-protein. In the case of pHC79-20, it was found that Tn5 had inserted at a site approximately 7 kb downstream from the A-protein structural gene, *vap*A. The location of the Tn5 insertion in pHC79-11 relative to *vap*A is unknown at this stage.

The insert in pHC79-20 was further subcloned to yield p2C1HE (Fig. 1). From this clone, which contains approximately 1 kb of Tn5 DNA, a 2.2 kb BamH1-Sal1 fragment was isolated from a gel and labeled with digoxygenin (Boehringer Mannheim) for use as a probe. This probe was hybridized against a number of *A. salmonicida* gene bank clones known to contain, or map close to, the *vap*A gene.

One positive clone, pHC79-42, was obtained and restriction analysis confirmed the presence and expected location of the 2.2 kb fragment. Based on the restriction map of pHC79-42, a 6 kb BamH1 fragment from this clone was ligated

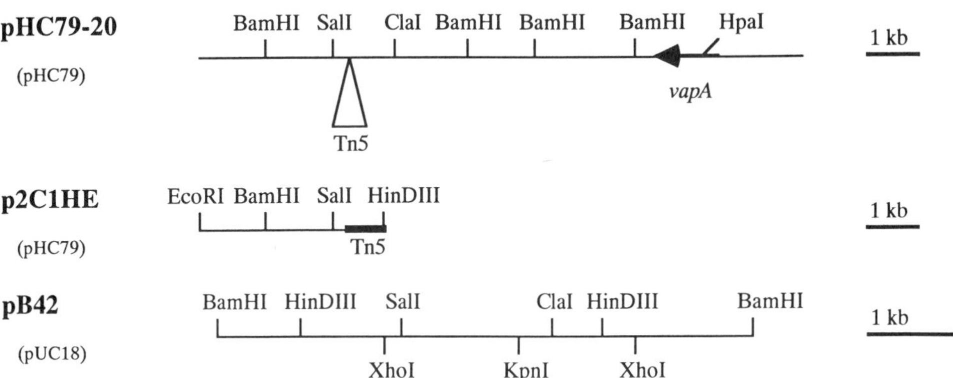

Figure 1. Genetic mapping and cloning of a gene involved in A-protein transport in *A. salmonicida*.

into pUC18 for further analysis. This clone was called pB42. Restriction analysis of the 6 kb insert showed that it contained the site of insertion of the Tn5 in TM2 (Fig. 1).

Further biochemical and genetic studies of the mutated genes in TM1 and TM2 should lead to a better understanding of the process of A-protein export in *A.salmonicida*. Of particular interest will be the apparent connection between LPS and the export of A-protein.

REFERENCES

Belland, R.J., and Trust, T.J., 1985, Synthesis, export and assembly of *Aeromonas salmonicida* A-layer analysed by transposon mutagenesis, *J. Bacteriol.* 163: 877.

Belland, R.J., and Trust, T.J., 1987, Cloning of the gene for the surface array protein of *Aeromonas salmonicida* and evidence linking loss of expression with genetic deletion, *J. Bacteriol.* 169: 4086.

Chu, S., Cavaignac, S., Feutrier, J., Phipps, B.M., Kostrzynska, M., Kay, W.W., and Trust, T.J., 1991, Structure of the tetragonal surface array protein and gene of *Aeromonas salmonicida*, *J. Biol. Chem.* 266: 15258-15265.

Dooley, J.S.G., Engelhardt, H., Baumeister, W., Kay, W.W., and Trust, T.J., 1989, Three-dimensional structure of an open form of the surface layer from the fish pathogen *Aeromonas salmonicida*, *J. Bacteriol.* 171: 190.

Hohn, B., and Collins, J., 1980, A small cosmid for efficient cloning of large DNA fragments, *Gene* 11: 291-298.

Hovömller, S., Sjogren, A., and Wang, D.N., 1988, The structure of crystalline bacterial surface layers, *Prog. Biophys. Molec. Biol.* 51: 131.

Ishiguro, E.E., Kay, W.W., Ainsworth, T., Chamberlain, J.B., Buckley, J.T., and Trust, T.J., 1981, Loss of virulence during culture of *Aeromonas salmonicida* at high temperature, *J. Bacteriol.* 148: 333.

Lory, S., 1992, Determinants of extracellular protein secretion in gram-negative bacteria, *J. Bacteriol.* 174: 3423.

Messner, P., and Sleytr, U.B., 1992, Crystalline bacterial cell-surface layers, *Adv. Microbial Physiol.* 33: 213.

Munn, C.B., Ishiguro, E.E., Kay, W.W., and Trust, T.J., 1982, Role of surface components in serum resistance of virulent *Aeromonas salmonicida*, *Infect. Immun.* 36: 1069.

Trust, T.J., Kay, W.W., and Ishiguro, E.E., 1983, Cell surface hydrophobicity and macrophage association of *Aeromonas salmonicida*, *Curr. Microbiol.* 9: 315.

Poole, K., and Hancock, R.E.W., 1983, Secretion of alkaline phosphatase and phospholipase C in *Pseudomonas aeruginosa* is specific and does not involve an increase in outer membrane permeability, *FEMS Lett.* 16: 25.

Wagner, W., Vogel, M., and Goebel, W., 1983, Transport of haemolysin across the outer membrane of *Escherichia coli* requires two functions, *J. Bacteriol.* 154: 200.

Chapter 36

GROWTH MEDIUM CONSIDERATIONS FOR THE SCALE-UP OF S-LAYER PROTEIN PRODUCTION BY *BACILLUS BREVIS* 47

Gordon K. Whitney, A. Wong, and A.J. Daugulis

Department of Chemical Engineering, Queen's University
Kingston, Ontario, Canada

B.N. White

Biology Department, McMaster University
Hamilton, Ontario, Canada

Bacillus brevis 47 is known to secrete 16 g/L of S-layer proteins (Wight et al., 1992) and therefore has great potential for the production of high levels of secreted heterologous proteins of commercial value. Investigations into the nutritional regulation of S-layer protein production should facilitate the production of foreign proteins whose genes are under the control of S-layer promoters. Polypeptone has been identified as the key nutrient for the stimulation of S-layer secretion (Shaku et al., 1980). Given the heterogeneous nature of this nutrient it should be possible to employ less expensive nitrogen sources, given medium cost considerations when scaling-up a commercial process.

Preliminary shake flask studies (10 mL of medium in 125 mL flasks at 37 °C and 250 rpm for 24 h) to test several nitrogen sources have been conducted in a semi-defined medium (10 g/L glucose, 0.3 g/L KH_2PO_4, 0.7 g/L K_2HPO_4, 0.1 g/L $MgSO_4 \cdot 7H_2O$, 10 mg/L $FeSO_4 \cdot 4H_2O$, 10 mg/L $MnSO_4 \cdot 4H_2O$, 1 mg/L $ZnSO_4 \cdot 7H_2O$), and have revealed that cultivation of *B. brevis* 47 with laboratory grade polypeptone (BBL), yeast extract (GIBCO) or peptone (Difco) resulted in comparable maximum secreted S-layer protein levels (4.5 to 5 g/L protein as determined by Bradford, 1976). Testing each ingredient at various concentrations (only the data for the yeast extract are shown) demonstrated that increasing the nitrogen supply (0 to 0.06 M) resulted in increased growth and S-layer protein levels of less than 1 g/L (Table 1). Nitrogen levels above 0.06 M resulted in no increase in growth but significantly increased S-layer protein secretion (4 to 5 g/L). It would appear that there is a level of nitrogen required for cell growth, and that nitrogen in excess of this value is channelled into S-layer protein production.

Prior to undertaking cell culture in bioreactors, two industrial grade nitrogen sources (yeast extract and Pancase S peptone, Champlain Industries, Cornwall,

Table 1. Growth and Secreted S-Layer Protein Production by *B. brevis* 47 in a Semi-Defined Growth Medium With Various Concentrations of Yeast Extract as the Sole Nitrogen Source

Moles/L Nitrogen as Yeast Extract	Cell Growth at OD660	Secreted S-Layer Protein (g/L)
0.00	0.3	0.0
0.02	4.3	0.1
0.04	7.7	0.2
0.06	9.6	0.9
0.08	8.9	4.4
0.10	9.6	4.5
0.20	8.9	4.6

Table 2. Evaluation of Industrial and Laboratory Nitrogen Sources for Cost and Ability to Support S-layer Protein Production

Nitrogen Sources	Cost in Dollars (Can) for 0.2 M Total Nitrogen	Maximum Secreted Protein (g/L)
Laboratory Yeast Extract	2.24	4.1
Laboratory Polypeptone	2.53	4.2
Industrial Yeast Extract	0.34	3.1
Industrial Peptone	0.56	4.0

Ontario, Canada) were compared with 2 laboratory grade nitrogen sources (BBL polypeptone and GIBCO yeast extract) for cost and the ability to support *B. brevis* 47 S-layer protein production. The industrial grade components were significantly less expensive, based upon 0.2 M total nitrogen, ($0.34 and $0.56/L) than the laboratory grade polypeptone ($2.53/L) or yeast extract ($2.24/L) (all costs are in 1992 Canadian dollars). However, only the Pancase S provided a suitable nitrogen source for S-layer protein production (Table 2) due to solubility problems with the industrial yeast extract. Using Pancase S as a nitrogen source it was possible to scale-up induction protocols developed in this laboratory (Wight et al.,1992; Daugulis et al., this book).

REFERENCES

Bradford, M., 1976, A rapid and sensitive method for the quantitation of microgram quantities of protein utilizing the principle of protein-dye binding, *Anal. Biochem.* 72:248.

Shaku, M., Koike, S., and Udaka, S., 1980, Cultural conditions for protein production by *Bacillus brevis* no. 47, *Agricult. Biol. Chem.* 44:99.

Wight, C.P., Whitney, G.K., White, B.N., and Daugulis, A.J., 1992, Enhancement and regulation of extracellular protein production by *Bacillus brevis* 47 through manipulation of cell culture conditions, *Biotechnol. Bioeng.* 40:46.

VIII. SUMMARY STATEMENTS

Chapter 37

SUMMARY STATEMENTS

Terry J. Beveridge

Department of Microbiology, College of Biological Science
University of Guelph
Guelph, Ontario, Canada

Susan F. Koval

Department of Microbiology and Immunology
University of Western Ontario
London, Ontario, Canada

Uwe B. Sleytr

Center for Ultrastructure Research and Ludwig Boltzman
Institute for Molecular Nanotechnology
University of Agriculture
Vienna, Austria

Helmut König

Department of Microbiology
University of Ulm
Ulm, Germany

Trevor J. Trust

Department of Biochemistry and Microbiology
University of Victoria
Victoria, British Columbia, Canada

In the early 1970's there were no more than three or four laboratories actively working with bacterial S-layers (Murray, Chapter 1). Few microbiologists recognized their existence as cell envelope components and the best possible electron microscopy was required for their detection. Then, they were not always called S-layers but were instead named paracrystalline arrays, regularly structured layers (RS-layers), planar

Advances in Bacterial Paracrystalline Surface Layers
Edited by T.J. Beveridge and S.F. Koval, Plenum Press, New York, 1993

crystalline layers, or surface layers (S-layers). They had been detected only on a few species of *Acinetobacter, Bacillus, Clostridium* and *Spirillum (Aquaspirillum)*. Yet, those few scientists who worked in the S-layer field not only recognized their fragile beauty and symmetry but also understood that, since they were metabolically expensive devices for each bacterium to synthesize, S-layers must have an important function.

A quarter of a century has passed since these halcyon days and most of the early researchers retain their enthusiasm for S-layers. Three international S-layer workshops have been held beginning in 1984 and since that year the ranks of interested scientists have increased ten times over so that books such as this present one contain articles written by approximately fifty senior researchers. These include structuralists, molecular biologists, chemists, ecologists, clinicians, biogeochemists, and biotechnologists. It is an interdisciplinary spectrum of scientists and it reflects the exponential growth of interest in S-layers amongst a broad array of researchers. This last workshop (the NATO-ARW) has reminded us that S-layer function for bacteria remains to be an elusive goal although, intuitively, several functions have been identified (i.e., S-layers protect against predacious bacteria, some bacteriophages, toxic heavy metals, and some degradative enzymes, they can be the only wall layer in some archaeobacteria (and therefore contribute cell shape and resist osmotic lysis), and they may be virulence factors on some pathogens). Yet, an undeniable encompassing role for these expensive prokaryotic corsets as a unifying feature of all S-layered bacteria has not been discovered and this is primarily because we are still in a descriptive and analytical phase of S-layer research. To aid functional studies we need new methods for their detection which do not rely on electron microscopy so that a wider array of laboratories can join in the search for function. Indeed, we may be too shallow in our insistance that an encompassing functional trait exists for each S-layer. When they exist, S-layers are an integral part of the cell envelope and it may be appropriate to think of them more in terms of <u>total cell wall function</u>.

S-layers are found in gram-positive and gram-negative eubacteria and in archaeobacteria (archeae). They can consist of proteins or glycoproteins. Generally, there seems to be no common denominator as to what bacterial genus, species or strain possess a S-layer, or as to why a carbohydrate substituent is added to one type of S-layer and not another. It is also impossible to determine at this point whether or not glycosylation of the protomeric unit makes certain S-layer symmetries preferred. For these reasons, the presence or absence of a S-layer has low taxonomic value.

Very little information is available about how the synthesis, transport and self-assembly of S-layers is controlled at the cellular level. From the available information (Chapter 12), it is possible that S-proteins require undecaprenyl phosphate as a membrane-carrier, whereas S-glycoproteins require dolichyl phosphate. Similar biosynthetic pathways are used for the S-glycoproteins and glycoflagellins in *Halobacterium*. For some S-layers, the structural genes have been identified, promoters suggested, and the conditions for self-assembly determined. But, it is more complex than this and must be interactive with several metabolic processes. For example, is there feedback control on S-layer synthesis, is there S-layer turnover and is this turnover integrated with the turnover of other surface components, and is the production of S-layers modulated by environmental factors? Pools of S-layer constituents seem to be common in gram-positive bacteria where protomers reside in the interstices of the polymeric network which make up the cell wall, but pools do not seem to be so essential for gram-negative types; no pool has been detected in *Aquaspirillum serpens* and *Campylobacter fetus* subsp. *fetus*. Since (during surface assembly) approximately 500 S-layer subunits are incorporated per second, it is remarkable that pools do not exist in all S-layered bacteria.

Some S-proteins are glycosylated and this carbohydrate moiety cannot only be a simple structural decoration. For *Halobacterium* the glycosubstituent substantially increases the charge density of the S-protein so that under high salt conditions the molecule can fold properly for assembly (Sumper, Chapter 11). Generally for biological systems, carbohydrate substituents are highly charged, are less metabolically expensive than proteins, and can be readily modified or tailored to suit a changeable environment; this may be the advantage for the glycosylation of select S-layers. Yet, for eubacterial glycosylated S-layers, the carbohydrate moieties are unchanged.

The NATO-ARW covered a wide range of S-layer topics and included both invited lectures and poster presentations. All of these are included as chapters in this book. The subjects ranged from the highest possible resolution analysis, to minute chemical and molecular biological analyses, to environmental effects, to medical implications, and to possible biotechnological uses. Three dimensional reconstructions and scanning probe microscopy are beginning to reveal the intricate folding patterns of the monomeric units within different lattice systems in both archaeobacterial (Hövmoller, Chapter 2; Phipps, Chapter 3; Kessel and Trachtenberg, Chapter 5; and Southam et al., Chapter 13) and eubacterial systems (Lounatmaa et al., Chapter 4; S. Lortal, Chapter 6; Šmarda and Komrska, Chapter 8; and Kay et al., Chapter 15). It is becoming apparent that surface topography (Firtel et al., Chapter 23) and its effect on the accessability of chemically reactive sites (Schultze-Lam and Beveridge, Chapter 7) are important factors for the interaction of S-layers with the external environment. In some cases, especially with those S-layers which are glycosylated, there is tantalizing evidence of the position of the carbohydrate moieties which seem to be situated close to the external surface of the S-layer (e.g., *Halobacterium halobium*, *Haloferax volcanii*, and *Bacillus stearothermophilus*; Kessel and Trachtenberg, Chapter 5; Messner et al., Chapter 10; and Sumper, Chapter 11). For these reasons, it is also possible for pathogens possessing S-layers which are prime antigenic determinants or virulence factors, that the molecular folding of S-proteins is of prime importance for the infectious process, for resistance to the host immune system (e.g., *A. salmonicida*; Kay et al., Chapter 15; and Garduño et al., Chapter 28), for antigenic variation (e.g., *C. fetus* subsp. *fetus*; Blaser, Chapter 17, and (possibly) for the delicate reconfiguration of the lattice system and exposed polypeptide sites (e.g., *A. salmonicida*; Kay et al., Chapter 15; and Garduño et al., Chapter 28).

High resolution spectroscopic and chromatographic methods are allowing some researchers to redefine the chemical nature of S-layers (Messner et al., Chapter 10). Sometimes, especially in eubacterial systems, carbohydrate substituents are a small percentage of the total S-layer mass and are, therefore, undetectable by more traditional chemical methods. These glycoproteins, and those with more apparent glycosylations (e.g., *H. halobium*), are becoming exquisitely defined by genetical analysis (Sumper et al., Chapter 11; and Kuen et al., Chapter 14). In fact, molecular biology has had a great impact on clarifying the chemical nature of a variety of S-layers (e.g., *Aeromonas* sp., *Bacillus* spp., *Caulobacter crescentus* and *C. fetus*; Trust, Chapter 16; Blaser, Chapter 17; Kuen et al., Chapter 14; and Bingle et al., Chapter 18). Now, the same molecular systems are defining genetic relationships between S-layers, translocation routes, and processing systems. Some S-layer genes are being genetically reworked by researchers so that the S-protein can carry peptide stretches of biotechnological importance (e.g., metallothionine has been incorporated into the *C. crescentus* S-protein; Bingle et al., Chapter 18).

Many archaeobacteria, especially the methanogens, have unusual cell wall formats which range from single S-layers (e.g., *Methanococcus*) to multilayered types (e.g., *Methanospirillum*); some possess the unique polymers, pseudomurein (e.g., *Methanobacterium*) and methanochondroitin (e.g., *Methanosarcina*). It is apparent that the synthetic steps towards their polymeric production and their transport across the plasma membrane have subtle differences to more usual eubacterial systems (König, Chapter 12). *M. hungatei* possesses an outer most sheath which possesses the most minute lattice system of any S-layer. Its structure and chemistry is so complex (Southam et al., Chapter 13) that it is difficult to still consider it a S-layer (at least under our current conception of S-layers as being planar paracrystalline self-assembled arrays composed of identical monomeric (glyco)proteins). As additional encompassing structures, *M. hungatei* possesses a bona fide S-layer surrounding each cell and multilamellar paracrystalline plugs; all of these have now been defined by image reconstructions and scanning probe microscopy (Southam et al., Chapter 13;and Firtel et al., Chapter 23).

Possibly, some of the most exciting reports of the NATO-ARW were those that dealt with environmental issues. Preliminary reports over a decade ago of *Aquaspirillum serpens* VHA surviving predacious *Bdellovibrio* attack because of its S-layer have been confirmed and expanded to a wider spectrum of S-layered bacteria (Koval, Chapter 9). This provides unequivocal evidence of at least one function for S-layers on bacteria exposed to the natural environment. In addition, research involving bacteria and ciliated protozoan "grazers" showed that there was no discrimination between the engulfment of naked versus S-layered bacteria (Koval, Chapter 9). It is still not clear whether or not S-layered bacteria can better survive phagosomal digestion than their naked counterparts.

Remarkably for the first time, an S-layered cyanobacterium (*Synechococcus* GL24) isolated from a groundwater-fed lake near Fayetteville, New York was shown to control fine-grain mineral development within the lake waters and sediments (Schultze-Lam and Beveridge, Chapter 7). This entire process of biomineralization is determined by the external physicochemistry of the S-layer and the regular exposure of surface functional anionic groups used to complex Ca^{2+}. This chemical complexation nucleates the growth of fine-grained minerals and relies on the abundance of electrolytes in the lake water. Photosynthesis is directly involved in determination of mineral type since it can increase the pH of the microenvironment surrounding each cell and therefore alter the geochemical "solid field". At circumneutral pH (the natural pH of the water) SO_4^{-2} complexes the Ca^{2+} to form gypsum, whereas at more alkaline pHs (during photosynthesis) HCO_3^- and Ca^{2+} form calcite; as the *Synechococcus* blooms in summer the solid field switches from gypsum to calcite and calcarious bioherms and a marl sediment is formed. This new functional trait of a S-layer may have tremendous implications for the global production of fine-grained minerals from "weathered" electrolytes.

The last day of the NATO-ARW was devoted to the biotechnological applications of S-layers and it was apparent that this is a very important focus for S-layer research. Not only can these microporous structures (SUMs) be used for ultrafiltration processes requiring sharp exclusion cutoffs, but they are exquisite templates for the addition of accessory molecules of industrial importance (Sára et al., Chapter 19). For example, enzymes, antibodies, or other reactive molecules can be bonded to the regular fabric of the S-layer and used to detect minute quantities of substrate. In fact, "sandwiches" of reactive molecules can be situated between a thin deposit of metal and a microfiltration membrane making possible the crafting of extremely small biosensors (Pum and Sleytr, Chapter 20). It was demonstrated, that S-layer proteins can be recrystallized into large-scale coherent monolayers at an air/water interface and on lipid films. These techniques appear of particular interest for studying the physicochemical surface properties of S-layers and to build composite structures which mimic the molecular architecture of those archaeobacterial cell envelopes which are exclusively composed of an S-layer and a plasma membrane. It is even possible that scanning probe microscopes could be used to move S-layer subunits around so that molecular messages could be written within the paracrystalline array. This has important ramifications for information storage and retrieval. These are exciting developments and, because they deal with minute manipulations, the term "nanotechnology" has been coined (Pum and Sleytr, Chapter 20).

So far there has been little attention paid to the stimulation or regulation of S-layer production for industrial purposes. Fermentor control of S-protein synthesis in *B. brevis* was outlined at the Workshop as well as the identification of cost effective nutrient sources for pilot scale production (Daugulis et al., Chapter 21). If substantial quantities of S-protein can be produced, highly antigenic substances chemically attached to them (e.g., *Streptococcus pneumoniae* oligosaccharide) and the conjugated S-protein assembled into microcarriers, a highly effective subunit vaccine is formed (Malcolm et al., Chapter 21). Another possibility as microcarriers uses the chemical resilience of archaeobacterial lipids, such as tetraether lipids, once they have been formed into unilamellar liposomes (Choquet et al., Chapter 24). The high organizational parameters of S-layers, their ability to be extracted and to self-assemble, their abundance as a high percentage of cellular protein, their easy regulation during pilot-scale bacterial growth, and their malleability by bioengineering techniques makes these paracrystalline surface arrays exciting devices for industrial exploitation. Certainly, since the last Workshop in Vienna, the area of S-layer biotechnology has grown in leaps and bounds.

The 1992 NATO-ARW in London, Canada was a successful workshop in every sense of the word. Scientists and their students from around the world gathered to exchange new ideas, data, and interpretations pertaining to S-layers. It was an intense meeting which emphasized the importance of these arrays and the necessity for their continued investigation. Clearly, there have been many advances in the study of bacterial paracrystalline surface layers.

October, 1992

CONTRIBUTORS

J.W. Austin, Department of Biochemistry and Microbiology, University of Victoria, Victoria, British Columbia, Canada V8W 2Y2

P. Awram, Department of Microbiology, University of British Columbia, Vancouver, British Columbia, Canada V6T 1Z3

D. Bayley, Department of Microbiology and Immunology, Queen's University, Kingston, Ontario, Canada K7L 3N6

M.W. Best, Alberta Research Council, P.O. Box 8330, Station F, 250 Karl Clark Road, Edmonton, Alberta, Canada T6H 5X2

T.J. Beveridge, Department of Microbiology, College of Biological Science, University of Guelph, Guelph, Ontario, Canada N1G 2W1

W.H. Bingle, Department of Microbiology, University of British Columbia, Vancouver, British, Columbia, Canada V6T 1Z3

B.L. Blackford, Department of Physics, Dalhousie University, Halifax, Nova Scotia, Canada B3H 3J5

M.J. Blaser, Division of Infectious Diseases, Department of Medicine, Vanderbilt University, School of Medicine, Nashville, Tennessee, 37232, USA

G. Bröckl, Angewandte Mikrobiologie, Universität Ulm, Oberer Eselsberg M23, D-7900 Ulm, Germany

S. Cavaignac, Department of Biochemistry and Microbiology, University of Victoria, Victoria, British Columbia, Canada V8W 2Y2

C.G. Choquet, National Research Council, Institute of Biological Sciences, 100 Sussex Drive, Ottawa, Ontario, Canada K1A 0R6

R. Christian, Scientific Software Company, A-1140 Wien, Austria

S. Chu, Department of Biochemistry and Microbiology, University of Victoria, Victoria, British Columbia, Canada V8W 2Y2

A. Daugulis, Department of Chemical Engineering, Queen's University, Kingston, Ontario, Canada K7L 3N6

E.M. Egelseer, Zentrum für Ultrastrukturforschung und Ludwig Boltzmann Institut für Molekulare Nanotechnologie, Universität für Bodenkultur, Gregor Mendel Strasse 33, A-1180 Wien, Austria

M. Firtel, Department of Microbiology, College of Biological Science, University of Guelph, Guelph, Ontario, Canada N1G 2W1

R.A. Garduño, Department of Biochemistry and Microbiology, University of Victoria, Victoria, British Columbia, Canada V8W 2Y2

L. Graham, Department of Microbiology, College of Biological Science, University of Guelph, Guelph, Ontario, Canada N1G 2W1

M. Haapasalo, Department of Cariology, University of Helsinki, Mannerheimintie 172, 00280 Helsinki 28, Finland

E. Hartmann, Angewandte Mikrobiologie, Universität Ulm, Oberer Eselsberg M23, D-7900 Ulm, Germany

S. Hovmöller, Structural Chemistry, University of Stockholm, S-10691 Stockholm, Sweden

M.H. Jericho, Department of Physics, Dalhousie University, Halifax, Nova Scotia, Canada B3H 3J5

H. Jousimies-Somer, National Public Health Institute, Helsinki, Finland

U. Kärcher, Angewandte Mikrobiologie, Universität Ulm, Oberer Eselsberg M23, D-7900 Ulm, Germany

W.W. Kay, Department of Biochemistry and Microbiology, University of Victoria, Victoria, British Columbia, Canada V8W 2Y2

E. Kerosuo, Department of Cariology, University of Helsinki, Mannerheimintie 172, 00280 Helsinki 28, Finland

M. Kessel, Department of Microbiology, Building 231, University of Maryland, College Park, Maryland 20742, USA

J. Komrska, Department of Physics, Faculty of Mechanical Engineering, Technical University, CS-616 69 Brno, Czechoslovakia

H. König, Angewandte Mikrobiologie, Universität Ulm, Oberer Eselsberg M23, D-7900 Ulm, Germany

S.F. Koval, Department of Microbiology and Immunology, Faculty of Medicine, University of Western Ontario, London, Ontario, Canada N6A 5C1

B. Kuen, Institut für Mikrobiologie und Genetik, Universität Wien, Althanstrasse 14, A-1090 Wien, Austria

S. Küpcü, Zentrum für Ultrastrukturforschung und Ludwig Boltzmann Institut für Molekulare Nanotechnologie, Universität für Bodenkultur, Gregor Mendel Strasse 33, A-1180 Wien, Austria

K. Leung, Department of Microbiology, University of British Columbia, Vancouver, British Columbia, Canada V6T 1Z3

S. Lortal, INRA, Technologie Laitière, 65, rue de Saint-Brieuce, 35042 Rennes Cedex, France

K. Lounatmaa, Department of Electron Microscopy, University of Helsinki, Mannerheimintie 172, 00280 Helsinki 28, Finland

W. Lubitz, Institut für Microbiologie und Genetik, Universität Wien, Althanstrasse 14, A-1090 Wien, Austria

A.J. Malcolm, Alberta Research Council, P.O. Box 8330, Station F, 250 Karl Clark Road, Edmonton, Alberta, Canada T6H 5X2

P. Messner, Zentrum für Ultrastrukturforschung und Ludwig Boltzmann Institut für Molekulare Nanotechnologie, Universität für Bodenkultur, Gregor Mendel Strasse 33, A-1180 Wien, Austria

A. Möschl, Zentrum für Ultrastrukturforschung und Ludwig Boltzmann Institut für Molekulare Nanotechnologie, Universität für Bodenkultur, Gregor Mendel Strasse 33, A-1180 Wien, Austria

Z. Mosleh, Alberta Research Council, P.O. Box 8330, Station F, 250 Karl Clark Road, Edmonton, Alberta, Canada T6H 5X2

P.J. Mulhern, Department of Physics, Dalhousie University, Halifax, Nova Scotia, Canada B3H 3J5

R.G.E. Murray, Department of Microbiology and Immunology, Faculty of Medicine, University of Western Ontario, London, Ontario, Canada N6A 5C1

B. Noonan, Department of Biochemistry and Microbiology, University of Victoria, Victoria, British Columbia, Canada V8W 2Y2

G.B. Patel, National Research Council, Institute of Biological Sciences, 100 Sussex Drive, Ottawa, Ontario, Canada K1A 0R6

B.M. Phipps, Protein Crystallography Group, Institute for Biological Sciences, National Research Council of Canada, Building M-54, Montreal Road, Ottawa, Ontario, Canada K1A 0R6

D. Pum, Zentrum für Ultrastrukturforschung und Ludwig Boltzmann Institut für Molekulare Nanotechnologie, Universität für Bodenkultur, Gregor Mendel Strasse 33, A-1180 Wien, Austria

M. Sára, Zentrum für Ultrastrukturforschung und Ludwig Boltzmann Institut für Molekulare Nanotechnologie, Universität für Bodenkultur, Gregor Mendel Strasse 33, A-1180 Wien, Austria

C. Schäffer, Zentrum für Ultrastrukturforshung und Ludwig Boltzmann Institut für Molekulare Nanotechnologie, Universität für Bodenkultur, Gregor Mendel Strasse 33, A-1180 Wien, Austria

S. Schultze-Lam, Department of Microbiology, College of Biological Science, University of Guelph, Guelph, Ontario, Canada N1G 2W1

J. Schuster-Kolbe, Zentrum für Ultrastrukturforschung und Ludwig Boltzmann Institut für Molekulare Nanotechnologie, Universität für Bodenkultur, Gregor Mendel Strasse 33, A-1180 Wien, Austria

U.B. Sleytr, Zentrum für Ultrastrukturforschung und Ludwig Boltzmann Institut für Molekulare Nanotechnologie, Universität für Bodenkultur, Gregor Mendel Strasse 33, A-1180 Wien, Austria

J. Šmarda, Department of Biology, Medical Faculty, Masaryk University, CS-662 44 Brno, Czechoslovakia

J. Smit, Department of Microbiology, University of British Columbia, Vancouver, British Columbia, Canada V6T 1Z3

G. Southam, Department of Microbiology, College of Biological Science, University of Guelph, Guelph, Ontario, Canada N1G 2W1

G.D. Sprott, National Research Council, Institute of Biological Sciences, 100 Sussex Drive, Ottawa, Ontario, Canada K1A 0R6

E. Sturm, Zentrum für Ultrastrukturforschung und Ludwig Boltzmann Institut für Molekulare Nanotechnologie, Universität für Bodenkultur, Gregor Mendel Strasse 33, A-1180, Wien, Austria

M. Sumper, Lehrstuhl Biochimie I, Universität Regensburg, Universitatsstrasse 31, 8400 Regensburg, Germany

R.J. Szarka, Alberta Research Council, P.O. Box 8330, Station F, 250 Karl Clark Road, Edmonton, Alberta, Canada T6H 5X2

S.R. Thomas, Department of Biochemistry and Microbiology, University of Victoria, Victoria, British Columbia, Canada V8W 2Y2

J.C. Thornton, Department of Biochemistry and Microbiology, University of Victoria, Victoria, British Columbia, Canada V8W 3P6

S. Trachtenberg, Department of Membrane and Ultrastructure Research, The Hebrew University-Hadassah Medical School, P.O. Box 1172, Jerusalem 91-010, Israel

T.J. Trust, Department of Biochemistry and Microbiology, University of Victoria, Victoria, British Columbia, Canada V8W 2Y2

F.M. Unger, Alberta Research Council, P.O. Box 8330, Station F, 250 Karl Clark Road, Edmonton, Alberta, Canada T6H 5X2

S.G. Walker, Department of Microbiology, University of British Columbia, Vancouver, British Columbia, Canada V6T 1Z3

S. Weigert, Zentrum für Ultrastrukturforschung und Ludwig Boltzmann Institut für Molekulare Nanotechnologie, Universität für Bodenkultur, Gregor Mendel Strasse 33, A-1180 Wien, Austria

C. Weiner, Zentrum für Ultrastrukturforschung und Ludwig Boltzmann Institut für Molekulare Nanotechnologie, Universität für Bodenkultur, Gregor Mendel Strasse 33, A-1180 Wien, Austria

B.N. White, Department of Biology, McMaster University, Hamilton, Ontario, Canada L8S 4L8

G.K. Whitney, Department of Chemical Engineering, Queen's University, Kingston, Ontario, Canada K7L 3N6

C.P. Wight, Department of Chemical Engineering, Queen's University, Kingston, Ontario, Canada K7L 3N6

A. Wong, Department of Chemical Engineering, Queen's University, Kingston, Ontario, Canada K7L 3N6

W. Xu, Department of Physics, Dalhousie University, Halifax, Nova Scotia, Canada B3H 3J5

INDEX